Applied Analysis of Composite Media

Woodhead Publishing Series in Composites Science and Engineering

Applied Analysis of Composite Media

Analytical and Computational Results for Materials Scientists and Engineers

Piotr Drygaś
Department of Mathematical Analysis
University of Rzeszow
Rzeszow, Poland

Simon Gluzman
Toronto, ON, Canada

Vladimir Mityushev
Institute of Computer Science
Pedagogical University of Cracow
Krakow, Poland

Wojciech Nawalaniec
Institute of Computer Science
Pedagogical University of Cracow
Krakow, Poland

WP
WOODHEAD
PUBLISHING
ELSEVIER   An imprint of Elsevier

Woodhead Publishing is an imprint of Elsevier
The Officers' Mess Business Centre, Royston Road, Duxford, CB22 4QH, United Kingdom
50 Hampshire Street, 5th Floor, Cambridge, MA 02139, United States
The Boulevard, Langford Lane, Kidlington, OX5 1GB, United Kingdom

Library of Congress Cataloging-in-Publication Data
A catalog record for this book is available from the Library of Congress

British Library Cataloguing-in-Publication Data
A catalogue record for this book is available from the British Library

ISBN: 978-0-08-102670-0

For information on all Woodhead Publishing publications
visit our website at https://www.elsevier.com/books-and-journals

Publisher: Matthew Deans
Acquisition Editor: Gwen Jones
Editorial Project Manager: Mariana Kuhl
Production Project Manager: Poulouse Joseph
Designer: Greg Harris

Typeset by VTeX

Working together
to grow libraries in
developing countries

www.elsevier.com • www.bookaid.org

To those who "live not by bread alone"

Contents

Biography

The authors belong to the research scientific group www.materialica.plus

Piotr Drygaś, is Assistant Professor at the Faculty of Mathematics and Natural Sciences at the University of Rzeszow. He is interested in Boundary Value Problems, describing conductivity of fibrous composites with non ideal contact, and 2D elastic composites of random structure with ideal contact between matrix and inclusions.

Simon Gluzman, PhD, is presently an Independent Researcher (Toronto, Canada) and formerly a Research Associate at PSU, and Researcher at UCLA. He is interested in algebraic renormalization method and theory of random and regular composites. Main achievement is development of novel, analytical Post-Padé techniques for re-summation of a divergent series, such as root, factor, additive and corrected Padé approximants. They are proven to achieve high accuracy in calculation of critical properties in a variety of physical and material science problems, see also S. Gluzman, V. Mityushev, W. Nawalaniec, Computational Analysis of Structured Media, Elsevier, 2017.

Vladimir Mityushev is Professor at the Institute of Computer Science of the Pedagogical University of Kraków, Poland. His academic and scientific activity is based on interdisciplinary international research devoted to applied mathematics and computer simulations. His main mathematical result is complete solution to the scalar Riemann–Hilbert problem for an arbitrary multiply connected domain. This result includes, for instance, the Schwarz–Christoffel mapping of multiply connected domains bounded by polygons onto circular domains, Poincaré series for the classical Schottky groups and other objects of the classical complex analysis constructed in other works in particular cases. The main scientific achievements of the group headed by VM concern application of the Riemann–Hilbert problem and of the generalized alternating method of Schwarz to 2D random composites and porous media, analytical formulas for the effective properties of regular and random composites, a constructive theory of RVE, pattern formation, and the collective behavior of bacteria. The developed approach revises the known analytical formulas and precisely answers the question how to determine macroscopic properties of random media.

Wojciech Nawalaniec is Assistant Professor at the Institute of Computer Science of the Pedagogical University of Kraków, Poland. WN is interested in simulation, classification and analysis of random structures. Also interested in application of symbolic-numerical calculations and machine learning to computational materials science. WN also develops algorithms and software modules providing solutions to scientific problems tackled by the Materialica+ Research Group.

Preface

In words of Sergei Prokofiev we define neoclassicism: "I thought that if Haydn were alive today he would compose just as he did before, but at the same time would include something new in his manner of composition. It seemed to me that had Haydn lived to our day he would have retained his own style while accepting something of the new at the same time. That was the kind of symphony I wanted to write, a symphony in classical style."

Classical theory of composites amounts to the celebrated Maxwell formula, also known as Clausius–Mossotti approximation. Actually, all modern self-consistent methods (SCM) perform elaborated variations on the theme, and are justified rigorously only for dilute composites when interactions among inclusions are neglected. In the same time, exact and high-order formulas for special regular composites which go beyond SCM were derived.

Let matrix conductivity be normalized to unity and σ denote the conductivity of inclusions. Introduce the contrast parameter $\varrho = \frac{\sigma - 1}{\sigma + 1}$. For many years it was thought that Maxwell's and Clausius–Mossotti approximation for the effective conductivity of 2D (3D) composites

$$\sigma_e = \frac{1 + \varrho f}{1 - \varrho f} + O(f^2) \tag{1}$$

can be systematically and rigorously extended to higher orders in f by taking into account interactions between pairs of spheres, triplets of spheres, and so on. However, it was recently demonstrated (Gluzman et al., 2017, Mityushev et al., 2018b, Mityushev, 2018) that the field around a finite cluster of inclusions can yield a correct formula for the effective conductivity only for non-interacting clusters. Rigorous justification of this fact is given in Appendix A.4 of the present book, based on the paper (Mityushev, 2018). The higher order term(s) can be properly found only after a subtle study of the conditionally convergent series.

The hard experimental evidence accumulated by material scientists and engineers begs for a constructive theory of random composites with explicit account for the geometry. The geometry is, de facto, another important *structural parameter*. As discussed in Nielsen (2005), "In itself the large number of completely different empirical stiffness expressions suggested for porous materials clearly indicates a need for a more rational research on composite properties versus composite geometry... Change of geometry will influence any mechanical/physical behavior of composites. Stiffness and viscoelasticity (creep and relaxation) will change. Shrinkage and eigenstress-strain (such as hydro-thermal properties, and heat conductivity are other examples of materials behavior, which will change with geometry. In order to cope rationally with such changes in composite analysis we must increase our freedom to choose other analytical models than the specific, non-variable ones most often used to day." Our new book is dedicated mainly to constructive topics of boundary value problems and their applications to macroscopic properties of composites and of porous media. Symbolic-numerical computations are widely used to deduce new formulas important for engineers and researchers. New formulas for the effective properties are deduced in the form customized for engineering applications.

The outline of typical exposition is given below for the case of 2D elastic composites. Composites with non-overlapping circular inclusions randomly embedded in matrix are investigated. Special attention is paid to critical regimes related to the optimal packing of inclusions and to extreme physical constants (rigid and soft inclusions). Investigation of regular and random structures is based on the general approach of the RVE (representative volume element) and the corresponding structural sums. The proposed method yields an effective algorithm in symbolic-numeric form to compute structural sums as discrete multiple convolutions. In this book, new algorithms are described systematically, codes or pseudo-codes are given, and complexity of computations is studied.

We present also modified averaging computational method applied to the local stresses and deformations. Properly constructed series are reduced to polynomials and rational functions depending on the concentration of inclusions f. But it is not a final solution to the problem. Furthermore, these functions are replaced by asymptotically equivalent expressions. Special methods of resummation suggested in the book and in (Gluzman et al., 2017), bring accurate and compact formulas for all concentrations. Accurate analytical formulas for deterministic and random composites and porous media can be derived employing approximants, when the low-concentration series are supplemented with information on the high-concentration regime. Typical problems we encounter are characterized by asymptotic power laws.

Our first book (Gluzman et al., 2017) may be considered as an neoclassical answer to the question associated to Fig. 0.1, why does James Bond prefer shaken, not stirred martini with ice? Highly accurate computational analysis of structural media allowed us to explain the difference between various types of random composite structures. It is strongly related to the critical exponent s in the asymptotic behavior of the effective conductivity. In the limiting case of a perfectly conducting inclusions, the effective conductivity σ_e is expected to tend to infinity as a power-law, as the concentration of inclusions f tends to $f_c = \frac{\pi}{\sqrt{12}}$, the maximal value in 2D

$$\sigma_e(f) \sim (f_c - f)^{-s}. \tag{2}$$

The dependence of the index s on the shaken-stirred regime of inclusions is displayed in Fig. 0.1. Similarly, one can consider different effective properties. Universality of the mathematical modeling implies that the same equations hold for the electric and thermal conductivity, magnetic permeability, anti-plane elastic strains and so on.

Scientific classicists are perpetually in search of universal answers to all questions concerning the transport properties of composite materials, which can be simply adjusted to any concrete case. Neoclassical approach includes their classical results as a

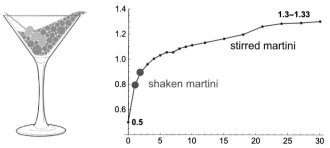

Figure 0.1 Why did James Bond prefer shaken, not stirred martini with ice? Because he sensed in martini the critical exponent s from formula (2). The dependence of s on the degree of disorder measured in steps of random walk is displayed in the graphics.

limit-case, complements it with another limit case of high-concentration percolating inclusions, but does not stop here. The problem now is getting shifted towards methodology (tool-box), concerned with properly matching the limit-cases. On the way one should develop a non-universal, structure-dependent expansions, as well as special approximation methods. From the warm and cosy world where universal answers exist, we are getting to the cold, unfriendly one. In such world each and every problem of the theory of composites should be studied in terms of the particular numerical set of structural sums, the particular large f limit found, and the particular approximation method carefully selected to receive formula for all f.

We are convinced that to derive a new formula, valid in the whole range of relevant variables, is not merely a mathematical exercise, or even a matter of convenience. It provides a fresh insight, since in the majority of cases realistic material sciences problems correspond neither to weak coupling (or low concentration) regime nor to strong coupling (high concentration) limit, but to the intermediate range of parameters. Such regime can be covered by some rather complex formula deduced from asymptotic regimes. It is quite handy for a scientist to possess a general mathematical toolbox to derive asymptotic, typically power laws, as well as explicit crossover formulas for a multitude of processes. Problems discussed in the book can be viewed as asymptotically classical, but in each of them, a neoclassical twist is supposed to make them more agreeable to the modern listener. Besides, they are all interesting and hold a promise in days to come.

Some authors equate an *approximate analytical formula* with a *model*. Such an approach is misleading, since a mathematical modeling involves fundamental governing equations, complemented by interface and boundary conditions. Different approximate formulas/solutions for the mathematical model hold under restrictions usually not discussed by authors. A serious methodological mistake may follow when intermediate manipulations are valid only within the precision $O(f)$, while the final formula is claimed to work with a higher precision, see an explicit example in Appendix A.4. In particular, it follows from our investigations that it is impossible to write a universal higher order formula independent on locations of inclusions. Such a universal formula holds only for a limited class of composites with non-interacting inclusions, e.g., for dilute composites and the Hashin-Shtrikman coated sphere assemblage (Cherkaev, 2009).

Since Einstein, the transport coefficients in random and regular media are expressed as expansions in concentration f. Nevertheless, despite persistent efforts of such outstanding researchers as Batchelor, Bergman, Brady, Jeffrey, Milton, McPhedran, Torquato, Wajnryb the problem still exists of finding correct numerical coefficients in expansions. Besides, the validity of such short series is very limited, and their true value is still remains to be seen.

Let us address the Hashin–Shtrikman bounds and their extensions (Milton, 2002). The dependence of the effective conductivity of a random composite on f corresponds to some monotonous curve, drawn between the bounds. Such a curve can be sketched arbitrarily, and it will correspond to some unspecified distribution of inclusions. Often, we deal with a uniform distribution corresponding to a stir-casting process described by the random sequential addition (RSA) model (Kurtyka and Rylko, 2013, 2017). Thence main theoretical requirement to the geometric model consists not only in writing a formula, but in a precise description of the geometrical conditions imposed on deterministic or random locations of inclusions. The rigorous statement and study of this theoretical problem is necessary for the proper approach to various applied problems, e.g., stir-casting process.

Our formulas for the effective properties of random composites are derived as the mathematical expectation of the effective conductivity over the independent and identically distributed (i.i.d.) non-overlapping inclusions. Even formulas obtained for uniformly distributed non-overlapping balls are not universal, because the effective conductivity depends on the protocol of computer simulations or experimental stirring, meaning that the very notion of randomness is non-universal (Torquato, 2002, Kurtyka and Rylko, 2013, Rylko, 2014, Gluzman et al., 2017).

Some famous formulas turn out to be questionable, in our opinion. The example, why various not rigorous, popular approaches can be questioned, is given in Chapter 6 based on the paper by Mityushev and Nawalaniec (2019). It is demonstrated that the classical Jeffrey formula contains wrong f^2 term. See Chapter 6, where the terms f^2 and f^3 are written explicitly! In particular, we demonstrate that the f^3 term depends on the deterministic and random locations of inclusions. A novel expansion for the random composite with superconducting (perfect conducting) inclusions is obtained, see (6.92) on page 226. It leads to a proper estimate for the critical index for superconductivity. The finding seems to justify the whole body of work on the short series for effective conductivity, and give a physical meaning to the series as a valuable source of estimating critical index s.

Piotr Drygaś
Simon Gluzman
Vladimir Mityushev
Wojciech Nawalaniec
Kraków, April 2019

Acknowledgments

PD, VM and WN were partially supported by National Science Centre, Poland, Research Project No. 2016/21/B/ST8/01181.

Nomenclature

Geometry

a_k	stands for center (complex coordinate) of the disk D_k
$\mathbf{a}_k$	stands for center of the sphere D_k
BCC	means body-centered cubic lattice
D	stands for a domain in $\mathbb{R}^2$ or in $\mathbb{R}^3$ or stands for circular multiply connected domain
D_k	stands for simply connected domain (inclusion) in $\mathbb{R}^2$ and in $\mathbb{R}^3$ or stands for circular inclusion in $\mathbb{R}^2$ and ball in $\mathbb{R}^3$
∂D	stands for boundary of D
$E_k(z)$	stands for Eisenstein function of order k
$E_m^{(j)}(z)$	stands for Eisenstein–Natanson–Filshtinsky functions
$e_{m_1,m_2,m_3,\ldots,m_n}$	stands for structural sum of the multi-order $\mathbf{m} = (m_1, \ldots, m_n)$ based on the Eisenstein functions
$e_{p_1,p_2,p_3,\ldots,p_n}^{f_0,f_1,f_2,\ldots,f_n}$	stands for structural sum of poly-disperse composite
$e_{\mathbf{n}}^{(\mathbf{j})} \equiv e_{n_1,\ldots,n_q}^{(j_1,\ldots,j_q)(l_1,\ldots,l_q)}$	stands for the elastic structural sum based on the Eisenstein–Natanzon functions
f	stands for volume fraction (concentration) of inclusions
FCC	means face-centered cubic lattice
HCP	means hexagonal close-packed lattice
$\mathbf{n} = (n_1, n_2)$	normal vector to a smooth oriented curve L
$n = n_1 + in_2$	complex form of the normal vector $\mathbf{n} = (n_1, n_2)$
$\mathbb{R}^d$	stands for the Euclidean space
$\mathbf{s} = (-n_2, n_1)$	tangent vector
SC	means simple cubic lattice
S_m	stands for Eisenstein–Rayleigh lattice sums
$S_p^{(q)}$	stands for Natanson–Filshtinsky lattice sums
$\mathbb{U}$	stands for unit disk on plane
ω_1 and ω_2	stand for the fundamental translation vectors on the complex plane

Conductivity

K	stands for permeability
s	stands for the critical index for superconductivity
t	stands for the critical index for conductivity
$\varrho = \frac{\sigma^+ - \sigma^-}{\sigma^+ + \sigma^-}$	stands for contrast parameter between media with the conductivities σ^+ and σ^-; for two-phase composites $\varrho = \frac{\sigma-1}{\sigma+1}$
$\sigma(\mathbf{x})$	stands for scalar local conductivity
σ_k	is used for conductivity of the kth inclusion
σ	is used for conductivity of inclusions of two-phase composites when the conductivity of matrix is normalized to unity
$\boldsymbol{\sigma}_e$	stands for effective conductivity tensor
σ_e	stands for scalar effective conductivity of the macroscopically isotropic composites when the conductivity of matrix is normalized to unity
σ_e^{ij}	stands for components of the effective conductivity tensor

Elasticity

G and G_1	stand for shear modulus of matrix and inclusions, respectively
k and k_1	stand for 2D bulk modulus of matrix and inclusions, respectively
k', k'_1, v' and v'_1	stand for the corresponding 3D elastic constants; the shear modulus is the same in 2D and 3D
$\mathcal{S}$	is the critical index for super-elasticity
$\mathcal{T}$	is the critical index for elasticity
ϵ_{ij}	stand for the components of the strain tensor
κ and κ_1	stand for Muskhelishvili's constant, $\kappa = 3 - 4v' = \frac{3-v}{1+v}$
μ and μ_1	stand for shear moduli of incompressible media and of viscosity
μ_e	stand for the effective viscosity
v and v_1	stand for 2D Poisson's ratio of matrix and inclusions, respectively
ϱ_j	stand for contrast parameters $\varrho_1 = \frac{\frac{G_1}{G}-1}{\frac{G_1}{G}+\kappa_1}$, $\varrho_2 = \frac{\kappa\frac{G_1}{G}-\kappa_1}{\kappa\frac{G_1}{G}+1}$, $\varrho_3 = \frac{\frac{G_1}{G}-1}{\kappa\frac{G_1}{G}+1}$
$\varrho_{jk,m}$	stand for contrast parameters of multi-phase composites $\varrho_{1k,m} = \frac{G_m-G}{G_k+\kappa_k G} \cdot \frac{G_k}{G_m}$, $\varrho_{2k,m} = \frac{\kappa G_m-\kappa_m G}{\kappa G_k+G} \cdot \frac{G_k}{G_m}$, $\varrho_{3k,m} = \frac{G_m-G}{\kappa G_k+G} \cdot \frac{G_k}{G_m}$, $\varrho_{ik,k} := \varrho_{ik}$, where G_m and κ_m denote the elastic constants of mth inclusion
σ_{ij}	stand for the components of the stress tensor

Other symbols and abbreviations

1D, 2D, and 3D	mean one-, two-, and three- dimensional
A and B	stand for the critical amplitudes
$\mathcal{A}^*$	stands for the additive approximants
$\mathbb{C}$	stands for the set of complex numbers
$\mathbf{C}$	stands for the operator of complex conjugation
$\mathbf{Cor}^*$, or $\mathbf{Cor}^*$	stand for Corrected Padé approximant. Sometimes, to avoid confusion, we write simply $\mathbf{Cor}$
$\mathcal{E}$	measures the respective relative error
EMA	means effective medium approximations
$\mathcal{F}^*$	stands for the factor approximants
HS	means the Hashin–Shtrikman lower and upper bounds
$\mathcal{H}^\alpha(L)$	stands for functions Hölder continuous on a simple smooth curve L
$\mathcal{H}^\alpha(D) \equiv \mathcal{H}(D)$	stands for functions analytic in D and Hölder continuous in its closure
Im	means the imaginary part of complex value
LHS and RHS	mean the expressions "left-hand side" and "right-hand side" (of an equation), respectively
$\mathcal{H}^{(2,2)}$	Hardy–Sobolev space
$PadeApproximant[F[z], n, m]$	stands for the Padé (n, m)-approximant of the function $F(z)$
$\mathbb{R}$	stands for the set of real numbers
$\mathcal{R}^*$	stands for the root approximants
$\mathcal{RA}^*$	stands for the $DLog$ additive recursive approximants
RCP	means random close packing
Re	means the real part of complex value
RS	means random shaking
RSA	means Random Sequential Addition
RVE	stands for representative volume element
RW	means random walks
SCM	means self-consistent methods
T	stands for temperature
$\mathcal{Z}$	is the compressibility factor
β	generally stands for the critical index in application to arbitrary case
$\varkappa$	is the fluid permeability exponent

ρ	stands for particles or cracks geometric density proportional to the number of elements per cell
$\wp(z)$	stands for the Weierstrass $\wp$-function
$\wp_j(z)$	stands for the Natanzon–Filshtinsky functions
$\zeta(z)$	stands for the Weierstrass ζ-function
$\sim$	indicates asymptotic equivalence between functions, i.e., indicates that functions are similar, of the same order
$\simeq$	indicates that the two functions are similar or equal asymptotically
$\approx$	is used as "approximately equal to", indicating that the number is acceptably close to an exact value
$*$	marks self-similar approximants

Introduction to computational methods and theory of composites

1

1.1 Traditional approaches

In many problems of material sciences one encounters the so-called crossover phenomena, when a physical quantity qualitatively changes its behavior in different domains of its variable. Let a function $\Phi(x)$ represent a physical quantity of interest, with a variable running through the interval $x_1 \leq x \leq x_2$. Let also the behavior of this function be essentially different near the boundary points x_1 and x_2. Assume that the function varies continuously from $\Phi(x_1)$ to $\Phi(x_2)$, as x changes from x_1 to x_2. Then we may say that the function in the interval $[x_1, x_2]$ undergoes a crossover between $\Phi(x_1)$ and $\Phi(x_2)$. Crossover behavior of different physical quantities is ubiquitous in nature. For instance, a number of physical quantities essentially change their behavior when passing from the weak-coupling to strong-coupling limit. Real physical systems are usually so complicated that governing equations almost never can be solved exactly. In reality, it is often possible to find asymptotic expansions of solutions in vicinity of boundary points.

The natural arising problem is how to construct a good approximation for the sought function, valid on the whole domain of its variable, knowing only its asymptotic behavior near the boundaries. This problem is complicated by the fact that only a few terms of the asymptotic expansions are usually available. Sometimes, only general structure of the expansion could be guessed. In such a case the problem looks very difficult. The crossover problem could be compared with general Weierstrass program aimed at extracting all properties of the global analytic function in terms of corresponding Taylor coefficients of power series.

There is a demand for an approach, which could overcome the discussed difficulties and would be applicable to a large variety of problems. It is important that such an approach would provide relatively simple formulas for the material properties of interest. The advantage of having analytical expressions, as compared to just numbers that could be obtained from a numerical procedure, is in the convenience of analysis of such expressions with respect to various parameters entering into them. Therefore, we keep in mind a relatively simple analytical method that would provide relatively compact and accurate representations for various material quantities.

To derive a new formula, valid in the whole range of physical variables, is not merely a mathematical exercise. It provides a fresh insight, since in the majority of cases realistic material sciences problems correspond neither to weak coupling (or low concentration) regime nor to strong coupling (high concentration) limit, but to the intermediate range of parameters (Li and Ostling, 2015), where only some rather complex formula can be deduced. It is quite handy to possess a general mathematical toolbox to derive asymptotic, typically power laws, as well as explicit crossover for-

Applied Analysis of Composite Media. https://doi.org/10.1016/B978-0-08-102670-0.00010-X

mulas for arbitrary phenomena. "Power laws are ubiquitous and should be exploited
for complete analysis of the system," rather than from imbuing them with a vague and
mistakenly mystical sense of universality" (Stumpf and Porter, 2012).

A numerical approach to the effective properties of elastic media, based on integral equations and series method, was presented in Grigolyuk and Filishtinskii (1970,
1992, 1994), Helsing (1995). Integral equations for 2D doubly periodic composites
first constructed in Grigolyuk and Filishtinskii (1970), are efficient for the numerical investigation of a non-dilute composites when the inclusion interactions are taken
into account. The method of boundary layer integral equations was developed and applied to various problems for composites in Ammari et al. (2018a,b). The numerical
method of finite elements is also widely applied to fibrous composites, see Eischen
and Torquato (1993), Barbero (2008), Zohdi and Wriggers (2008), Selvadurai and
Nikopour (2012), Devireddy and Biswas (2014), Majewski et al. (2017). The developed numerical methods yield an effective way to determine elastic properties of
composites with a fixed geometry, for given values of shear moduli G_1 and G, bulk
moduli k_1 and k, for the inclusions and matrix, respectively.

We leave aside many other purely numerical methods, since their place in material
science was already discussed in Gluzman et al. (2017). We just note that optimally
designed technologies rely on explicit dependencies of the effective constants on the
elastic constants of constituents and on the concentration of inclusions f. In the case
of two-phase composites with isotropic components we have four elastic constants
G_1, G, k_1, k that complicates the complete numerical investigation. Moreover, the
limit percolation regimes when $\frac{G_1}{G}$ is close to zero or to infinity, and f tends to the 2D
maximal packing fraction $\frac{\pi}{\sqrt{12}}$, can be hardly investigated numerically.

1.1.1 Self-consistent methods

Analytical methods are applied to tackle various multiparametric problems. From
reading the titles and abstracts to many papers one can have an impression that a
wide variety of a diverse analytical formula was found as the final solutions. And
the problem has been "completely solved", or "almost solved" in closed form! However, an attempt to extract from these papers the concrete analytical formula, typically
show its validity up to $O(f^3)$ at most. In fact, all such formulas without exception,
are asymptotically equivalent to the same, lowest-order approximation. Some authors
put forward an "explicit" procedures which include solution to implicitly infinite order
systems, integral equations (see the discussion in Andrianov and Mityushev (2018)),
or absolutely divergent series. We outline below the typical results originating from
various analytical approaches. Nevertheless, some of them are worth of recognition
as the first fundamental results for the effective material constants. Such formulas are
widely applied in material sciences.

The dilute elastic composites were studied by solving exactly a simple, single–
inclusion problem (Eshelby, 1957, Hill, 1964). The obtained formula is asymptotically
equivalent to the Hashin–Shtrikman bounds (3.69) and (3.76) valid up to $O(f^2)$. This
is a unique pair of analytical formula known in the theory of random 2D macroscopically isotropic elastic composites with circular section of fibers. Their followers

frequently perform a Sisyphean job of sophisticated manipulations actually masking the preceding results of Giants. Under the umbrella of self-consistent methods (SCM), we find various effective medium approximations, Mori–Tanaka method, difference method, finite cluster method, reiterative homogenization and so forth. All are frequently declared "a rigorous method of micromechanics" (Kushch and Sevostianov, 2016) valid also for high concentrations. However, it was explicitly demonstrated by Mityushev and Rylko (2013) that the self-consistent methods can produce formula for the effective conductivity only to $O(f^2)$ for macroscopically isotropic composites.

The improvement claimed by SCM actually yields results within the linear precision in f. It is usually based on asymptotic manipulations within $O(f)$, followed typically by declaration of its universal validity. It is hard to argue with adherents of SCM, since they typically avoid any discourse. But it was shown in Mityushev and Rylko (2013), Drygaś and Mityushev (2016) that already the terms on f^2 in the 2D effective constants include a conditionally convergent sums. The feature is completely ignored in theory and application of self-consistent methods. Maxwell's approach yields the effective conductivity of dilute clusters (Mityushev, 2018), but not of a composite considered as a geometric limit of cluster constructions. We explain the proper treatment of the cluster method and give its revision in Appendix A.4, where subtle convergence questions neglected by some authors will be discussed.

What are the reasons for reasonably accurate numerical results obtained for the hexagonal array by the self-consistent methods (except the high concentration, percolation regime (Gluzman et al., 2016a))? In (Wall, 1997) it was noted that the hexagonal array can be approximated by a coated structure similar to lubrication approximation (Andrianov et al., 1999, Gluzman et al., 2016a). Moreover, the coated structure is nothing else but the famous Hashin's coated disks assemblage (Hashin, 1962), which attains the bounds for macroscopically isotropic composites. This geometric observation yields similar effective elastic properties for moderate concentrations for both structures. The same arguments explain why the effective elastic constants for the hexagonal array and the corresponding Hashin–Shtrikman lower bounds are in a close agreement for the wide range of material parameters and f, namely for low contrasts or moderate concentrations as in examples illustrated in Chapter 3.

Regular composites with deterministic structure are sufficiently well studied, for instance in the book (Chatterjee, 2008), devoted to regular geometric structures used in crystallography. Sometimes, deterministic structures and their weakly perturbed variations, e.g., the hexagonal array, are considered as a faithful representation for random composites[1]. However, accurate analytical formulas demonstrate the essential difference between regular and random composites (Ryan et al., 2015, Gluzman and Mityushev, 2015, Gluzman et al., 2016a,b). These papers show that the hexagonal array is an exceptional geometrical structure. This statement becomes clear after analysis of analytical formulas and figures generated in the present book in Chapter 3.

---

[1] See Fig 9.2 on page 197 in the book (Mityushev et al., 2018a).

1.1.2 Series method

Few attempts to find the effective elastic properties of regular arrays were based on the fundamental works (Natanson, 1935), (Grigolyuk and Filishtinskii, 1970, 1992)[2]. The complex potentials were represented in the form of series in the Weierstrass elliptic functions with coefficients satisfying an infinite system of linear algebraic equations. As in Rayleigh (1892), the truncation method was applied to the infinite system. The local elastic fields were sufficiently well described in Natanson (1935), Grigolyuk and Filishtinskii (1970, 1992). The effective elastic constants were computed by averaging of the local fields by the method described in Grigolyuk and Filishtinskii (1992, Chapter 4) for regular arrays. De facto, the same series method for the effective elastic constants for the hexagonal array was reported in Guinovart-Diaz et al. (2001), where the numerical truncation method for infinite systems was referred to as "closed-form" (see discussion in Andrianov and Mityushev (2018)).

The next step of importance was made in Mityushev (2000), Drygaś and Mityushev (2016), where the method of functional equations was developed for 2D elastic problems with circular inclusions. The method of functional equations can be considered as a continuous form of the series method. The functional equations can be solved by the series method. The equivalence of these methods was demonstrated by Rylko (2000). The method of functional equations in combination with the cluster method includes the conditionally convergent sums. Analytical formulas up to $O(f^4)$ for arbitrary locations of disks were presented.

However, the Supplement to Drygaś and Mityushev (2016) contains the correct high-order in f formulas for the local fields within the hexagonal array, and also contains incorrect formulas for the effective constants. The error arises because of a surprising property of the conditionally convergent series (3.78), while taking into account the induced moment at infinity. It was proved in Drygaś and Mityushev (2017), Mityushev (2018), Mityushev and Drygaś (2019) that the induced moment vanishes for the local fields, if the conditionally convergent series are defined by the Eisenstein summation (3.79). In the same time, computation of the effective constants by the cluster extension of the Maxwell self-consistent approach via dipoles must be defined by the symmetric summation (3.80), in order to get the zero induced moment.

1.1.3 Getting from traditional approach to neoclassical illustrated by example

We proceed now to illustrate both classical and neoclassical thinking, as discussed in Preface. The best way to do so is just to give an example. This subsection presents an application of the approach coined as neoclassical. It will be applied along the similar

---

[2] To the best of our knowledge only a single paper by Natanson (1935) was published. His results were developed by Filshtinsky, as early as 1964. The seminal paper (Natanson, 1935) can be considered as an extension of the method of Rayleigh (1892). An independent extension as a multipole method was discussed in McPhedran and Movchan (1994) and Movchan et al. (1997), were the series method was applied for the local elastic fields in doubly periodic composites.

lines in the bulk of the book, in particular to the more complicated case of elastic composites.

Consider a classical problem of the effective conductivity (thermal, electric etc.) of a 2D regular composite. An accurate approximate formula can be deduced for a 2D, two-component composite made from a collection of non-overlapping, identical, ideally conducting circular disks, embedded regularly in an otherwise uniform, locally isotropic host. Let σ denote the ratio of the conductivity of inclusions to the matrix conductivity. Usually, the conductivity of the matrix is normalized to unity. Introduce the contrast parameter

$$\varrho = \frac{\sigma - 1}{\sigma + 1}, \tag{1.1}$$

so that $|\varrho| \leq 1$. The exact formula for the effective conductivity tensor of an arbitrary regular array was written in the most general form in Gluzman et al. (2017, eq. (4.2.28)) as a power series in ϱ and f with exactly written all the coefficients. In the case of a square array of inclusions this series diverges as $f \to f_c = \frac{\pi}{4} \approx 0.7854$ and $\varrho \to 1$. It is difficult to analytically investigate its singular behavior. However, we know it from other asymptotic formulas (Keller, 1963), (Gluzman et al., 2017, Chapter 6). Consider the truncated expansion for the square array

$$\sigma_e \approx \tfrac{1+f\varrho}{1-f\varrho} + 0.611654 f^5 \varrho^3 + 1.22331 f^6 \varrho^4 + 1.83496 f^7 \varrho^5 + 2.44662 f^8 \varrho^6. \tag{1.2}$$

It respects the phase interchange symmetry (Keller, 1964) $[\sigma_e(\sigma)]^{-1} = \sigma_e(\sigma^{-1})$. One should check their method's compliance with the symmetry, and the proper steps has to be taken to guarantee corresponding critical properties. The latter can be accomplished from scaling considerations, in the vicinity of f_c. For small σ the form of a correction to generic power-law may be also found as a power-law. We are going to look for an approximate analytical solution in the model with two critical exponents. There are two limit-cases. For non-conducting disks, as $\sigma = 0$,

$$\sigma_e \simeq \sqrt{\frac{f_c - f}{f_c}}, \tag{1.3}$$

and for weakly-conducting disks, as $f = f_c$, with $\sigma \neq 0$,

$$\sigma_e \approx \sigma^{1/2}, \tag{1.4}$$

following (Efros and Shklovskii, 1976). The simplest solution satisfying both limits can be constructed in additive form

$$\sigma_e \approx \sqrt{\sigma} + \frac{2\sqrt{\frac{\pi}{4} - f}}{\sqrt{\pi}}, \tag{1.5}$$

with a subsequent correction in the form of a diagonal Padé approximant, achieving asymptotic equivalence with (1.2).

In high orders one can still obtain closed-form expressions and manage them with *Mathematica*. They are too long to be brought up here. But for concrete parameters their derivation and final form are pretty simple. Assuming the form $P_{4,4}$ for the correcting Padé approximant, we obtain an accurate formula for $\sigma = 0.02$ (compared with numerical results of (Perrins et al., 1979))

$$\sigma_e \approx \frac{0.817042(f(f(f(f-0.13819)-0.034015)+2.33313)-3.22041)\left(\sqrt{0.785398-f}+0.125331\right)}{f(f(f(f+0.226828)+1.01293)-1.70169)-2.66162}. \tag{1.6}$$

For further comparison we also constructed the Padé approximant for conductivity

$$\sigma_e \approx \frac{(f-0.871396)(f+1.75656)\left(f^2-0.88516f+2.31417\right)}{(f-1.75656)(f+0.871396)\left(f^2+0.88516f+2.31417\right)}. \tag{1.7}$$

In Figs. 1.1 and 1.2 it is clearly seen a linear behavior of conductivity in the intermediate region of f. Thus, there are three characteristic parts described by the formula (1.6). For small f there is a diluted situation, also well covered by the Clausius–Mossotti formula (1). For intermediate f the conductivity can be approximated by linear behavior, also well covered by the Padé approximant. The critical region close to f_c is described only by our formula (1.6). Various approximations are compared in Fig. 1.1. Overall, only neoclassical formula (1.6) can cover all three situations.

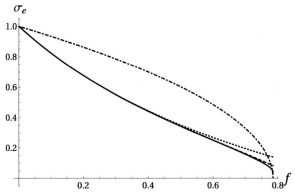

Figure 1.1 Our suggestion (1.6) is shown with solid line, and the Padé approximant (1.7) is shown with dashed line. The Clausius–Mossotti approximation (1) is shown with dotted line. Naive power-law extrapolation of the formula (1.3) to the whole region is shown with dot–dashed line.

The form (1.5) is particularly suitable to include the critical behavior as $\rho \to -1$, and could be adapted to the case $\rho \to 1$, respectively. The phase-interchange symmetry is preserved, when analogous calculations are performed for highly conducting inclusions, $\sigma \gg 1$, with the following "symmetric" choice of the approximation for the critical behavior, $\sigma_e \approx \left(\frac{2\sqrt{\frac{\pi}{4}-f}}{\sqrt{\pi}}+\frac{1}{\sqrt{\sigma}}\right)^{-1}.$

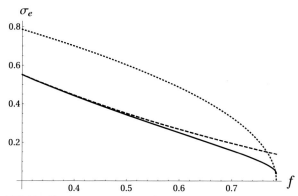

Figure 1.2 $\sigma = 1/50$. The Clausius–Mossotti approximation (1) is shown with dashed line. Our suggestion (1.6) is shown with solid line. Naive power-law extrapolation of the formula (1.3) to the whole region is shown with dotted line.

1.2 Mathematical foundations of constructive homogenization

1.2.1 Different approaches to heterogeneity in material sciences and in statistical mechanics

Statistical mechanics studies the behavior of macroscopic systems by investigation of their microscopic structures. Every structure, for instance an ensemble of particles, is considered as a probabilistic event and its macroscopic property as the mathematical expectation over the configuration space. The pairs $(\mathbf{x}, \mathbf{p})$ form a position and momentum space of particles. This yields a phase space which includes all possible states of a dynamic system. Formalism of statistical mechanics can be extended to various problems of physics, biology and geometry to study heterogeneous structures.

Self-assembly is the fundamental part of such a study common to static and dynamic systems of nature and technology (Whitesides and Grzybowski, 2002). Heterogeneity and self-assembly are discussed here from the viewpoint of composites, having in mind applications to material sciences.

Our approach is much different from the traditional methods used in statistical mechanics and material science. First, the methodological difference between statistical mechanics and material science should be outlined. Next, we explain the efficiency of our new method and unveil its potential. We do not discuss the dynamic mechanisms of the self-assembly and investigate only static "frozen" pictures to qualitatively analyze the degree of regularity and to classify heterogeneous structures. The main attention is paid to positions $\mathbf{x}$ and momentum $\mathbf{p}$ is neglected (with some exception). It is not an oversimplification of the physical theory, since another object, not a set of particles, but a structured continuum is discussed. We have to distinguish the meaning of *energy* in statistical mechanics and in material science. Some simplifications will be used below in order to explain the main idea of the theoretical distinction between these two topics. Statistical mechanics begins with chaotic motion of balls and their kinetic energy. The interaction energy is introduced through a potential function ϕ as

$$E_s = \frac{1}{2} \sum_{\mathbf{x} \neq \mathbf{y}} \phi(\|\mathbf{x} - \mathbf{y}\|), \tag{1.8}$$

where $\mathbf{x}$ and $\mathbf{y}$ run over the coordinates of particles. Consider a disperse composite with hard inclusions considered as the identical balls. In material sciences, the dissipation of energy E_c of medium around inclusions has to be determined. The quantity E_c determines the effective constants via Dirichlet's integral (Cherkaev, 2009) under normalization of the applied external field. One can see here the essential difference between mathematical problems of statistical mechanics and of heterogeneous media. The energy in statistical mechanics is given and the joint motion of particles is investigated by means of dynamic systems. The dissipation energy in heterogeneous media determines its macroscopic properties.

As an example, consider a simple sedimentation problem (Guazzelli and Morris, 2011) when a hard n balls of mass m fall along the axis x_3 under gravity in viscous fluid. In statistical mechanics, the energy of balls $E_s = nmgx_3$ is considered in weakly viscous medium when $E_s \gg E_c$. We stress that the energy is given here and the trajectory (simple trajectory along the axis x_3) has to be found. Next, the corresponding phase space can be studied in the framework of statistical mechanics. The theory of composites sets out to estimate the dissipation of energy E_c in the medium for any relation between E_s and E_c. The motion of balls is considered as an accompanying problem. In the case $E_s \ll E_c$, the particles are frozen in medium and a static boundary value problem is considered to determine the macroscopic behavior of suspension. The ultimate result in the theory of composites is an estimation of the quantity E_c.

In the present book, we estimate the dissipation of energy E_c for conducting, elastic and viscous media for dispersed composites. The classic boundary value problems are applied to stationary and dynamic processes following the traditional approach of stationary and quasi-stationary statements. This, in particular, means that a boundary value problem for every "frozen" geometry is considered independently of the previous ones. However, a set of such frames can be useful to study the dynamics of the process. Though problems of statistical mechanics and the theory of composites are different, their geometrical structure is a joint attribute of both theories. Some ideas originating in statistical mechanics reoccur throughout the book.

1.2.2 Random geometry of composites

The first step in material sciences is experimental. It is based on preliminary classification of observed structures. The simple classification divides composites on laminates, fibrous and dispersed. Such division corresponds to mathematical 1D, 2D and 3D problems. We follow such a mathematical simplification and set aside some interesting complex structures such as polycrystals, short fiber and multi-ply fibrous composites, etc. (Cherkaev, 2000, Milton, 2002). It follows from 1D homogenization that 1D randomness does not impact the effective properties. For instance, the longitudinal and transverse effective conductivity of laminates are calculated by the arithmetic and harmonic means, respectively, independently on the order and the widths of layers with prescribed phase concentrations. Randomness becomes essential in 2D (unidirectional fiber composites) and 3D cases (dispersed composites).

Usually, geometrical analysis in material sciences is reduced to observation and measurement of the "first-order" parameters such as the concentration of inclusions f, description of their shapes; and of the "second-order" parameters such as the correlation function describing the distribution of distances between inclusions. It is not sufficient to establish influence of randomness on the macroscopic behavior through the lower order parameters. The limitations of SCM to the first- and second-order approximations in f have been already discussed in Section 1.1.1.

Typically, generality of the description of random composites is declared and a universal formula is derived by SCM. Randomness in SCM is reflected in *existence of the concentration* f over the whole space which can be not properly defined from mathematical point of view. Frequently, the condition of *macroscopic isotropy* or of *local anisotropy* is added to the assumptions. These two axioms form the foundation of the SCM. Of course, as the consequence such too general assumptions lead to "discovery" of the universal properties of composites. The main universal property is that the coefficients in the expansion of the effective conductivity of macroscopically isotropic composites do not depend on distribution of inclusions. This is true, but only up to $O(f^3)$. The term f^3 does depend on geometry, in particular, on the joint probabilistic distributions of inclusions. Sometimes, higher order terms are estimated empirically, as some correction coefficients.

The triumphantly declared SCM formulas deceptively demonstrate universality of random composites. Frequently, they lead to misleading conclusions. As an example, we recall that the percolation threshold f_c of randomly packed inclusions was considered as the universal constant until it was shown the dependence of f_c on the protocol employed to produce the random packing (Torquato et al., 2000, Torquato and Stillinger, 2010).

In the same time, straightforward computations of the local physical fields for fixed geometries carry more geometrical information than observations. Specially fixed geometries have to be considered as a set of statistical realizations according to a given theoretical probabilistic distribution. However, presently it is impossible to carry out such straightforward computations based on the geometric configurations as outlined in Krauth (2006). Numerical examples given in literature represent some information about the variety of structures, but still not all of them. Frequently, study of random composites in material sciences begins with a specific picture with supposedly "randomly distributed" inclusions. Neglect of proper geometrical simulations can distort the final results in the numerical homogenization. It is similar to random number simulations when only proper generators yield correct sequences of random numbers.

The theory of spatial correlation functions can be considered as the theoretical basis of random composites (Torquato, 2002), since any random geometry of measured sets can be completely described by the infinite set of n-point correlation functions for $n = 1, 2, \ldots$. Consider, for instance, a two-phase composite with the concentrations of inclusions f. Its effective properties can be approximately found through the two 1-point correlation functions equal to constants f and $1 - f$. The 2-point correlation functions (auto-correlation functions) are used to construct the next approximation in contrast parameters. The integral moments of the 3-point correlation functions known as the Milton–Torquato parameters (Milton, 2002, Torquato, 2002) are used in the next approximations. Calculation of the n-point correlation functions for $n \geq 3$ is confronted with a hard computational problem.

1.2.3 The method of structural sums

The main goal of our approach is a decomposition of the dissipation energy E_c discussed in Section 1.2.1 onto the terms including only physical and only geometric parameters. It is equivalent to the decomposition of the effective properties tensor (Cherkaev, 2009), (Gluzman et al., 2017, Chapter 3) under a normalization of the external field. For simplicity, consider a macroscopically isotropic 2D two-phase dispersed conducting composites with the normalized matrix and inclusions scalar conductivity 1 and σ, respectively. Introduce the concentration of inclusions f and the contrast parameter (1.1). The normalized effective conductivity σ_e, the ratio of the effective conductivity to the matrix conductivity, can be written in the form

$$\sigma_e = \sum_{k=0}^{\infty} c_k(\mathbf{x}, \mathbf{r}, f)\varrho^k, \tag{1.9}$$

where the coefficients $c_k(\mathbf{x}, \mathbf{r}, f)$ depend on the location of inclusions $\mathbf{x}$, on their concentration f and do not depend on ϱ. The size parameter $\mathbf{r}$ is introduced here in order to take into account dispersity. For instance, $\mathbf{x}$ denotes the set of the disks centers and $\mathbf{r}$ the set of their radii for circular inclusions. Such a decomposition is called in the theory of composites the contrast expansion (Milton, 2002, Torquato, 2002). The convergence of the series (1.9) had been proved for $|\varrho| \leq 1$ (Mityushev, 1993) (see also (Gluzman et al., 2017, Chapter 2)). Let the coefficients $c_k(\mathbf{x}, \mathbf{r}, f)$ be expanded onto a series in f and the obtained double series is summed in ϱ. Then, we arrive at the concentration expansion

$$\sigma_e = \sum_{k=0}^{\infty} C_k(\mathbf{x}, \mathbf{r}, \varrho) f^k. \tag{1.10}$$

Sometimes, the concentration expansion is called the cluster expansion associated with the cluster method used to obtain explicitly the coefficients $C_k(\mathbf{x}, \mathbf{r}, \varrho)$. However, the cluster method frequently leads to misleading results. Its revision is outlined in Appendix A.4.

The expansions (1.9) and (1.10) can be obtained by the generalized alternating method of Schwarz and corresponding method of integral equations (Mityushev and Rogosin, 2000, Gluzman et al., 2017). The integral equations solved by the explicit iterative scheme yield the contrast expansion and by an implicit scheme the concentration expansion, for details see Mityushev (2015), Gluzman et al. (2017). Here, we want to stress that the method of Schwarz is computationally efficient for dispersed composites when interactions between inclusions can be exactly or approximately determined. It is the case when a single inclusion problem corresponding to an iteration in the method of Schwarz has a simple form solution. This is the reason why the method is effective, for instance, for circular and elliptic shapes of inclusions (Mityushev et al., 2018b).

The structure of the coefficients $C_k(\mathbf{x}, \mathbf{r}, \varrho)$ is much simpler than of $c_k(\mathbf{x}, \mathbf{r}, f)$. More precisely, every coefficient $C_k(\mathbf{x}, \mathbf{r}, \varrho)$ is a polynomial in ϱ

$$C_k(\mathbf{x}, \mathbf{r}, \varrho) = \sum_{j=1}^{k} e'_{kj}(\mathbf{x}, \mathbf{r})\varrho^j, \tag{1.11}$$

where the coefficients $e'_{kj}(\mathbf{x}, \mathbf{r})$ $(j = 1, 2, \ldots, k, k = 0, 1, 2, \ldots)$ depend only on the location of inclusions. Therefore, these coefficients completely determine the geometric factors. Geometric factors, in turn, determine the effective properties of composites.

The values $e'_{kj}(\mathbf{x}, \mathbf{r})$ are normalized and transformed into the multiple convolution sums called the structural sums (basic sums). The structural sums were introduced by Mityushev (2006). We now outline their theory following (Gluzman et al., 2017, Chapter 4). The computational algorithms will be described, following (Nawalaniec, 2016, 2017), in Chapter 2 of the present book. Moreover, the application of basic sums can be extended to other objects (shapes) by covering the considered shape with disks which form the cluster approximating the desired shape (Czapla, 2018a), (Czapla, 2018b, PhD Thesis), (Nawalaniec, 2019c).

In the beginning, we consider monodisperse composites with N identical circular non-overlapping inclusions arbitrary distributed in the periodicity cell. Let a_k denote the complex coordinate of the kth disk; q be a natural number; k_s run over 1 to N, $m_q = 2, 3, \ldots$. Let $\mathbf{C}$ denote the operator of complex conjugation. Introduce the multiple convolution sums called the structural sums based on the Eisenstein functions

$$\begin{aligned} e_{m_1,\ldots,m_q} &= \frac{1}{N^{1+\frac{1}{2}(m_1+\cdots+m_q)}} \\ &\times \sum_{k_0,k_1,\ldots,k_n} E_{m_1}(a_{k_0} - a_{k_1})\overline{E_{m_2}(a_{k_1} - a_{k_2})}\ldots \mathbf{C}^{q+1} E_{m_q}(a_{k_{q-1}} - a_{k_q}), \end{aligned} \tag{1.12}$$

where the Eisenstein functions $E_m(z)$ are introduced in Appendix A.2. Mind that the definition (A.32) is used for shortness.

For instance, the structural sums e_2 and e_{22} take the following form,

$$\begin{aligned} e_2 &= \frac{1}{N^2} \sum_{k_0=1}^{N} \sum_{k_1=1}^{N} E_2(a_{k_0} - a_{k_1}), \\ e_{22} &= \frac{1}{N^3} \sum_{k_0=1}^{N} \sum_{k_1=1}^{N} \sum_{k_2=1}^{N} E_2(a_{k_0} - a_{k_1})\overline{E_2(a_{k_1} - a_{k_2})}. \end{aligned} \tag{1.13}$$

It is worth noting that the triple sum e_{22} can be written as the double sum, see (2.18) for $p = 2$.

We now proceed to discuss polydisperse composites. Assume that inclusions are modeled by N non-overlapping disks of different radii r_j $(j = 1, 2, 3, \ldots, N)$ (see Fig. 1.3). Thus, the total concentration of inclusions equals

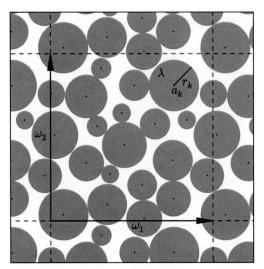

Figure 1.3 Polydisperse two-dimensional composite modeled as a two-periodic cell $Q_{(0,0)}$.

$$f = \pi \sum_{j=1}^{N} r_j^2.$$

Let $r = \max\limits_{1 \leq j \leq N} r_j$. Introduce the constants

$$f_j = \left(\frac{r_j}{r}\right)^2, \quad j = 1, 2, 3, \ldots, N, \tag{1.14}$$

describing polydispersity of inclusions. Then, the corresponding structural sums of the multi-order $\mathbf{p} = (p_1, \ldots, p_n)$ are defined by the following formulas

$$
\begin{aligned}
e_{p_1, p_2, p_3, \ldots, p_n}^{f_0, f_1, f_2, \ldots, f_n} = {} & \frac{1}{\eta^{1 + \frac{1}{2}(p_1 + \cdots + p_n)}} \sum_{k_0, k_1, \ldots, k_n} f_{k_0}^{t_0} f_{k_1}^{t_1} f_{k_2}^{t_2} \cdots f_{k_n}^{t_n} E_{p_1}(a_{k_0} - a_{k_1}) \\
& \times \overline{E_{p_2}(a_{k_1} - a_{k_2})} E_{p_3}(a_{k_2} - a_{k_3}) \cdots \mathbf{C}^{n+1} E_{p_n}(a_{k_{n-1}} - a_{k_n}),
\end{aligned}
\tag{1.15}
$$

where $\eta = \sum_{j=1}^{N} f_j$. The superscripts t_j $(j = 0, 1, 2 \ldots, n)$ are given by the recurrence relations

$$t_0 = 1,$$
$$t_j = p_j - t_{j-1}, \quad (j = 1, 2, \ldots, n) \tag{1.16}$$
$$t_n = 1. \tag{1.17}$$

In the case of equal disks, (1.15) becomes (1.12).

The structural e-sums (1.12) and (1.15) are expressed by means of the Eisenstein functions[3]. The e-sums (1.12) and (1.15) can be treated as basic because the 2D effective conductivity tensor of any composites with circular inclusions is a linear combination of the e-sums. The 2D elasticity and viscosity problems require introduction of the additional structural sums described below.

Besides the structural sums (1.15) based on the Eisenstein functions we need structural sums constructed from the Eisenstein–Natanson–Filshtinsky functions. Let j_s take the values 0 or 1, $k_s = 1, 2, \ldots, N$, $s = 1, 2, \ldots, q$, $\alpha = \sum_{s=1}^{q}(m_s - j_s)$ and $m_s = 2, 3, \ldots$. Introduce the structural sums

$$e_{\mathbf{m}}^{(\mathbf{j})(\mathbf{l})} \equiv e_{m_1,\ldots,m_q}^{(j_1,\ldots,j_q)(l_1,\ldots,l_q)} = \frac{1}{N^{\frac{\alpha}{2}+1}} \sum_{k_0,k_1,\ldots,k_q} \prod_{s=1}^{q} \mathbf{C}^{l_s-1} E_{m_s}^{(j_s)}(a_{k_{s-1}} - a_{k_s}), \quad (1.18)$$

where $E_m^{(j)}(z)$ denote the Eisenstein–Natanson–Filshtinsky functions discussed in Appendix A.3.

The notation (1.18) includes the structural sum (1.12) as the particular case when $j_1 = \ldots = j_q = 0$ and $l_s = 0$ for even s, $l_s = 1$ for odd s ($s = 1, 2, \ldots, q$). For instance, for even q we have

$$e_{m_1,\ldots,m_q} = e_{m_1,\ldots,m_q}^{(0,0,\ldots,0)(0,1,\ldots,0,1)}. \quad (1.19)$$

The structural sums (1.15) are used for conductivity problems governed by the Laplace equation; sums (1.15) and (1.18) are used for elasticity problems, governed by the bi-Laplacian equation (biharmonic functions). The systematic study of structural sums and the corresponding effective symbolic-numerical algorithms were accomplished in Czapla et al. (2012a,b), Mityushev and Nawalaniec (2015). We describe this computational approach in Chapter 2.

1.2.4 Random homogenization and ergodicity

It is worth to briefly review here the notion of *ergodicity*, used in various physical situations. Boltzmann's Ergodic Hypothesis central for statistical mechanics (Szász, 2000) declares that *for large systems of interacting particles in equilibrium time averages are close to the ensemble, or equilibrium average*. In the theory of composites, the term ergodicity is associated with random composite considered as a statistically homogeneous medium (Telega, 2004). In other words, the medium whose geometric statistical properties are invariant under translations (Telega, 2004), (Gluzman et al., 2017, Chapter 3).

Metastable phases and glasses defy ergodicity over long observation timescales. "Lars Onsager used to tell a story about a glycerin factory somewhere in Canada. One winter it was so cold that the glycerin froze, and from then on no matter how thoroughly the place was cleaned, it was impossible to get rid of all the nuclei of

---

[3] The notation e is hommage to Eisenstein.

solid glycerin. As a result, it was no longer possible to produce glycerin in its usual (metastable) liquid phase. They had to close the factory, he would say with his impish grin." (Penrose, 1995). Penrose interpreted Maxwell's approach to the theory of metastability as saying that thermodynamic functions for metastable phases are analytic continuations of those of stable phases. Some other method of extrapolation such as based on asymptotic expansions can be used as well. Since analytical continuation is a form of extrapolation, particular care should be taken of the uncontrolled errors introduced by extrapolation on the sought continued quantities (Penrose, 1995). It is our understanding that the crossover approach wherever possible would help to achieve the latter goal.

It is possible (but not rigorous), both in Monte-Carlo simulations and theoretically, to extend the density expansion valid for the stable liquid phase, to the metastable region of concentrations till the very end at random close packing (RCP). It is widely believed that at many of the RCP protocols at $f = f_c \approx 0.64$, there is a simple pole in the compressibility factor $\mathcal{Z}(f)$ of hard sphere liquid. Many different algorithms that yield the RCP pole around 0.64, were all designed to avoid crystallization and follow a metastable branch (Brady, 1993, Speedy, 1994, Wu and Sadus, 2005, Kamien and Liu, 2007, Torquato, 2018). Thus, there is an understanding that the hard spheres equation of state may be extrapolated smoothly from the expansions at small densities, avoiding the first-order freezing transition, to describe the metastable fluid branch, and end at the first-order pole at RCP.

There is also an alternative approach to RCP through jamming phenomena, based on studying the ensemble of states that are at the threshold of jamming. For a given packing protocol it aims to understand typical configurations and their frequency of occurrence. Both the bulk and shear moduli are shown to have their thresholds at the same packing density. It places the maximally random jammed packing fraction to be where the highest number of initial states have final states at the jamming threshold. In the thermodynamic limit it appears that the width of the distribution of jamming thresholds approaches zero. And all of the configurations, which were sampled randomly, are jammed at the same packing fraction. The value of this packing fraction is associated with random close-packing (O' Hern et al., 2003, 2006). RCP is useful for practical purposes and is widely used, but maybe ill-defined. Instead of RCP it is suggested to consider "maximally random jammed state". For protocols with varying compression rate, the most rapid compression of the liquid would end in mechanically stable packing, which is considered as the maximally random jammed state (Torquato, 2002).

In the theory of composites, ergodicity is restricted to homogenization of random media when the average over the whole space is reduced to the average over a *representative volume element (RVE)* . In the context of the constructive homogenization, ergodicity was discussed in Mityushev (1999), where it was shown that the correctly defined concentration does not guarantee the existence of the effective conductivity. Its coefficient in f^2 depends on a series[4] which can be not properly defined. The theory of homogenization discusses only statistically homogeneous media (Jikov et al.,

---

[4] Associated to Rayleigh's type series $S_2 = \sum_{k=1}^{\infty} a_k^{-2}$ for non-regular locations of inclusions.

1994, Telega, 2004). In such medium, there exists an RVE and the effective constants are determined via solution to a boundary value problem for the RVE. The theory of homogenization is consistent with the physical theory, e.g., the MMM principle by Hashin (1983) outlined in Gluzman et al. (2017, Chapter 3).

When the mathematical theory of existence is formulated and uniqueness is established, one can develop some constructive procedure to describe the RVE and to calculate the effective properties of composites. Discussion of extensive, purely numerical methods is outside of our current scope. The current book, as well as its predecessor (Gluzman et al., 2017), are devoted to a constructive analytical approach. It includes as the foundation the powerful method of structural sums for classification of dispersed composites and computation of their effective constants.

The essence of our approach to introduce randomness into geometry, consists in the "delayed" introduction of the random variables. Almost all studies, introduce first a random representative cell, e.g., Ostoja-Starzewski (2008), and deal with random variables from the beginning. We introduce the set $(\mathbf{x}, \mathbf{r})$ per periodicity cell in a symbolic form. One can assume that the set $(\mathbf{x}, \mathbf{r})$ describes a deterministic geometry or a random geometry obeying a prescribed probabilistic joint distribution of locations and sizes. If we work with symbols and do not use the probabilistic operations like the mathematical expectation, it does not matter whether $(\mathbf{x}, \mathbf{r})$ is deterministic or probabilistic variables.

Let the elements $e'_{kj}(\mathbf{x}, \mathbf{r})$ be constructed in a symbolic form. Only then we introduce the probabilistic distribution of $(\mathbf{x}, \mathbf{r})$, and calculate the mathematical expectation $e''_{kj} = \langle e'_{kj}(\mathbf{x}, \mathbf{r}) \rangle$. The elements e''_{kj} are written in a dimensionless form in Chapter 2 in terms of the structural sums $e_{\mathbf{m}}$ with the multi-index $\mathbf{m}$. The infinite set $\{e_{\mathbf{m}}\}$ completely determines the random structure of the considered composites. Hence, their effective properties can be computed.

Practical applications of the proposed method begins with the solution to the Riemann–Hilbert and $\mathbb{R}$-linear problem for an arbitrary multiply circular domain (Mityushev, 1993, Mityushev and Rogosin, 2000), see Gluzman et al. (2017, Chapter 2). The next step consists in extension of the method to arbitrary shapes. Such an extension is possible by covering the considered domain by an appropriate cluster of disks, i.e., by circle packing.

The approximation of different objects by configurations of circles is useful in analysis of biophysical data, see examples of modeling a collective behavior of bacteria by stadiums and segments in Czapla (2018b). An efficient algorithm was developed to compute $e_{\mathbf{m}}$ of the almost linear complexity in practice, and the quadratic complexity in theory, see Chapter 2. Of course, the set $\{e_{\mathbf{m}}\}$ is truncated in computations to reach the precision $O(f^M)$ in (1.10), where M is usually taken from 3 to 26 in various random models. An alternative way to introduce and study the shape, is the method of conformal mapping (Pesetskaya et al., 2018) not discussed in the present book.

Even consistently derived power series in concentration and contrast parameters may be not sufficient, because most often the series are too short and unstable. In other cases they do not converge fast enough, or even diverge in the most interesting regimes. The typical answer to the challenge is to apply additional methods, powerful enough to extract information from the series. In addition to a traditional Padé approximants applied in such cases (Awrejcewicz et al., 2018), we would need a post-Padé approximants and techniques for analysis of the divergent or poorly convergent series. Thus, the next important step is asymptotically equivalent transformation of the obtained polynomial

$$\langle \sigma_e \rangle = \sum_{k=0}^{M} \langle C_k(\mathbf{x}, \mathbf{r}, \varrho) \rangle f^k + O(f^{M+1}). \tag{1.20}$$

The method of structural sums was applied to construction of the RVE following the scheme of Mityushev (2006). For a given finite set of $e_\mathbf{m}$ ($|\mathbf{m}| \leq M$) constructed for a composite, one has to find the location and size, i.e., the set $(\mathbf{x}, \mathbf{r})$ with the minimal inclusions per cell for which the structural sums $\widetilde{e}_\mathbf{m}$ coincide with a prescribed accuracy with the corresponding given values $e_\mathbf{m}$ ($|\mathbf{m}| \leq M$). Any two given sets of structural sums can be compared to establish whether the considered two geometric composites refer to the same random structure, see examples (Czapla and Mityushev, 2017, Nawalaniec, 2019c).

In order to demonstrate the superior efficiency of our method for random composites, we have to compare it with the previously applied methods. As it is noted above any self-consistent method holds at most up to $O(f^3)$. Though SCM method can be applied to study general physical fields in composites, its geometrical limitations do not allow to apply the effective properties formulas to the non-dilute random composites.

The method of structural sums is related to the method of multi-point correlation functions (Torquato, 2002) in the following way. Assume that we do possess such multi-point correlation functions. Then, it is possible to plug them into integrals entering in the contrast expansions of the effective constants (see for instance Section 20.1.2 from Torquato (2002)). The structural sums are linear combinations of these integrals. Hence, the hard computations needed to determine high order correlation functions and to calculate the corresponding integrals may be skipped if we know structural sums!

Besides computational advantages, our method gives simple and quick answer to the question about macroscopic isotropy of composites. In order to establish the isotropy by means of the two-point correlation functions one has to find that these functions depend only on the distance between points. The same task when addressed by means of structural sums is reduced to checking the relation $e_2 = \pi$, see Chapter 2.

The expansions (1.9)–(1.11) are written for 2D conductivity with the contrast parameter (1.1). They were systematically discussed in (Gluzman et al., 2017). Analogous expansions for the 2D elastic composites with two contrast parameters are discussed in present book. The development of the mathematical method of functional equations for 2D elastic composites is presented in Chapters 3 and 4.

We now proceed to discuss applications of stochastic geometry to random composites. The general geometrical theory is based on theories of measure and random fields (Adler and Taylor, 2008). Spatial point processes (Illian et al., 2008, Chiu et al., 2013) can be considered as a constructive computational method to describe the arrangement of randomly distributed objects.

Stochastic geometry incorporates a lot of interesting models. However, only part of them can be used in computational material science. We believe that the only applicable patterns constitute samples of *hard-core point processes* in which the constituent points are forbidden to get closer than a certain positive minimum distance (Chiu et al., 2013, section 5.4). Application of non-hard-core models may result in appearance of relatively small inclusions. Taking the interactions with such inclusions into account may introduce computational difficulties or unnecessary increase of the complexity of calculations with a negligible influence on the effective properties.

Another computational model is presented in Krauth (2006), where hard disks are applied to simulations of systems based on Newtonian deterministic mechanics and, in general, on the method of molecular dynamics. Though the computer simulations (Krauth, 2006) were discussed in context of statistical mechanics, the described geometrical distributions perfectly fit the homogenization theory, see brief discussion below.

A simple computational methodology to determine representative ensembles of particles in composites has to follow the general method of statistical mechanics (Krauth, 2006), but neglect mechanical (elastic) collisions between hard particles. A sufficient approach may incorporate a purely geometrical random walk algorithms representing mechanical stirring of inclusions, see for instance algorithm presented in Chapter 2. It is crucial to perform such Monte Carlo simulations in the parallelogram with boundary conditions providing periodicity of the cell, i.e. in torus topology.

In order to achieve high concentrations of inclusions, one can apply protocols being combination of packing algorithms and random walks, with initial configurations of smaller density being either a regular arrays of inclusions or a result of various protocols, e.g., Random Sequential Adsorption (RSA) and others discussed in Section 2.6 in Chapter 2. To the best of our knowledge the pure geometric simulations were not properly applied to random composites in practice. Frequently, engineers do not apply proper random geometrical simulations taking only single "random picture". Such a naive approach fails to recognize the diversity of random structures.

1.3   Asymptotic methods. Critical index

We are interested in the behavior of a real function $\Phi(x)$ of a real variable $x \in [0, \infty)$. Let this function be defined by a complicated problem that does not allow for an explicit derivation of the function form. But what can be done is only the use of some kind of perturbation theory yielding asymptotic expansions representing the function. Note that exact solutions are exceptionally rare and even then mostly represented as series. Our task is to obtain an accurate approximate solution represented as series, and recast them into convergent expressions by means of analytically expressed approximants.

The simplest way to make sense of such perturbative results or extrapolate (Baker and Graves-Moris, 1996), is to apply the Padé approximants $P_{n,m}(x)$. However, solutions to many problems, e.g., nonlinear equations exhibit irrational functional behavior, which cannot be well described by Padé approximants or, sometimes, these are not applicable at all. Approximate solutions in the class of irrational functions can be constructed by invoking such irrational functional forms as self-similar root approximants (Gluzman et al., 2017). In principle, it could be possible to employ the additive self-similar approximants as such. But it would be unreasonable to forget the well developed techniques of Padé approximants. So, the question has been advanced whether it would be possible to modify the use of Padé approximants in such a way that to extend their applicability to the class of irrational functions. The desired modification can be performed by splitting the sought solution into two factors, one, represented by iterated root approximants or factor approximants (Gluzman et al., 2017), taking care of the irrational part of the solution, and the other being a Padé approximant, characterizing the rational part of the solution. The so corrected Padé approximants are applicable to a larger class of problems and are well defined even for those cases, where the standard Padé approximants cannot be used (Gluzman and Yukalov, 2016, Gluzman et al., 2017).

The key to the success is the choice of the control function as the self-similar approximant, which not only should capture the essence of the irrational part of the given function, but must also secure convergence of the standard Padé approximants for the remaining rational part of the function. Moreover, both the boundary conditions at small and large values of the independent variable must be satisfied. When detailed information on large x behavior is available one can employ as control function more sophisticated self-similar roots, or additive self-similar approximants (Gluzman et al., 2017). Sometimes, instead of the multiplicative corrections an additive corrections ansatz will be employed, in particular for calculation of the higher-order critical amplitudes. The idea of "correction" can be adapted for calculating the critical index and threshold. Introduction of the additive approximants was motivated by the problems appearing in theory of regular composite materials. But they appeared useful for the most typical field-theoretical problems.

Extrapolation methods attempt to, from the knowledge of several terms of an asymptotic expansion (1.21), as $x \to 0$, to find the limit corresponding to $x \to \infty$. The variable $x > 0$ can represent, e.g., a coupling constant or concentration properly transformed. Interpolation problem consists in constructing such a representation for the sought function $\Phi(x)$ that would reproduce the small-variable, as well as large-variable expansions, providing an accurate approximation for the whole domain $[0, \infty]$.

The Padé approximants $P_{n,m}$ are nothing else but ratio of the two polynomials $Q_n(x)$ and $P_m(x)$ of the order n and m, respectively. If $m = n$ the approximant is called a diagonal Padé approximant of order n. Usually, $P_m(0) = 1$. The coefficients of the polynomials are derived directly from the coefficients of the given power series (Baker and Graves-Moris, 1996) from the requirement of asymptotic equivalence to the given series or function $\Phi(x)$. When there is a need to stress the last point, we simply write $Pade Approximant [\Phi[x], n, m]$, adopting a reduced notation from *Mathematica*.

The Padé approximants locally are the best rational approximations of power series. Their poles determine singular points of the approximated functions (Baker and Graves-Moris, 1996). But the functions in the vicinity of their critical points, in general, are non-rational. Therefore the direct use of Padé approximants for functions exhibiting critical behavior is impossible. A Padé approximant can have a pole that could be associated with a finite critical point, but the related critical index would be an integer, while usually critical indices are not integers. The same concerns the large-variable behavior where the power of x is always an integer. Another benefit of calculations with Padé approximants is that they can be performed in very high orders with *Mathematica*. The holomorphy of diagonal Padé approximants in a given domain implies their uniform convergence inside this domain (Gonchar, 2011).

When the character of the large-variable limit is known, one can invoke the two-point Padé approximants . Two-point Padé are applied when in addition to the expansion about $x_0 = 0$,

$$\Phi(x) \sim \sum_{n=0}^{\infty} c_n x^n, \tag{1.21}$$

an additional information is available and contained in the expansion about $x = \infty$,

$$\Phi(x) \sim \sum_{n=0}^{\infty} b_n x^{-n}. \tag{1.22}$$

A two-point Padé approximant to $\Phi(x)$ is a rational function

$$F(x) = \frac{Q_n(x)}{P_m(x)}. \tag{1.23}$$

The polynomials of degrees n and m have their coefficients chosen arbitrarily to make the first J terms of the Taylor series of $F(x)$ around $x = 0$ agree with the expansion (1.21), and the first L terms of the expansion of $F(x)$ about $x = \infty$ to agree with the given expansion (1.23), so that also $J + L = n + m + 1$.

The self-similar approximants allow to extrapolate and interpolate between the small-variable and large-variable expansions. In particular, based on root approximants one can construct an analytical expression uniformly approximating the sought function in the whole domain $[0, \infty]$ and reproducing both the small-variable and large-variable expansions (Gluzman and Yukalov, 2015, Gluzman et al., 2017). Importantly the uniqueness conditions are satisfied. The method of root approximants is more general than that of standard Padé approximants as well as the Baker–Gammel method of fractional Padé approximants. By construction, the accuracy of the method is not worse, and often better, than that of the latter methods.

In what follows we would also attempt to compute the coefficients in the expansion using various techniques motivated by the renormalization group. We are inspired here by Richard Feynman, who hoped that in perturbative theories it would be possible to estimate the result for the coefficient in a given order, without the brute force evaluation of all the diagrams contributing in this order. The complexity of calculations increases very rapidly in high orders, and even a way of determining the sign of the contribution would be useful (Samuel and Li, 1994).

1.3.1   Method of self-similar renormalization

The description of the method with the corresponding mathematical details and complete referencing can be found in Gluzman et al. (2017). In this section we provide the ingredients which are necessary for a self-contained understanding of this book (Gluzman and Yukalov, 1997, Yukalov and Gluzman, 1997b, 1998).

Let us consider a function $\Phi(x)$ of the variable $x \in [0, \infty)$. Let this function satisfy a complicated equation that cannot be solved exactly. Assume that by means of perturbation theory we can get a sequence of perturbative approximations $p_k(x)$, where $k = 0, 1, 2...$, enumerates the approximation order. Usually, perturbation sequences are divergent. To extract a meaningful result from a divergent sequence, one has to use so-called resummation techniques. In the method of self-similar renormalization, a divergent sequence can be made convergent by introducing additional functions ensuring convergence. Because of their role, these functions are called control functions. Let m be a set of such control functions entering into a sequence $\{F_k(x, m)\}$ obtained by a perturbation algorithm. In addition to introducing the control functions, the main idea of the method of self-similar renormalization is to treat the passage from one approximation to another as a motion with respect to the approximation number $k = 0, 1, 2, ...$ considered as an effective time variable. This motion is realized in the functional space of the considered function.

Let us define the initial approximation

$$F_0(x, m) = \phi \qquad (1.24)$$

as an equation for the expansion function $x = x(\phi, m)$. Substitute the latter back to F_k, so that

$$y_k(\phi, m) \equiv F_k(x(\phi, m), m). \qquad (1.25)$$

The relation inverse to (1.25) is

$$F_k(x, m) = y_k(F_0(x, m), m). \qquad (1.26)$$

Let $\{y_k\}$ form a group of transformations with respect to $k = 0, 1, 2....$ Then, the trajectory $\{y_k(f, m)\}$ of this dynamical system, according to definitions (1.25) and (1.26), is in one-to-one correspondence to the approximation sequence $\{F_k(x, m)\}$. This dynamical system with discrete time k is called the approximation cascade. The attracting fixed point of the cascade trajectory is, by construction, is in a one-to-one correspondence to the limit of the approximation sequence $\{F_k(x, m)\}$, and thus corresponds to the sought function.

Technically is easier to deal with continuous than with discrete time. Let us embed the approximation cascade $\{y_k\}$ into an approximation flow $\{y(\tau, ...)\}$ with continuous time $\tau \geq 0$. This implies that the trajectory $\{y(\tau, \phi, m)\}$ of the flow passes through all the points of the cascade trajectory at the integer times $\tau = k = 0, 1, 2, ...$,

$$y(\tau, \phi, m) = y_k(\phi, m). \qquad (1.27)$$

The evolution equation for the flow reads

$$\frac{\partial}{\partial \tau} y(\tau, \phi, m) = v(y(\tau, \phi, m)), \tag{1.28}$$

with the right-hand side being the velocity field. The latter, in the language of renormalization-group theory, is called the β-function. Integrating the evolution equation (1.28) from $\tau = k$ to $\tau = k^*$, we get the evolution integral

$$\int_{y_k}^{y_{k+1}^*} \frac{d\phi}{v(\phi, m)} = k^* - k, \tag{1.29}$$

in which $y_k = y(k, \phi, m)$ and $y_{k+1}^* = y(k^*, \phi, m)$. Before specifying the numbers k and k^* in the limits of the evolution integral, let us note that the differential form (1.28) of the evolution equation, or its integral form (1.29), are equivalent to the functional relation

$$y \left(\tau + \tau', \phi, m \right) = y(\tau, y(\tau', \phi, m), m). \tag{1.30}$$

The latter is called the self-similarity relation, which explains our terminology. In general, the self-similarity can occur with respect to motion over different parameters. In our case, this is the motion over the steps of a computational procedure, the number of steps playing the role of effective time.

If there exists an attractive fixed point of the approximation-flow trajectory, then it is always possible to find a number k^* in the evolution integral (1.29) such that the upper limit y_k^* would correspond to an expression

$$F_k^*(x, m) \equiv y \left(k^*, F_0(x, m), m \right), \tag{1.31}$$

representing, with the desired accuracy, the sought function $\Phi(x)$. If y_k^* was an exact fixed point, then (1.31) would give an exact answer to the problem. However, a fixed point can be reached only after infinite number of steps $k \to \infty$. For a finite number k, the limit y_k^* may represent the fixed point only approximately, and is thus named the quasi-fixed point. Our aim is to reach the latter during the minimal time

$$\tau_k^* = \min(k^* - k), \tag{1.32}$$

or the minimal number of steps. When there are no additional restrictions, the minimal number of steps counted by k is 1, so that

$$|\{\min\}\tau_k^*| = 1. \tag{1.33}$$

In the case where some constraints are imposed on the motion, the minimal time (1.32) should correspond to a conditional minimum. In practice, if a value $\Phi_0 \equiv \Phi(x_0)$ of the sought function $\Phi(x)$ is given for some x_0, then we can find τ_k^* by requiring that the trajectory of the approximation cascade should pass through the given point Φ_0.

To calculate the evolution integral (1.29), we need to know explicitly the velocity field. This can be done approximately, by the Euler discretization of (1.28) yielding

$$v_k(\phi, m) = y_k(\phi, m) - y_{k-1}(\phi, m). \tag{1.34}$$

Substituting (1.34) into (1.29), and using (1.26), we come to the representation

$$\int_{F_k}^{F_{k+1}^*} \frac{d\phi}{v_{k+1}(\phi, m)} = \tau_k^*, \tag{1.35}$$

for the evolution integral (1.29), where $F_k = F_k(x, m)$, $F_{k+1}^* = F_{k+1}^*(x, m)$.

Finally, we have to define the set m of control functions, whose role is to govern the convergence of the approximation sequence. This convergence can be expressed, in the language of dynamical theory, as the stability of the cascade trajectory. A useful tool for analyzing stability is the set $\{\mu_k\}$ of the local multipliers

$$\mu_k(\phi, m) = \frac{\partial}{\partial \phi} y_k(\phi, m). \tag{1.36}$$

The inequality

$$|\mu_k(\phi, m)| < 1 \tag{1.37}$$

is the condition of local stability at the step k with respect to the variation of an initial point ϕ. The equality $|\mu_k(\phi, m)| = 1$ implies local neutral stability. For a convergent sequence corresponding to a contracting mapping, the condition of asymptotic stability is

$$|\mu_k(\phi, m)| \to 0, \text{ as } k \to \infty. \tag{1.38}$$

The approximation cascade $\{y_k\}$ describes the motion in the functional space $\{\phi\}$. To return to the domain $\{x\}$, we must use the inverse transformation (1.26) which allows to pass from the multiplier (1.36) given on the functional space $\{\phi\}$ to its image

$$\mu_k(x, m) = \mu_k(F_0(x, m), m) \tag{1.39}$$

being a function of x. For the image (1.39), the same stability condition as in (1.37) can be written,

$$|\mu_k(x, m)| < 1. \tag{1.40}$$

According to (1.38), the local multipliers decrease when approaching an attracting fixed point. This means that the variation of the initial condition ϕ produces weaker effects on the trajectory as the attractor becomes closer. In other words, the smaller are the absolute values of the multipliers, the more stable is the trajectory. Therefore, it is reasonable to define the control functions as those minimizing the absolute values of the local multipliers, ensuring that the trajectory becomes more stable at each step k. In this way, a set m of control functions is defined by the principle of maximal stability written as

$$|\mu_k(x, m_k(x))| = \min_m |\mu_k(x, m)|. \tag{1.41}$$

The control functions $m_k(x)$ defined by the principle (1.41) may be called the stabilizing functions or stabilizers.

After the stabilizers are defined, we have to substitute them into the corresponding approximations $F_k(x, m)$ getting

$$\Phi_k(x) \equiv F_k(x, m_k(x)).\tag{1.42}$$

This stage can be called the stabilizing renormalization of a perturbative sequence. Then, considering the motion near the renormalized quantity (1.42) by means of the evolution integral (1.81), we obtain

$$\Phi_k^*(x) \equiv F_k^*(x, m_k(x)).\tag{1.43}$$

This step can be called the dynamical renormalization. And the whole procedure of the double renormalization (1.42) and (1.43) is named the self-similar renormalization.

The evolution equation (1.28) is generally nonlinear and can have several different solutions leading to different self-similar approximations (1.43). In such a case, to select a physically meaningful solution, we need to involve additional conditions as constraints. The constraints can involve the properties of symmetry, the asymptotic properties as $x \to 0$ or (and) $x \to \infty$, sum rules or other relations containing some known information on the character of the sought solution. Such additional constraints narrow down the set of possible solutions to a class with desired properties.

Since our goal is to obtain a good accuracy for the sought function from just a few available perturbative terms, one is obliged to invent some tricks which amount effectively to increase the perturbation order. Here, we present the simplest and natural algebraic method. In a nutshell, we style the renormalization as a specific critical phenomenon, where control functions are introduced to mimic critical indices and amplitudes. In turn, when applied to the real critical phenomena, we are able to calculate the indices and amplitudes! Significant novelty is in equating critical index with the control function. The name algebraic self-similar renormalization method has been used to refer to this choice wherein the control functions are introduced in the exponents of perturbative polynomials.

Suppose that there is a sequence of approximations $\{p_k(x)\}$ having polynomial structure (1.21), k showing the order of the polynomial. This order can be effectively increased by means of the multiplicative transformation

$$P_k(x, m) = x^m\, p_k(x), \quad m \geq 0.\tag{1.44}$$

Then, the order of the expression (1.44) becomes $k + m$. The transformation inverse to (1.44) is

$$p_k(x) = x^{-m}\, P_k(x, m).\tag{1.45}$$

Following the technique described above, we consider the sequence $\{P_k(x, m)\}$ and construct an approximation cascade $\{y_k\}$ whose trajectory $\{y_k(\phi, m)\}$ is bijective to $\{P_k(x, m)\}$. Solving the evolution integral (1.35), we have $P_k^*(x, m)$. From the principle of maximal stability (1.41) we define the stabilizers $m_k(x)$. Substituting these into

$P_k^*(x, m)$ and applying the inverse transformation (1.45), we obtain the self-similar approximation

$$\Phi_k^*(x) = x^{-m_k(x)} P_k^*(x, m_k(x)). \tag{1.46}$$

The multiplicative transformation (1.44) is natural one in the case when the perturbative approximations $p_k(x)$ have the form of polynomials or series of a variable with not necessarily integer powers. The factor x^m effectively increases the approximation order, and m plays simultaneously the role of a stabilizer. The power or effective order m can be dictated by the principle of maximal stability selecting the most stable trajectory of the approximations cascade.

In particular, it may happen that $m = 0$, and we do not need to proceed further, or we may have to go to the limit of $m \to \infty$, thus allowing for all approximation orders. In each concrete case, the effective order we need to reach depends on how good is the perturbative sequence $\{p_k(x)\}$ we start with and, how much information can be extracted from its truncations by means of the renormalization (1.42) and (1.43). The principle of maximal stability (1.41) defines the stabilizers $m_k(x)$, whose explicit expressions are expressed only through the coefficients c_n.

The procedure works as follows for the polynomials of the order k,

$$p_k(x) = \sum_{n=0}^{k} c_n x^n, \quad c_n \neq 0. \tag{1.47}$$

Following (1.44), define

$$P_k(x, m) = \sum_{n=0}^{k} c_n x^{n+m}. \tag{1.48}$$

Similarly to (1.24), we have

$$P_0(x, m) = c_0 x^m = \phi, \tag{1.49}$$

from which the expansion function is

$$x(\phi, m) = \left(\frac{\phi}{c_0}\right)^{1/m}. \tag{1.50}$$

The definition (1.25) yields the points

$$y_k(\phi, m) = \sum_{n=0}^{k} c_n \left(\frac{\phi}{c_0}\right)^{n/m+1} \tag{1.51}$$

of the approximation-cascade trajectory. For the velocity field (1.34), we get

$$v_{k+1}(\phi, m) = c_{k+1} \left(\frac{\phi}{c_0}\right)^{\frac{k+1}{m}+1}. \tag{1.52}$$

From the evolution integral (1.35), we find the final approximation

$$P_{k+1}^* = \frac{P_k}{\left(1 - \frac{(k+1)\, c_{k+1}\, \tau_k^*}{m\, c_0^{\frac{k+1}{m}+1}}\, P_k^{\frac{k+1}{m}}\right)^{\frac{m}{k+1}}}. \tag{1.53}$$

The multiplier (1.36) becomes

$$\mu_k(\phi, m) = \sum_{n=0}^{k} \frac{c_n}{c_0} \left(1 + \frac{n}{m}\right) \left(\frac{\phi}{c_0}\right)^{\frac{n}{m}}, \tag{1.54}$$

and its image (1.39) reads

$$\mu_k(x, m) = \sum_{n=0}^{k} \frac{c_n}{c_0} \left(1 + \frac{n}{m}\right) x^n. \tag{1.55}$$

According to the transformations (1.44)–(1.46), we obtain from (1.53)

$$\Phi_{k+1}^*(x) = \frac{p_k(x)}{\left(1 - \frac{(k+1)\, c_{k+1}\, \tau_k^*}{m\, c_0^{\frac{k+1}{m}+1}}\, x^{k+1} p_k(x)^{\frac{k+1}{m}}\right)^{\frac{m}{k+1}}}, \tag{1.56}$$

where $m \equiv m_k(x)$ defines the most stable trajectory. When there are no additional conditions, one can set $\tau_k^* = 1$, as in (1.33). Formula (1.56) is generic. The beauty and power of the algebraic renormalization is precisely due to the possibility to perform all necessary transformations explicitly and find the final formula analytically.

It may happen that the most stable trajectory corresponds to $m \to \infty$. It is straightforward to check that the limit of the right side in (1.56), as $m \to \infty$, leads to

$$\Phi_{k+1}^*(x) = p_k(x) \exp\left(\frac{c_{k+1}}{c_0} x^{k+1}\right). \tag{1.57}$$

One may notice that, renormalizing $p_k(x)$ in (1.57), we obtain the recurrence relation

$$\Phi_{k+1}^*(x) = \Phi_k^*(x) \exp\left(\frac{c_{k+1}}{c_0} x^{k+1}\right). \tag{1.58}$$

It is possible then to repeat the self-similar renormalization several times, which is useful when working with high-order terms. Comparing (1.56) with (1.57), we see that the self-similar renormalization can yield quite different expressions, from the fractional form to exponential one. Accomplishing exponential renormalization of all sums appearing in expression of type (1.58), we follow the bootstrap procedure (Yukalov and Gluzman, 1997b) according. When all polynomial expressions emerging on the way are transformed following the rule expressed by (1.58), we arrive to the self-similar exponential approximants discovered by Euler (Bender and Vinson, 1996, Yukalov and Gluzman, 1997b). Euler exponential approximants appear to be deeply connected with the self-similar approximation theory, algebraic renormalization and bootstrap. When the subsequent self-similar approximants $\Phi_k^*(x)$ display substantial scatter it is proved effective to introduce the average from (Yukalov and Gluzman, 1999b)

$$\langle \Phi_k^*(x) \rangle = \sum_{i=1}^{k} \mathsf{p}_i(x) \Phi_i^*(x), \qquad (1.59)$$

where $\Phi_i^*(x)$ are weighted with probabilities

$$\mathsf{p}_i(x) = \frac{|\mathcal{M}_i(x)|^{-1}}{\sum_{j=1}^{N} |\mathcal{M}_j(x)|^{-1}},$$

in which N is the number of approximants involved and

$$\mathcal{M}_i(x) = \frac{\frac{\partial}{\partial x} \Phi_i^*(x)}{\frac{\partial}{\partial x} \Phi_0^*(x)},$$

are the mapping multipliers (Yukalov and Gluzman, 1999b). The result of averaging does not depend on the "initial condition" $\Phi_0^*(x)$. It can be chosen simply as a linear function. Also, there is no principal difference between multipliers introduced for polynomial and non-polynomial expressions, and averaging may be considered over solutions given by the non-renormalized, original series (with varying order) and renormalized expressions obtained from the series.

The technique based on multipliers and stability is quite general, but also cumbersome. To simplify and standardize calculations some more powerful than exponential or Padé constructs, the so-called self-similar factor approximants have been based on the general ideas presented above by Gluzman et al. (2003). The k-th order self-similar factor approximant reads as

$$\mathcal{F}_k^*(x) = c_0 \prod_{i=1}^{N_k} (1 + \mathcal{P}_i x)^{m_i}, \qquad (1.60)$$

where

$$N_k = \frac{k}{2}, \quad k = 2, 4, \ldots; \quad N_k = \frac{k+1}{2}, \quad k = 3, 5, \ldots \qquad (1.61)$$

and parameters $\mathcal{P}_i$ and m_i are defined typically from the standard procedure, solely from various constraints on the sought function, e.g., the asymptotic properties at $x \to 0$ or (and) $x \to \infty$.

For example, parameters $\mathcal{P}_i$ and m_i can be found by expanding expression (1.60) in powers of x, comparing the latter expansion with the given power series for small x, and equating the like terms in these expansions. When the approximation order $k = 2p$ is even, the above procedure uniquely defines all $2p$ parameters. When the approximation order $k = 2p + 1$ is odd, one sets some additional condition dictated by the problem specifics, and uniquely defining all other parameters. Fixing the parameters to some known or plausible values can be very helpful for short series, as is often the case. The factor approximants (1.60) may have singularities when some $\mathcal{P}_i$ and m_i are negative. Different from factors, self-similar approximants are available, see (Gluzman et al., 2017) and next subsections.

1.3.2   Critical index. Direct methods

The method of self-similar root approximants (Gluzman and Yukalov, 1998, Gluzman et al., 2017), provides accurate interpolating formulas for functions for which small-variable expansions are given and the behavior of the functions at large variables is known (Gluzman and Yukalov, 2015). This method can be generalized for the purpose of extrapolating small-variable expansions to the region of finite and large variables, where the sought function exhibits critical behavior (Gluzman and Yukalov, 2017b). The method leads to the procedure of calculating critical indices as formulated below.

A real function $\Phi(x)$ of a real variable x is said to exhibit critical behavior, with a critical index β[5], at a finite critical point x_c, when in the vicinity of this point it behaves as

$$\Phi(x) \simeq A(x_c - x)^\beta, \text{ as } x \to x_c - 0. \tag{1.62}$$

The function can tend to infinity, if the critical index β is negative, or to zero, if this index is positive. When the value of critical index is known in advance, the problem consists in finding the critical amplitude A. In Gluzman et al. (2017), various ways to calculate A were discussed. In the present book, we pay more attention to calculation of the critical index.

The critical behavior can also occur at infinity, where the function behaves as

$$\Phi(x) \simeq Ax^\beta, \text{ as } x \to \infty, \tag{1.63}$$

with the critical index β. Respectively, the function can tend to infinity, if β is positive and to zero, if β is negative. The critical behavior at infinity can formally be interpreted as the case, where the critical point is located at infinity.

Critical phenomena are widespread in physics and material science. And it is an important problem of defining the related critical indices. However, for realistic systems one can only resort to perturbation theory for obtaining the behavior of the sought function at small variable,

$$\Phi(x) \simeq \Phi_k(x), \text{ as } x \to 0, \tag{1.64}$$

where the function is approximated by an expansion

$$\Phi_k(x) = 1 + \sum_{n=1}^{k} c_n x^n. \tag{1.65}$$

---

[5] Sometimes, for convenience, the critical index could be defined as $-\beta$, when β is negative.

Such expansions are usually asymptotic and strongly divergent, not allowing for their direct use at finite values of the variable.

Let us discuss how the critical exponents can be found by using standard definition of the critical index. The techniques employed here express critical index directly, as the limit of explicitly calculable approximants. Critical index can be estimated from a standard representation called here a "single pole" approximation for the following derivative (or else called a $DLog$ transformation for the series $\Phi(x)$),

$$\mathcal{B}_a(x) = \partial_x \log(\Phi(x)) \simeq -\frac{\beta}{x_c - x}, \tag{1.66}$$

as $x \to x_c$, thus defining critical index β as the residue in the corresponding single pole. The pole here defines the critical point x_c, while the critical index is given by the residue

$$\beta = \lim_{x \to x_c} (x - x_c)\mathcal{B}_a(x).$$

To the $DLog$-transformed series one would consider applying the Padé approximation. As is known (Baker and Graves-Moris, 1996), for a given expansion of order k, one can construct the whole table of Padé approximants. This means that defining the critical indices through the $DLog$ Padé method is not a uniquely defined procedure. And different Padé approximants can result in basically different values. Then it is not clear which of these quantities to prefer. Usually, outside of the immediate vicinity of the critical point a diagonal Padé approximant is assumed for the residue estimation.

One can also use another representation for $\mathcal{B}_a(x)$, e.g., in the form of a factor approximant, so that

$$\mathcal{B}_a(x) = -\frac{(\mathcal{P}_2 x + 1)^{m_2}}{1 - \frac{x}{x_c}}, \tag{1.67}$$

with the values for parameters $\mathcal{P}_2$, m_2, found for the series (1.65). Formula (1.67) leads to the simple expression for the critical index

$$\beta = x_c (\mathcal{P}_2 x_c + 1)^{m_2}.$$

The functional dependence can be reconstructed by proper integration, and the result could be expressed through hypergeometric functions.

When a function, at asymptotically large variable, behaves as in (1.63),

$$\Phi(x) \simeq A x^\beta, \; as \; x \to \infty,$$

then the critical exponent can be defined similarly, by means of the $DLog$ transformation. It is represented by the limit

$$\beta = \lim_{x \to \infty} x \mathcal{B}_a(x). \tag{1.68}$$

Assume that the small-variable expansion for the function $\mathcal{B}_a(x)$ is given. In order that the critical index be finite it is necessary to take the asymptotically equivalent approximants behaving as x^{-1} as $x \to \infty$. It leaves us no choice but to select the non-diagonal Padé $P_{n,n+1}(x)$ approximants, so that the corresponding approximation β_n is finite.

In the particular case of $\beta = 0$, the $P_{n,n+1}(x)$-type approximants would not do. Then, different approximants of $P_{n,n+n'}(x)$-type, with integer $n' \geq 2$ will bring the correct asymptotic behavior at ∞. Once again we are confronted with multiple choice among all such non-diagonal Padé approximants.

Consideration of the factor approximants for all these problems helps to avoid the selection. The factor approximants (1.60) in all the cases discussed above remain defined uniquely. For instance, in the last case we simply have to set $\sum_{i=1}^{N_k} n_i = \beta = 0$, and perform standard calculations. Mind that factor approximants were derived in Gluzman et al. (2003) without invoking Padé approximants directly, but they do give different, more compact and convenient representation for some $DLog$ non-diagonal Padé approximants.

One can also apply, in place of Padé, some different approximants. For example, one can apply iterated roots $\mathcal{R}_n^*(x)$, behaving as x^{-1} as $x \to \infty$. The self-similar root approximant of general type, based on the small-variable expansion (1.65), has the form

$$\mathcal{R}_k^*(x, m_k) = \left(\left((1 + \mathcal{P}_1 x)^{m_1} + \mathcal{P}_2 x^2 \right)^{m_2} + \ldots + \mathcal{P}_k x^k \right)^{m_k}, \qquad (1.69)$$

in which all parameters $\mathcal{P}_j$ may be found from comparing the like orders of the expansion of Eq. (1.69) with the given expansion (1.65). The internal powers for the iterated roots are defined as $m_j = \frac{j+1}{j}$ ($j = 1, 2, \ldots, k-1$), while the external power m_k plays the role of a control parameter to be found from the described condition at infinity.

Calculations with iterated roots are really easy to perform, and find the critical index in n-th order according to the following formula (Gluzman et al., 2017),

$$\beta_n = \lim_{x \to \infty} \left(x \, \mathcal{R}_n^*(x) \right). \qquad (1.70)$$

Looking for novel methods is not just a fancy, but harsh necessity. Consider several examples which illustrate the failure of $DLog$ Padé approximants in evaluation of the critical index. We also give below a constructive solution to the problem invoking iterated roots. And consider the following function (see e.g., Gluzman et al. (2017)),

$$\Phi(x) = \left(\sqrt{x^2 + 1} + x \right)^{\beta},$$

with arbitrary positive β, and the coefficients in the expansion at small $x > 0$,

$$c_n = \frac{2^n \left(\frac{\beta}{2} - \frac{n}{2} + 1 \right)^{\bar{n}}}{n! \left(\frac{n}{\beta} + 1 \right)},$$

while for large x,

$$\Phi(x) \simeq 2^\beta x^\beta.$$

Here $m^{\bar{k}}$ means $m(m+1)\ldots(k+m-1)$. Consider the case of arbitrary β. The following expansion is valid at small $x > 0$,

$$\Phi(x) = 1 + \beta x + \frac{\beta^2 x^2}{2} + \frac{1}{6}\left(\beta^3 - \beta\right)x^3 + \frac{1}{24}\left(\beta^4 - 4\beta^2\right)x^4 + O(x^5). \quad (1.71)$$

Formal application of the $DLog$ Padé approximants to the expansion for small x brings meaningless results for the critical index, as its value alternates between 0 and ∞, with increasing approximation order. Thus, we are confronted with the indeterminate problem (Gluzman and Yukalov, 2016, Gluzman et al., 2017). Therefore, it is simply unavoidable to go beyond the standard approach and develop some form of the corrected Padé approximants (Gluzman et al., 2017).

After the $DLog$ transformation application to the original series we have the transformed series $L(x)$, which reads

$$L(x) \simeq \beta - \frac{\beta x^2}{2} + \frac{3\beta x^4}{8} - \frac{5\beta x^6}{16}, \text{ as } x \to 0.$$

To the series $L(x)$ we apply the formula (1.70). The few starting roots are shown below,

$$\mathcal{R}_2^*(x) = \frac{\beta}{\sqrt{x^2+1}}, \quad \mathcal{R}_3^*(x) \equiv \mathcal{R}_2^*(x),$$

and there is no change in the sequence with further increasing approximation number. One can invert the $DLog$ transformation and find an approximation Φ^* for the sought function Φ,

$$\Phi^*(x) = \exp\left(\int_0^x \mathcal{R}_2^*(X)\,dX\right) = e^{\beta \sinh^{-1}(x)},$$

which appears to be an equivalent representation of the original function, meaning its exact reconstruction!

Consider one more example. The following functional form arises in the calculation of the ground state energy of a quantum particle in a one-dimensional box (Kastening, 2006),

$$\Phi(g) = \frac{1}{16}\pi^4\left(\frac{1}{2}\sqrt{\frac{64}{\pi^4 g^2} + 1} + \frac{16}{\pi^4 g^2} + \frac{1}{2}\right)g^2.$$

As $g \to 0$, this function possesses an expansion of the type,

$$\Phi(g) \simeq \sum_{n=0} a_n g^n,$$

with

$$c_0 = 1, \quad c_1 = \frac{\pi^2}{4}, \quad c_2 = \frac{\pi^4}{32}, \quad c_3 = \frac{\pi^6}{512},$$

$$c_4 = 0, \quad c_5 = -\frac{\pi^{10}}{131072}, \quad c_6 = 0, \dots.$$

We shall be interested in finding the critical index, knowing that as $g \to \infty$

$$\Phi(g) \simeq \frac{\pi^4}{16} g^2,$$

and the exact value of the index $\beta = 2$. The standard $DLog$ Padé scheme fails again, leading to rapid oscillations between the values of 0 and ∞ with changing approximation number. We should just follow the same scheme as in previous case to get an exact solution. The expansion for the transformed series $L(g)$ is expressed in the same form as, namely

$$L(g) \simeq \frac{\pi^2}{4} - \frac{\pi^6 g^2}{512} + \frac{3\pi^{10} g^4}{131072} - \frac{5\pi^{14} g^6}{16777216}, \quad \text{as } g \to 0.$$

The control function for the method of corrected Padé approximants appears to be simple, second-order iterated root approximant

$$\mathcal{R}_2^*(g) = \frac{\pi^2}{4\sqrt{\frac{\pi^4 g^2}{64} + 1}}.$$

The final result

$$\Phi^*(g) = e^{2 \sinh^{-1}\left(\frac{\pi^2 g}{8}\right)},$$

is exact reconstruction of the sought function. In both examples the final solutions can be obtained from the trial form

$$AP(g) = e^{s \sinh^{-1}(Ag)},$$

with the parameters s, A defined from asymptotic equivalence with small g expansion. It appears that just a three non-trivial terms are needed to reconstruct the exact solution. The second-order coefficient is not independent, and is expressed only through the first-order coefficient. Now, avoiding intermediate considerations, we can show directly that yet another function

$$\Phi(x) = \frac{1}{2}\left(\sqrt{x^2 + 4} - x\right),$$

can be exactly reconstructed as the approximant $AP(x) = e^{\sinh^{-1}\left(\frac{-x}{2}\right)}$.

Most important, one can also combine calculations with Padé and roots, using low-order iterated roots as the control function, and not to be restricted to the simple examples just solved. Let us define the new series $L_1(x) = \frac{L(x)}{\mathcal{R}_2^*(x)}$, and apply the

technique of diagonal Padé approximants to satisfy the new series asymptotically. The
following expression for the critical index follows,

$$\beta_n = \lim_{x \to \infty} (x \, \mathcal{R}_2^*(x)) \lim_{x \to \infty} (PadeApproximant[L_1[x], n, n]). \tag{1.72}$$

For example, correlation function of the Gaussian polymer is given in the closed
form by Debye–Hukel function (Grosberg and Khokhlov, 1994),

$$\Phi(x) = \frac{2}{x} - \frac{2(1 - \exp(-x))}{x^2}.$$

For small $x > 0$,

$$\Phi(x) = 1 - \frac{x}{3} + \frac{x^2}{12} - \frac{x^3}{60} + \frac{x^4}{360} + O(x^5),$$

and

$$\Phi(x) \simeq 2x^{-1},$$

as $x \to +\infty$. The standard $DLog$ Padé scheme fails in this case, as it gives the solution alternating (oscillating) between the values of -2 and 0. The $DLog$-transformed
series

$$L(x) \simeq -\frac{1}{3} + \frac{x}{18} - \frac{x^2}{270} - \frac{x^3}{3240} + \frac{x^4}{13608},$$

has more general form than in previous examples. On the other hand, method of corrected Padé approximants expressed by (1.72), with rather simple control function

$$\mathcal{R}_2^*(x) = -\frac{1}{3\sqrt{\frac{x^2}{30} + \left(\frac{x}{6} + 1\right)^2}},$$

demonstrates very good convergence, and relative error for the critical index is negligible for n around 40, as results of computations are shown in Fig. 1.4.

Consider the popular example of quartic anharmonic oscillator, described by the
Hamiltonian

$$\hat{H} = -\frac{1}{2}\frac{d^2}{dx^2} + \frac{1}{2}x^2 + gx^4,$$

where $x \in (-\infty, \infty)$ and with the anharmonicity strength $g \in [0, \infty)$. The ground-
state energy is given by the lowest eigenvalue of this Hamiltonian. By perturbation
theory (Bender and Wu, 1969, Hioe et al., 1978) with respect to the parameter g, one
has got for the ground-state energy the following expansion

$$e(g) = \sum_{n=0}^{k} c_n g^n,$$

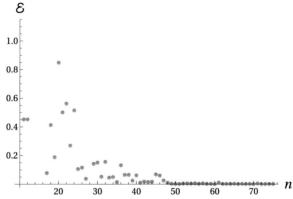

Figure 1.4 Debye–Hukel function. The relative error for the critical index with increasing approximation order obtained with corrected approximants formula (1.72). The relative error is negligible for n around 40.

with the starting coefficients

$$c_0 = \frac{1}{2}, \quad c_1 = \frac{3}{4}, \quad c_2 = -\frac{21}{8}, \quad c_3 = \frac{333}{16},$$

while much more terms are available for computations. The value of these coefficients quickly increases signifying strong divergence of the expansion. In the large anharmonicity limit, the finite series $e(g)$ have fractional powers, with the leading term

$$e(g) \simeq 0.667986 g^{1/3},$$

as $g \to \infty$. The standard method of $DLog$ Padé approximants seems to work, but the convergence is slow, and accuracy is limited to 20% for n around 35. The control function can be written as in the cases considered above,

$$\mathcal{R}_2^*(g) = \frac{3}{2\sqrt{\left(\frac{17g}{2} + 1\right)^2 - 58g^2}},$$

and calculations, according to (1.72), demonstrate a good convergence. The relative error is less than 0.5% for n around 35, as shown in Fig. 1.5.

The $N = 4$ Super Yang–Mills Circular Wilson Loop (see, e.g., Banks and Torres (2013)) is given by the following expression

$$\Phi(y) = \frac{2 \exp\left(-\sqrt{y}\right) I_1\left(\sqrt{y}\right)}{\sqrt{y}},$$

where I_1 is a modified Bessel function of the first kind. Let us set $\sqrt{y} = x$. For small $x > 0$,

$$\Phi(x) = 1 - x + \frac{5x^2}{8} - \frac{7x^3}{24} + \frac{7x^4}{64} + O(x^5),$$

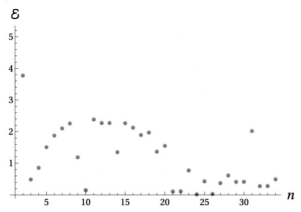

Figure 1.5 Quartic anharmonic oscillator. The relative error for the critical index with increasing approximation order obtained with formula (1.72). The relative error is less than 0.5% for n around 35.

and

$$\Phi(x) \simeq \sqrt{\frac{2}{\pi}} x^{-\frac{3}{2}}, \text{ as } x \to \infty.$$

Application of the standard $DLog$ Padé approximants brings divergent results. Mind that for the critical behavior

$$\Phi \sim x^{\beta}, \text{ as } x \to \infty,$$

we usually consider the following definition of the critical index, when the $DLog$ transformation is made explicit,

$$\beta = \lim_{x \to \infty} x \frac{\Phi'(x)}{\Phi(x)}.$$

In the case of Wilson loop the series $L(x)$ miss the quadratic term, which plays important part in definition of the control function. But we can use the another definition, through the higher-order derivatives

$$\beta = 1 + \lim_{x \to \infty} x \frac{\Phi''(x)}{\Phi'(x)}.$$

In such definition the sought series do have the quadratic term,

$$\frac{\Phi''(x)}{\Phi'(x)} = M(x) \simeq -\frac{5}{4} + \frac{3x}{16} + \frac{x^2}{64} - \frac{x^3}{256} - \frac{x^4}{1024},$$

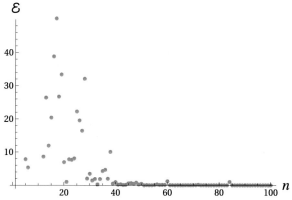

Figure 1.6 Wilson loop function. Critical index with increasing approximation order obtained with corrected approximants. The relative error for the critical index with increasing approximation order obtained with corrected approximants formula. The relative error becomes negligible for n around 40.

and the technique of control functions is again applicable. The control function is found from the asymptotic equivalence with the series $M(x)$,

$$\mathcal{R}_2^*(x) = -\frac{5}{4\sqrt{\frac{7x^2}{100} + \left(\frac{3x}{20} + 1\right)^2}},$$

and we can again construct corrected Padé approximants. To this end, define the new series

$$M_1(x) = \frac{M(x)}{\mathcal{R}_2^*(x)},$$

and apply the technique of diagonal Padé approximants to satisfy the new series asymptotically, order-by-order. The following expression for the critical index follows,

$$\beta_n = 1 + \lim_{x \to \infty} (x \, \mathcal{R}_2^*(x)) * \lim_{x \to \infty} (PadeApproximant[M_1[x], n, n]).$$

With such modified formula, we obtain rather good numerical results for the critical index, shown in Fig. 1.6.

Sometimes, it is better to avoid the $DLog$ (and alike) transformation altogether, and develop an approach with simpler, Log Padé approximants to be introduced. Mind that the critical index can be defined without a differentiation operation as follows:

$$\beta = \lim_{g \to \infty} \frac{\log(\Phi(g))}{\log(g)}. \qquad (1.73)$$

The definition (1.73) can be turned into the method of calculating critical index as suggested below. Indeed, let us look for the sought function in the following form,

$$\Phi^*(g) = \left(1 + \frac{g}{g_0}\right)^{\beta(g)}, \tag{1.74}$$

where g_0 is always positive. As $g \to \infty$, $\beta(g) \to \beta$, supposedly the correct value of the index. The function $\beta(g)$ will be designed in such a way, that it smoothly interpolates between some initial value $\beta(0)$ and the sought value $\beta(\infty) = \beta$. From (1.74) one can express $\beta(g)$, but only formally since $\Phi^*(g)$ is not really known. But we can use its asymptotic form $\Phi(g)$ at small g, express $\beta(g)$ as a power series and apply to the series some resummation procedure (e.g. Padé technique) to find an improved approximation for $\beta(g)$ with a finite value at infinity. Finally, we have to calculate the limit of the approximants as $g \to \infty$ to find the sought β. Indeed, as $g \to 0$ we find the following representation,

$$\beta(g) \simeq \frac{\log(\Phi(g))}{\log\left(1 + \frac{g}{g_0}\right)}, \tag{1.75}$$

which can be easily expanded in powers g, around the value of $\beta(0)$. The simple choice of $g_0 = 1$, corresponds to $\beta(0) = a_1$.

Consider another popular touchstone, the integral

$$Z(g) = \frac{1}{\sqrt{\pi}} \int_{-\infty}^{\infty} \exp\left(-z^2 - gz^4\right) dz$$

which is typical for numerous problems in quantum chemistry, field theory, statistical mechanics, and condensed-matter physics dealing with the calculation of partition functions. Here $g \in [0, \infty)$ plays the role of coupling parameter (Kleinert, 2006). The integral can be expanded at small $g \to 0$, yielding strongly divergent series, with the k-th order sums

$$Z_k(g) = \sum_{n=0}^{k} c_n g^n,$$

whose coefficients are

$$c_n = \frac{(-1)^n}{\sqrt{\pi}\, n!} \Gamma\left(2n + \frac{1}{2}\right).$$

For example, the low order

$$Z_3(g) = 1 - \frac{3}{4}g + \frac{105}{32}g^2 - \frac{3465}{128}g^3.$$

The coefficients c_n quickly grow with increasing n, tending to infinity, as n^n for $n \gg 1$, meaning that the weak-coupling expansion is strongly divergent. At strong coupling, we have in the leading order

$$Z(g) \simeq 1.022765g^{-1/4}, \text{ as } g \to \infty.$$

The standard method of $DLog$ Padé approximants seems to work, but its convergence is slow, and accuracy is limited to 7% for n around 100. From the knowledge of the expansion of $Z(g)$ for small g, the following expansion for $\beta(g)$ can be found,

$$\beta(g) = -\frac{3}{4} + \frac{21g}{8} - \frac{371g^2}{16} + \frac{9387g^3}{32} + \dots.$$

From the series for $\beta(g)$, one can construct the diagonal Padé approximants $P_{n,n}(g)$, and find their corresponding limit as $g \to \infty$. Thus found values will be our estimates for the critical index. With the $2n$-terms from the expansion for $\beta(g)$ being employed, we can find for the critical index,

$$\beta_n = \lim_{g \to \infty} P_{n,n}(g). \tag{1.76}$$

The corresponding Padé approximants need to appear as holomorphic functions. In such case they represent not only the critical index, but the whole "index" function $\beta_n(g)$ which is just the Padé approximant $P_{n,n}(g)$. The functional dependence in n-th order

$$\Phi_n^*(g) = \left(1 + \frac{g}{g_0}\right)^{\beta_n(g)}, \tag{1.77}$$

is reconstructed easily from the "index" function in the whole region of couplings, and the final expression will be called the Log Padé approximant of the n-th order. Calculation of the critical index according to the Log Padé approximation (1.76), demonstrates good, monotonous convergence, with the relative error less than 0.2% for n around 100, as shown in Fig. 1.7. Thus, given practically unlimited supply of c_n, one can accurately calculate critical indices[6].

1.3.3 Critical indices from self-similar root approximants. Examples

In order to understand the function's critical behavior, it is necessary to extrapolate the asymptotic expansion (1.65) to finite and even large values of the variable. Such an extrapolation can be accomplished by means of techniques just discussed above. But their successful application requires multi-termed expansions. How would it be possible to attempt and obtain reliable results for the critical indices employing a small number of terms in the asymptotic expansion? To this end we can employ the self-similar root approximants given by (1.69), where the external power m_k and the critical index by itself, is to be determined from additional conditions. If the large-variable

---

[6] Of course, in lieu of the Padé approximants one can use factor approximants conditioned on a zero-power at infinity.

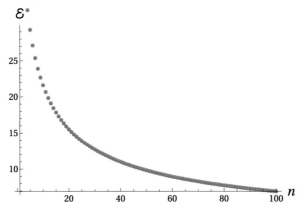

Figure 1.7 Anharmonic partition integral. The relative percentage error for the critical index with increasing approximation order obtained with the *Log* Padé approximants. The relative percentage error is less than 0.2% for n around 100.

power β in Eq. (1.63) were known, then we could compare it with the large-variable behavior of the root approximant (1.69),

$$\mathcal{R}_k^*(x, m_k) \simeq A_k x^{k m_k}, \tag{1.78}$$

where

$$A_k = \left(\left(\left(\mathcal{P}_1^{m_1} + \mathcal{P}_2 \right)^{m_2} + \mathcal{P}_3 \right)^{m_3} + \ldots + \mathcal{P}_k \right)^{m_k}. \tag{1.79}$$

This comparison yields the relation $k m_k = \beta$, defining the external power $m_k = \frac{\beta}{k}$, when β is known. This way of defining the external power is used when the root approximants are applied for interpolation. When several terms in the large-variable behavior are known, then the related powers define the values for the corresponding powers m_j.

Consider the situation of exceptional difficulty when the large-variable behavior of the function is not known and β is not given. Moreover, the critical behavior can happen at a finite value x_c of the variable x. The direct method for calculating the critical index β by employing the technique of self-similar root approximants was developed by Gluzman and Yukalov (2017b).

Suppose we construct several root approximants $R_k^*(x, m_k)$, in which the external power m_k plays the role of a control function. It is possible to treat the sequence $\{R_k^*(x, m_k)\}$ as a trajectory of a dynamical system, with the approximation order k playing the role of discrete time. A discrete-time dynamical system is called cascade, termed the approximation cascade, and its trajectory consists of the sequence of approximants. The cascade velocity is defined by the general form of Euler discretization formula

$$V_k(x, m_k) = \mathcal{R}_{k+1}^*(x, m_k) - \mathcal{R}_k^*(x, m_k) + (m_{k+1} - m_k) \frac{\partial}{\partial m_k} \mathcal{R}_k^*(x, m_k). \tag{1.80}$$

The effective limit of the sequence $\{\mathcal{R}_k^*(x, m_k)\}$ corresponds to the fixed point of the cascade, where the cascade velocity tends to zero, as k tends to infinity. Having a finite number of approximants, the cascade velocity is not necessarily tending to zero, but certainly has to diminish for the sequence being convergent. Thus, the control functions $m_k = m_k(x)$, controlling convergence, are defined as the minimizers of the absolute value of the cascade velocity

$$|V_k(x, m_k(x))| = \min_{m_k} |V_k(x, m_k)| . \tag{1.81}$$

A finite critical point x_k^c, in the k-th approximation, exists if the equation

$$[\mathcal{R}_k^*(x_k^c, m_k)]^{1/m_k} = 0 \qquad (0 < x_k^c < \infty) \tag{1.82}$$

enjoys a finite solution for $x_k^c = x_k^c(m_k)$. Then the critical index in the k-th approximation is

$$\beta_k = \lim_{x \to x_k^c} m_k(x).$$

When we are studying the critical behavior at infinity, which we denote as $x_c \sim \infty$, keeping in mind that this case formally corresponds to the critical point at infinity, then the critical index is

$$\beta = k \lim_{x \to \infty} m_k(x), \text{ near } x_c \sim \infty . \tag{1.83}$$

Thus, the critical indices are defined, provided the control functions $m_k(x)$ are found. However, the minimization of the cascade velocity (1.80) poses some problems. First, Eq. (1.81) contains two control functions, m_{k+1} and m_k. Hence it is impossible to find two solutions from one equation. But it is possible to simplify the problem, when minimizing velocity (1.80), so that to get one equation for one control function. This can be done in two ways.

For instance, keeping in mind that m_{k+1} is close to m_k, Eq. (1.81) reduces to the minimal difference condition

$$\min_{m_k} |\mathcal{R}_{k+1}^*(x, m_k) - \mathcal{R}_k^*(x, m_k)| \qquad (k = 1, 2, \dots) . \tag{1.84}$$

In particular, one can look for a solution $m_k = m_k(x)$ of the equation

$$\mathcal{R}_{k+1}^*(x, m_k) - \mathcal{R}_k^*(x, m_k) = 0 . \tag{1.85}$$

If the latter does not possess a solution for m_k, one has to return to (1.84). In general, when nothing is known on the form of the sought function $\Phi(x)$, the control functions m_k, depend on the variable x. But when we are looking for a function $\Phi(x)$ in the vicinity of its critical point x_c, where the function $\Phi(x)$ acquires critical form, the control functions are to be treated as the limits of $m_k(x)$ for $x \to x_c$. The control functions m_k, characterizing the critical behavior of $\Phi(x)$ become the numbers $m_k(x_c)$. In what follows, writing for short m_k, we assume $m_k = m_k(x_c)$.[7]

In the vicinity of a finite critical point, the function R_k^* behaves as

$$R_k^*(x, m_k) \simeq \left(1 - \frac{x}{x_k^c}\right)^{m_k}, \text{ as } x \to x_k^c - 0 . \tag{1.86}$$

Then condition (1.85) becomes

$$x_{k+1}^c(m_k) - x_k^c(m_k) = 0 \qquad (0 < x_k^c < \infty) . \tag{1.87}$$

When the critical behavior occurs at infinity, then it is convenient to introduce the control function

$$s_k = k m_k , \tag{1.88}$$

so that the large-variable behavior of the root approximants reads as

$$R_k^*(x, s_k) \simeq A_k(s_k) x^{s_k}, \text{ as } x \to \infty . \tag{1.89}$$

As a result, the minimal difference condition

$$R_{k+1}^*(x, s_k) - R_k^*(x, s_k) = 0 \tag{1.90}$$

leads to the equation

$$A_{k+1}(s_k) - A_k(s_k) = 0, \text{ as } x_k^c \sim \infty . \tag{1.91}$$

The other equation for defining control functions follows from the minimal velocity condition (1.81), by remembering that R_{k+1}^* is close to R_k^*, but usually m_{k+1} does not exactly coincide with m_k. Therefore, one has to consider the minimal derivative condition

$$\min_k \left| \frac{\partial}{\partial m_k} R_k^*(x, m_k) \right| \qquad (k = 1, 2, \ldots) . \tag{1.92}$$

Specifically one can look for the solution of the equation

$$\frac{\partial}{\partial m_k} R_k^*(x, m_k) = 0 . \tag{1.93}$$

However, the minimal derivative condition cannot be applied directly, when the sought function exhibits critical behavior, where the function either diverges or tends to zero. To apply this condition, it is necessary to extract from the function non-divergent parts.

---

[7] Later in the book, we also consider several examples when the complete control function $m_k(x)$, could be obtained analytically.

For example, if the critical point is finite, one can study the residue of the function $\partial \log \mathcal{R}_k^* / \partial m_k$, for which we have

$$\lim_{x \to x_k^c} (x_k^c - x) \frac{\partial}{\partial m_k} \log \mathcal{R}_k^*(x, m_k) = m_k \frac{\partial x_k^c}{\partial m_k}.$$

Thus, instead of Eq. (1.93), we get the condition

$$\frac{\partial x_k^c}{\partial m_k} = 0 \qquad (0 < x_k^c < \infty). \tag{1.94}$$

When the critical behavior occurs at infinity, then we can consider the limit

$$\lim_{x \to \infty} \frac{\mathcal{R}_k^*(x, s_k)}{x^{s_k}} = A_k(s_k),$$

for which Eq. (1.93), defining control functions, reduces to the form

$$\frac{\partial A_k(s_k)}{\partial s_k} = 0, \text{ near } x_k^c \sim \infty. \tag{1.95}$$

Let us consider a simple but generic example, when the sought function leads to the small-variable expansion

$$\Phi(x) \simeq 1 + c_1 x + c_2 x^2, \text{ as } x \to 0. \tag{1.96}$$

The first-order root approximant is

$$\mathcal{R}_1^*(x, m_1) = (1 + \mathcal{P}x)^{m_1}, \tag{1.97}$$

with

$$\mathcal{P} = \mathcal{P}(m_1) = \frac{c_1}{m_1}$$

found from asymptotic equivalence procedure. Expression (1.97) possesses a finite critical point, provided there exists a finite positive value

$$x_1^c(m_1) = -\frac{1}{\mathcal{P}(m_1)} = -\frac{m_1}{c_1}.$$

The second-order root approximant reads as

$$\mathcal{R}_2^*(x, m_2) = \left((1 + \mathcal{P}_1 x)^2 + \mathcal{P}_2 x^2 \right)^{m_2}, \tag{1.98}$$

with the parameters

$$\mathcal{P}_1 = \mathcal{P}_1(m_2) = \frac{c_1}{2m_2}, \qquad \mathcal{P}_2 = \mathcal{P}_2(m_2) = \frac{c_1^2(1 - 2m_2) + 4c_2 m_2}{4m_2^2}.$$

Now, the finite critical point is given by a positive value

$$x_2^c(m_2) = \frac{-\mathcal{P}_1(m_2) \pm \sqrt{-\mathcal{P}_1(m_2)}}{\mathcal{P}_1^2(m_2) + \mathcal{P}_2(m_2)} .$$

The minimal difference condition (1.85), in the form $\mathcal{R}_2^*(x, m_1) - \mathcal{R}_1(x, m_1) = 0$ is equivalent to the condition $x_2^c(m_1) - x_1^c(m_1) = 0$, which yields $m_1 = \frac{c_1^2}{c_1^2 - 2c_2}$. Then, the first-order critical point is $x_1^c(m_1) = -\frac{m_1}{c_1} = \frac{c_1}{2c_2 - c_1^2}$. In what follows, keeping in mind the minimal difference condition, we shall write,

$$x_1^c(m_1) = x_2^c(m_1) \equiv x_c.$$

The corresponding root approximant (1.97) has the form

$$\mathcal{R}_1^*(x) = \left(1 - \frac{x}{x_c}\right)^{\beta_1} \qquad (0 \le x \le x_c) ,$$

with the first-order critical index $\beta_1 = m_1 = \frac{c_1^2}{c_1^2 - 2c_2}$. The second-order critical index $\beta_2 = m_2$ is defined by the condition

$$\frac{\partial}{\partial m_2} x_2^c(m_2) = 0.$$

When the critical behavior happens at infinity, then we have to consider the large-variable asymptotic forms for the root approximants, with the substitution (1.88). The first-order approximant gives

$$\mathcal{R}_1^*(x, s_1) \simeq A_1(s_1) x^{s_1}, \; as \; x \to \infty ,$$

where $A_1(s_1) = \mathcal{P}(s_1)^{s_1}$. While the second-order approximant leads to

$$\mathcal{R}_2^*(x, s_2) \simeq A_2(s_2) x^{s_2}, \; as \; x \to \infty ,$$

with $A_2(s_2) = \left[\mathcal{P}_1(s_2)^2 + \mathcal{P}_2(s_2)\right]^{s_2/2}$. The first-order critical index $\beta_1 = s_1$ is found from the condition

$$A_2(s_1) - A_1(s_1) = 0 \qquad (\beta_1 = s_1) .$$

And the second-order critical index $\beta_2 = s_2$ can be obtained from the condition

$$\frac{\partial A_2(s_2)}{\partial s_2} = 0 \qquad (\beta_2 = s_2) .$$

The final answer can be presented as the average

$$\beta^* = \frac{1}{2}(\beta_1 + \beta_2) \pm \frac{1}{2}|\beta_1 - \beta_2| .$$

In the similar way, one can proceed to higher orders. For many important cases, even in the lower orders we can get rather accurate critical indices, as demonstrated by Gluzman and Yukalov (2017b).

1.3.3.1 Conductivity in two-dimensional site percolation

The problem of conductivity for site percolation is studied within the framework of a minimal model for transport of classical particles through a random medium (Nieuwenhuizen et al., 1986). This minimal model, known as the Lorenz 2D gas, is a particularly simple statistical hopping model allowing both for analytical consideration and numerical simulations (Nieuwenhuizen et al., 1986, Frenkel, 1987). It can be realized on a square lattice with a fraction of sites being excluded at random. The test particle, or tracer, walks randomly with Poisson-distributed waiting times between the moves. At every move the tracer attempts to jump on to one of the neighboring sites also selected at random. The move is accepted if the site is not excluded. Through the diffusion coefficient for the tracer one can express the macroscopic conductivity (Nieuwenhuizen et al., 1986). The diffusion ceases to exist at the critical density of the excluded sites. If f stands for the concentration of conducting or not excluded sites in the Lorenz model, then $x = 1 - f$ is the concentration of excluded sites. In the vicinity of the site percolation threshold the conductivity behaves as

$$\sigma(x) \sim (x_c - x)^t, \text{ as } x \to x_c - 0,\tag{1.99}$$

with $x_c = 0.4073$, $t = 1.310$ (Grassberger, 1999, Ziff and Torquato, 2017). Perturbation theory in powers of the variable $x = 1 - f$ gives (Nieuwenhuizen et al., 1986) for the two-dimensional square lattice the expansion

$$\sigma(x) \simeq 1 - \pi x + 1.28588x^2, \text{ as } x \to 0.\tag{1.100}$$

In our approach, we obtain the percolation threshold $x_c = 0.4305$ and the critical index $t_1 = 1.352$, with an error 3% and $t_2 = 1.423$, with an error 8.6%. The final answer is $t^* = 1.388 \pm 0.036$.

1.3.3.2 Conductivity in three-dimensional site percolation

The three-dimensional site percolation conductivity problem, similar to the two-dimensional one, exhibits the critical behavior (Kirkpatrick, 1973, Hofling et al., 2006, Bauer et al., 2010) as in Eq. (1.99), with $x_c = 0.688$, $t = 1.9$ (Clerc et al., 1990, Ziff and Torquato, 2017). Perturbation theory gives for the effective conductivity (conductance)

$$\sigma(x) \simeq 1 - 2.52x + 1.52x^2, \text{ as } x \to 0.\tag{1.101}$$

Using our method, we get for the critical index $t_1 = 1.918$, with an error 0.9% and $t_2 = 1.855$, with an error 2%. The answer is $t^* = 1.887 \pm 0.032$.

The critical index for superconductivity is considered in Section 6.6.

1.3.3.3 Permeability of two-dimensional channel

Let us consider the widely studied case of the two-dimensional channel bounded by the surfaces $z = \pm b(1 + \varepsilon \cos x)$, where ε is termed *waviness*. The permeability possesses the critical behavior (Adler, 1992), when (in the case of $b = 0.5$) it tends to zero as

$$K(\varepsilon) \sim (\varepsilon_c - \varepsilon)^{\varkappa}, \text{ as } \varepsilon \to \varepsilon_c - 0, \tag{1.102}$$

with $\varepsilon_c = 1$, $\varkappa = \frac{5}{2}$. An expression for permeability as a function of the waviness parameter can be derived by perturbation theory in the form of an expansion in powers of the waviness (Adler, 1992, Malevich et al., 2006). The permeability, for $b = 0.5$, has the expansion

$$K(\varepsilon) \simeq 1 - 3.14963\,\varepsilon^2 + 4.08109\,\varepsilon^4, \text{ as } \epsilon \to 0. \tag{1.103}$$

With our method, setting $\varepsilon = 1$, and using the change of the variable $y = \frac{\varepsilon^2}{1-\varepsilon^2}$, one shifts the critical point to infinity. Then, the critical index is calculated as explained above, giving $\varkappa_1 = 2.184$, with an error 12.6% and $\varkappa_2 = 2.559$, with an error 2.37%. Finally, we have $\varkappa^* = 2.372 \pm 0.19$.

The permeability, for $b = 0.25$, has the expansion

$$K(\varepsilon) \simeq 1 - 3.03748\,\varepsilon^2 + 3.54570\,\varepsilon^4, \text{ as } \varepsilon \to 0. \tag{1.104}$$

Setting $\varepsilon = 1$, and using the same change of the variable as above, the critical index is found, $\varkappa_1 = 2.342$, and $\varkappa_2 = 2.743$. Thus, we have $\varkappa^* = 2.543 \pm 0.2$. We also illustrate below the numerical convergence of root approximants in high-orders, when applied for calculating critical indices.

As an example to be widely referred to in the book, we also consider the same two cases of permeability $K(\varepsilon)$, but with higher-order terms included, limited to the 16th order terms inclusively. The numerical values of the corresponding coefficients can be found in Chapter 7, Eq. (7.45) on page 252, and Eq. (7.53) on page 255. Namely, we construct the root approximants

$$\mathcal{R}_k^*(y) = \left(\left(\left((1 + \mathcal{P}_1 y)^2 + \mathcal{P}_2 y^2 \right)^{3/2} + \mathcal{P}_3 y^3 \right)^{4/3} + \ldots + \mathcal{P}_k y^k \right)^{\beta_k/k}, \tag{1.105}$$

defining the parameters $\mathcal{P}_j$ from the asymptotic equivalence procedure. This gives the permeability the following asymptotic forms

$$\mathcal{R}_k^*(y) \simeq A_k y^{\beta_k}, \text{ as } y \to \infty, \tag{1.106}$$

where the amplitudes $A_k = A_k(\beta_k)$ are

$$A_k = \left(\left((\mathcal{P}_1^2 + \mathcal{P}_2)^{3/2} + \mathcal{P}_3 \right)^{4/3} + \ldots + \mathcal{P}_k \right)^{\beta_k/k}. \tag{1.107}$$

In order to define the critical index β_k, we analyze the differences

$$\Delta_{kn}(\beta_k) = A_k(\beta_k) - A_n(\beta_k). \tag{1.108}$$

Composing the sequences $\Delta_{kn} = 0$, we find the related approximate values β_k for the critical indices. It is possible to investigate different sequences of the conditions $\Delta_{kn} = 0$, the most logical from which are the sequences of $\Delta_{k\,k+1} = 0$ and of $\Delta_{k\,8} = 0$, with $k = 1, 2, 3, 4, 5, 6, 7$. The results, presented in Table 1.1 (for $b = \frac{1}{2}$), show good numerical convergence of the approximate critical indices $\beta_k \equiv \varkappa_k$ in our case, to the exact value $\varkappa = \frac{5}{2}$. Similarly, the results, presented in Table 1.2 (for $b = \frac{1}{4}$), again demonstrate rather good numerical convergence of the approximate critical indices to the exact value. The value of critical index does not depend on parameter b.

The sequence, based on the $DLog$ Padé method, is convergent as well, and results in consistent physical solutions, see Chapter 7.

Consider yet different case of permeability $K(\varepsilon)$, see Chapter 7, subsection 7.4.3. For the parallel sinusoidal two-dimensional channel there is no possibility of the walls touching, and permeability remains finite, but is expected to decay as a power-law as ε becomes large. In the expansion of $K(\varepsilon)$ in small parameter ε^2 we retain the terms up the 16th order terms inclusively. The numerical values of the corresponding coefficients can be found in Chapter 7, see (7.56) on page 257. In this case, the permeability decays as

$$K(\varepsilon) \sim \varepsilon^{\varkappa}, \text{ as } \varepsilon \to \infty,$$

Table 1.1 Case of $b = 1/2$. Critical indices for the permeability $\varkappa_k$ obtained from the optimization conditions $\Delta_{kn}(\varkappa_k) = 0$. The sequences demonstrate numerical convergence to the exact value $\varkappa = 5/2$.

$\varkappa_k$	$\Delta_{k\,k+1}(\varkappa_k) = 0$	$\Delta_{k\,8}(\varkappa_k) = 0$
$\varkappa_1$	2.18445	2.39678
$\varkappa_2$	2.68311	2.52028
$\varkappa_3$	2.48138	2.49208
$\varkappa_4$	2.49096	2.49692
$\varkappa_5$	2.5012	2.49982
$\varkappa_6$	2.49935	2.499
$\varkappa_7$	2.49861	2.49861

Table 1.2 Case of $b = 1/4$. Critical indices for the permeability $\varkappa_k$ obtained from the optimization conditions (1.108), $\Delta_{kn}(\varkappa_k) = 0$. The sequences of β_k demonstrate good numerical convergence to the exact value $\varkappa = 5/2$.

$\varkappa_k$	$\Delta_{k\,k+1}(\varkappa_k) = 0$	$\Delta_{k\,8}(\varkappa_k) = 0$
$\varkappa_1$	2.34165	2.452
$\varkappa_2$	2.52463	2.50542
$\varkappa_3$	2.4976	2.49933
$\varkappa_4$	2.49941	2.50004
$\varkappa_5$	2.50028	2.50033
$\varkappa_6$	2.50032	2.50036
$\varkappa_7$	2.50041	2.50041

Table 1.3 Case of $b = 1/2$. Walls can not touch. Critical indices for the permeability $\varkappa_k$ obtained from the optimization conditions $\Delta_{kn}(\varkappa_k) = 0$. The sequences demonstrate reasonably good numerical convergence to the value $\varkappa = -4$.

$\varkappa_k$	$\Delta_{k\,k+1}(\varkappa_k) = 0$	$\Delta_{k\,8}(\varkappa_k) = 0$
$\varkappa_1$	-6	-4.36
$\varkappa_2$	-4.04	-4.1
$\varkappa_3$	n.a.	-4.13
$\varkappa_4$	-4.09	-4.05
$\varkappa_5$	-3.97	-4.03
$\varkappa_6$	n.a.	-4.08
$\varkappa_7$	-3.94	-3.94

with negative index $\varkappa$. Results of calculations are presented in Table 1.3 (for $b = \frac{1}{2}$), and show good numerical convergence, especially in the last column, to the conjectured value -4. The sequence, based on the $DLog$ Padé method, is convergent as well, see Chapter 7, subsection 7.4.3.

1.3.3.4 Expansion factor of three-dimensional polymer chain

A polymer is a chain of molecular monomers attached at random angles to the end of the chain. A monomer cannot attach at an already occupied spot. For many monomers, the molecular chain will be modeled as a random walk, which cannot cross its own path. This is applicable to macromolecules such as proteins or DNA. Note that here the monomers are not identical. Polymer can serve as an example of self-organized criticality, since it demonstrates power-law feature without any external tuning. The expansion (swelling) factor of 3D polymer chain (see the discussion and definitions in Chapter 8, subsection 8.7 on page 307), as a function of the dimensionless coupling parameter g, can be expressed with sufficient accuracy by the phenomenological equation (Muthukumar and Nickel, 1984, 1987)

$$\Upsilon(g) = (1 + 7.52g + 11.06g^2)^{0.1772}. \tag{1.109}$$

At large g, this gives

$$\Upsilon(g) \simeq 1.531g^\beta, \; as \; g \to \infty, \tag{1.110}$$

with the critical index at infinity $\beta = 0.3544$. In the theory of polymers, one also considers the critical index

$$\nu \equiv \frac{1}{2}\left(1 + \frac{\beta}{2}\right), \tag{1.111}$$

which characterizes power-law increase

$$\sqrt{\langle R^2 \rangle} \sim N^\nu,$$

of the typical chain radius $\sqrt{\langle R^2 \rangle}$ with number of monomers N composing the chain (see Section 8.7), so that $\nu = 0.5886$. Other numerical calculations (Li et al., 1995) give slightly lower results, $\nu = 0.5877$, or 0.5876 (Clisby, 2010). At small g, perturbation theory yields (Muthukumar and Nickel, 1984, 1987) the expansion

$$\Upsilon(g) \simeq 1 + \frac{4}{3} g - 2.075385 g^2, \text{ as } g \to \infty. \tag{1.112}$$

By the method of Subsection 1.3.3, we obtain the critical indices $\beta_1 = 0.2999$, $\nu_1 = 0.5750$, the error being just 0.023%. And the error of $\nu_2 = 0.5878$ is only 0.0013%. The latter estimate is very good. In this way, $\nu^* = 0.5814 \pm 0.006$. Atomic force microscopy over long DNA molecules adsorbed on a surface, measures the average end-to-end distance as a function of the DNA length, and its full distribution function. Measured scaling exponent ν equals 0.589 (0.006) (Valle et al., 2005).

1.3.3.5 Expansion factor of two-dimensional polymer chain

At small g, perturbation theory yields (Muthukumar and Nickel, 1984) the expansion for the swelling factor of the two-dimensional polymer coil,

$$\Upsilon(g) \simeq 1 + \frac{1}{2} g - 0.12154525 \, g^2. \tag{1.113}$$

The critical index at infinity is known exactly (Grosberg and Khokhlov, 1994, Pelisetto and Vicari, 2002), $\beta = 1/2$, so that the critical index

$$\nu \equiv \frac{1}{2}(1 + \beta) = \frac{3}{4}. \tag{1.114}$$

This result, conjecture in fact, is reliably supported by various simulations (Pelisetto and Vicari, 2002). Again, let us apply the method of Subsection 1.3.3. We obtain the critical indices $\beta_1 = 0.507006$, $\nu_1 = 0.753503$, the error being just 0.467%. And the error of $\beta_2 = 0.593995$, $\nu_2 = 0.796997$, is 6.266%. Their average gives, $\nu^* = 0.77525 \pm 0.021747$. There are several experimental estimates for ν obtained for 2D monolayers. For polyvinylacetate $\nu = 0.79(1)$, for atactic polymethyl acrylate $\nu = 0.78(1)$, and for DNA molecules electrostatically bound to cationic liquid bilayers, $\nu = 0.79(1)$ (Vilanove and Rondelez, 1980, Vilanove et al., 1988, Maier and Rädler, 1999, Pelisetto and Vicari, 2002).

1.3.4 Factor approximants and critical index. Example

The factor approximants are given by the expression (1.60). Their singularities are associated with critical points and phase transitions (Gluzman et al., 2003), see also Section 8.8 for very important physical example. Singularity can be also located at ∞. For long series it is typical that their accuracy improves with inclusion of larger and larger number of terms from available expansions. But it is not clear how even attempt to improve factor approximants when the series are short. Below, we suggest the way to accelerate convergence of factors.

Without losing generality, let us consider the so-called $(2 + 1)$-dimensional Ising model given by the Hamiltonian

$$\hat{H} = \sum_i \left(1 - \sigma_i^z\right) - g \sum_{\langle ij \rangle} \sigma_i^x \sigma_j^x - h \sum_i \sigma_i^x,$$

on where i and j enumerate sites on the two-dimensional triangular lattice, $\langle ij \rangle$ denotes nearest-neighbor pairs, σ_i^α are the Pauli matrices, g corresponds to the dimensionless inverse temperature in the Euclidean formulation, and h is the magnetic field variable. The second derivative of the ground state energy with respect to magnetic field, as the magnetic field goes to zero, is called susceptibility. The susceptibility $\chi(g)$ is known (He et al., 1990) to diverge at a critical point $g_c \approx 0.2097$, as

$$\chi(g) \sim (g_c - g)^{-\gamma}, \tag{1.115}$$

with the critical index $\gamma \approx 1.243$ (He et al., 1990). In order to generalize the approach suggested below, one can recall the notation $\chi(g) \sim (g_c - g)^\beta$, where $\beta \equiv -\gamma$.

The weak-interaction or high-temperature expansion of the susceptibility yields (He et al., 1990) the series in powers of g,

$$\chi(g) \simeq 1 + c_1 g + c_2 g^2 \quad (g \to 0),$$
$$c_1 = 6, \qquad c_2 = 32.95. \tag{1.116}$$

Based on the available information, we can construct just a simple factor approximant, based on the threshold g_c and two known coefficients in the expansion,

$$\chi^*(g) = \mathcal{F}_3^*(g) = (1 - 4.76804g)^{-1.30535}(1 + g)^{-0.223957},$$

and reasonably well estimate the critical index, as $\gamma \approx 1.3$. Such estimate looks encouraging, but can not be understood or extended further without knowing precisely the higher order coefficients, or without some additional information from the critical region. We would like to get a better idea why the low-order approximants can produce such reasonable value of the index, around 1.3.

For the factor approximants we have to develop some special techniques for accelerating their convergence, based on mimicking addition of a higher-order coefficient, different from methods developed for root approximants, routinely applied in the book. The problem can be also understood in terms of accelerating convergence of the *DLog* Padé approximants, closely related to the factor approximants, see Subsection 1.3.2. Primarily interested in the case of very short series, we would like to anticipate what would happen to the results if some more information, e.g., higher-order terms are added.

Consider, in general form, the following two approximations of two different types, $\Phi(g)$ and $\Psi(g)$, devised to solve (approximately) the problem at hand. Both approximations respect asymptotically, as $g \to 0$, the same asymptotic expansion with n terms available, and also demonstrate critical behavior as $g \to g_c$. The critical point g_c is known, but the value of critical index γ_Φ and γ_Ψ are different, but may be close.

How to increase accuracy further with no exact, additional terms available? Consider now introducing a higher-order term to the expansion $\chi_n(g)$ at small g, namely

$$\chi_{n+1}(g) = \chi_n(g) + ag^{n+1},$$

with unknown trial parameter a, introduced to accelerate convergence of the series. To find a explicitly we have to formulate an additional condition. It seems natural to expect that introducing a will change the value of either of critical indices,

$$\gamma_\Phi \to \gamma_{\Phi,n+1}(a), \ \gamma_\Psi \to \gamma_{\Psi,n+1}(a).$$

The value of $\gamma_{n+1}(a)$ should be calculated in analytical form by two particular algorithms, leading to some expressions for $\Phi_{n+1}(g,a)$ and $\Psi_{n+1}(g,a)$. Then, the parameter can be found from minimizing the difference

$$|\gamma_{\Phi,n+1}(a) - \gamma_{\Psi,n+1}(a)|,$$

with respect to a. Ideally, introduction of the free parameter a will move the indices to the same unknown value, and we would have the equality

$$\gamma_{\Phi,n+1}(a) = \gamma_{\Psi,n+1}(a).$$

But one would settle for the minimal difference condition, when the equality is not possible. The formulas presented below are of general nature. Let us select $\Phi(g)$ and $\Psi(g)$ as two different, root and factor approximants. After introduction of the free parameter a, they can be written explicitly

$$\Phi_3(g,a) = 1 + c_1 g \left(1 + p(a)\frac{g}{g_c - g}\right)^{\gamma_{\Phi,3}(a)},$$

$$\gamma_{\Phi,3}(a) = \frac{c_2^2 g_c}{-2ac_1 g_c + 2c_1 c_2 + c_2^2 g_c},$$

$$(1.117)$$

and

$$\Psi_3(g,a) = (1 + \mathcal{P}_1(a)g)^{c_1(a)}\left(1 - \frac{g}{g_c}\right)^{-\gamma_{\Psi,3}(a)},$$

$$\gamma_{\Psi,3}(a) = \frac{g_c^3\left(3ac_1 + c_2\left(c_1^2 - 4c_2\right)\right)}{g_c^2\left(3a + c_1^3 - 3c_1 c_2\right) + 2g_c\left(c_1^2 - 2c_2\right) + c_1}.$$

$$(1.118)$$

The rest of parameters from (1.117), (1.118) can be expressed in closed form, not shown here. The equality leads in the case of susceptibility to a quadratic equation with two complex-conjugated solutions for a. But turning to the minimization condition gives $a \approx 172.58$. Then there are following two values for the index,

$$\gamma_{\Phi,3}(a) \approx 1.206, \ \gamma_{\Psi,3}(a) \approx 1.302.$$

The root approximant gives lower bound and the factor approximant supplies upper bound for the critical index. Their average gives $\gamma \approx 1.25 \pm 0.05$. We can also state that estimated a is in a good agreement with exact $a = 166.5$ (He et al., 1990). In the case of quadratic lattice (He et al., 1990, Yukalov and Gluzman, 1997a), similar accuracy can be found and analogous interpretation given.

Generally speaking, the methodology requires any two explicit expressions for the critical index β, obtained from two different approximants, based on the same asymptotic input. The critical index β could be expressed equivalently as[8]

$$\beta = \lim_{g \to g_c} \frac{(g - g_c)\chi'(g)}{\chi(g)}, \text{ or } \beta = 1 + \lim_{g \to g_c} \frac{(g - g_c)\chi''(g)}{\chi'(g)}. \quad (1.119)$$

The two expressions, $\frac{\chi'(g)}{\chi(g)}$, $\frac{\chi''(g)}{\chi'(g)}$, can be expanded as $g \to 0$

$$\frac{\chi'(g)}{\chi(g)} \simeq c_1 + g\left(2c_2 - c_1^2\right) + g^2\left(3a + c_1^3 - 3c_1c_2\right),$$
$$\frac{\chi''(g)}{\chi'(g)} \simeq \frac{2c_2}{c_1} + g\left(\frac{6a}{c_1} - \frac{4c_2^2}{c_1^2}\right). \quad (1.120)$$

Then such approximants to (1.120) should be constructed which would satisfy asymptotic expressions (1.120) and (1.119). From the first (longer) series from (1.120), we construct the following simple Padé approximant

$$P_{1,2}(g,a) = \frac{p_1 + p_2(a)f}{(1 - p_3(a)g)(g_c - g)},$$
$$p_1 = c_1g_c, \quad p_2(a) = \frac{c_1g_c(3ag_c + 2c_2) - c_1^3g_c + c_1^2\left(c_2g_c^2 - 1\right) - 4c_2^2g_c^2}{c_1^2g_c + c_1 - 2c_2g_c},$$
$$p_3(a) = -\frac{3af_c + c_1^3g_c + c_1^2 - 3c_1c_2g_c - 2c_2}{c_1^2g_c + c_1 - 2c_2g_c}. \quad (1.121)$$

From the second (shorter) series from (1.120), we construct the following simplest diagonal Padé approximant

$$P_{1,1}(g,a) = q_1(a) + \frac{q_2(a)}{g_c - g},$$
$$q_1(a) = \frac{-6ac_1g_c + 2c_1c_2 + 4c_2^2g_c}{c_1^2}, \quad q_2(a) = \frac{2g_c^2\left(3ac_1 - 2c_2^2\right)}{c_1^2}. \quad (1.122)$$

Corresponding limits (1.119), expressing the critical index, should be found explicitly. The value of parameter a found simply by equating (or minimizing the difference) between the two expressions for the index. In the case of susceptibility, from the equation

$$q_2(a) - 1 = \frac{p_1 + p_2(a)g_c}{1 - p_3(a)g_c},$$

[8] See also (Stevenson, 2016).

we estimate $a = 173.365$. Thus, we obtain the following index, $\gamma \approx 1.32$, agreeing with previous estimate. After proper integration, the approximant (1.121) results in a factor approximant. Thus, we conclude that the factor approximants give robust upper bound for the index, even without knowing the exact cubic term. Factors are able to evaluate it rather well, already from the available information. In this sense, the cubic term is not required to be known in advance to plausibly evaluate the index.

Of course, some middle-of-the road approach is possible. One should come up with two different $DLog$ approximants, expected to give equal value to the critical index expressed conventionally as $\gamma = \lim_{g \to g_c} \frac{(g_c - g)\chi'(g)}{\chi(g)}$. To such end, in addition to (1.121), we can construct also the second-order iterated root approximant. The corresponding limit for the two approximants could be found explicitly, and the value of parameter a found by equating the two limiting expressions for the critical index.

1.3.5 Additive self-similar approximants and $DLog$ additive recursive approximants

Suppose that for the sought function we find its asymptotic expansion near one of the domain boundaries, say, for asymptotically small $x > 0$, with the k-th order finite series

$$\Phi_k(x) = \sum_{n=0}^{k} c_n x^n .$$

Also the large-variable expansion is available, given by the finite series

$$\Phi^{(p)}(x) = \sum_{n=1}^{p} b_n x^{\beta_n}.$$

The powers in the series are arranged in the ascending order. The coefficients b_n may be unknown and can be calculated approximately from the known coefficients c_n, β_n. The standard situation corresponds to the uniform power decrease with the constant difference $\Delta\beta \equiv \beta_n - \beta_{n+1}$ ($n = 1, 2, \ldots$). Only the main steps of the procedure leading to additive approximants (Gluzman and Yukalov, 2017a), are stressed below. First, subject the variable x to the affine transformation $x \mapsto \mathcal{P}(1 + \lambda x)$, consisting of a scaling and shift. Then, the self-similar transformation of the power series is the affine transformation of its terms, which yields

$$A_k^*(x) = \sum_i \mathcal{P}_i (1 + \lambda x)^{m_i} . \tag{1.123}$$

The powers of the first k terms of this series correspond to the powers of series in the strong-coupling limit $m_i = \beta_i$ ($i = 1, 2, \ldots, k$), All coefficients $\mathcal{P}_i$ and λ if needed, can be found from asymptotic equivalence, expanding the approximant in powers of x and equating it asymptotically to the weak-coupling expansion. Thus we arrive to the additive self-similar approximant (Gluzman and Yukalov, 2017a). In the large-variable limit, the approximant will reproduce the terms with the powers of series borrowed from the strong-coupling expansion. Except the terms with the correct powers β_i, there appear the terms with the powers $\beta_i - 1$. In case that the powers $\beta_i - 1$ do not pertain to the set of the powers β_i, the terms with the incorrect powers should be canceled by including in approximant correcting terms (counter-terms). Their powers $m_j = \beta_j - 1$ ($j = 1, 2, \ldots, q$) and the coefficients (amplitudes) for such approximants are defined by the cancellation of the terms with incorrect powers in the large-variable expansion.

Consider now a given interval $[0, f_c]$. By a change of variables it is practically always possible to reduce a given interval to the ray $[0, \infty)$, and vice versa. Naturally, we would like to extend the region of applicability for our low-concentration approximations in principle, without even invoking the detail expression from high-concentration limit. Only the k coefficients c_n in the expansion of the effective quantity (modulus)

$$\mu_k(f) = 1 + \sum_{n=1}^{k} c_n f^n,$$

are given. Consider an alternative approach, based on the high-concentration expansions written in general form. By analogy to the conductivity problem (Gluzman et al., 2017), one can consider the following additive ansatz for the effective quantity (such as effective shear modulus) valid in the vicinity of f_c,

$$\mu^{(p)}(f) = b_1(f_c - f)^{\beta}\left(1 + \tfrac{b_2}{b_1}(f_c - f)^{\gamma} + \tfrac{b_3}{b_1}(f_c - f)^{2\gamma} + \ldots\right), \qquad (1.124)$$

and $p \geq 1$. The critical index β can be positive or negative, and the correction index γ is always positive.

The expression (1.124) extends the ansatz considered for the problem of effective conductivity, with $\beta = -\tfrac{1}{2}$, $\gamma = \tfrac{1}{2}$. Generally, such form describes the case, when in addition to the leading singularity there are correction terms, called the confluent singularities. They tend to zero at the critical point, but can not be ignored in away from the critical point.

It turns out resummation with additive approximants works better, when the following feature is added. It consists in modification to the standard $DLog$ technique for critical index calculation (Baker and Graves-Moris, 1996). Specifically, it incorporates the known value of the index and takes into account confluent corrections (Czaplinski et al., 2018). Mind that the critical index β can be estimated from the standard representation in terms of the derivative

$$\mathcal{B}_a(f) = \frac{d}{df}\log(\mu(f)) \simeq -\frac{\beta}{f_c - f}, \qquad (1.125)$$

as $f \to f_c$, thus defining critical index β as the residue in the corresponding pole. For all f, well outside of the immediate vicinity of the critical point, a diagonal Padé approximant is assumed for the residue estimation (Baker and Graves-Moris, 1996). But it is not able to take into account the confluent singularities (1.124).

Let us define a corrected in the form of a specific additive approximant, also vaild for all f, specifically constructed to take into account confluent additive singularities, which can be a called a modified "single-pole" approximation,

$$\mathcal{B}_a(f) = \frac{d}{df} \log\left(\mu\left(f\right)\right) \simeq \frac{-\beta}{f_c - f} + \sum_{k=1}^{n} \alpha_k \left(f_c - f\right)^{\gamma k - 1}, \tag{1.126}$$

and $n \geq 1$. In (Czaplinski et al., 2018) the critical index was fixed to the value $\beta = -\frac{1}{2}$, and $\gamma = \frac{1}{2}$, while here we consider general case. In such formulation we can derive formulas satisfying both the low-concentration and high-concentrations expansions. The complete expression for the effective modulus $\mu(f)$ can be written as quadrature,

$$\mu\left(f\right) = \exp\left(\int_0^f \mathcal{B}_a(F)\,dF\right). \tag{1.127}$$

In principle, an explicit formula can be found for $\mu\left(f\right)$ from (1.127) and (1.126) in arbitrary order, since the integration can be performed explicitly.

As the result, for all $n \geq 1$ one can find the following recursive *DLog* additive approximants $\mathcal{RA}^*(f)$ for the quantity $\mu(f)$

$$\mathcal{RA}_n^*(f) = \mathcal{RA}_{n-1}^*(f)e^{\frac{\alpha_n\left(f_c^{\gamma n} - (f_c - f)^{\gamma n}\right)}{\gamma n}} \tag{1.128}$$

For convenience, let us also consider

$$\mathcal{RA}_0^*\left(f\right) = \left(1 - \frac{f}{f_c}\right)^{\beta}.$$

This result corresponds to approximation (1.125) per se, and just expresses it differently. $\mathcal{RA}_0^*$ is present as a factor in all higher-order approximations guaranteeing the correct form of the leading power-law, while the confluent singularities are treated with recursively emerging exponential terms. For example in low-orders they are given as follows:

$$\mathcal{RA}_1^*(f) = \exp\frac{\alpha_1\left(f_c^{\gamma} - (f_c - f)^{\gamma}\right)}{\gamma}\left(1 - \frac{f}{f_c}\right)^{\beta},$$

$$\mathcal{RA}_2^*(f) = \exp\left(\frac{\alpha_1\left(f_c^{\gamma} - (f_c - f)^{\gamma}\right)}{\gamma} + \frac{\alpha_2\left(f_c^{2\gamma} - (f_c - f)^{2\gamma}\right)}{2\gamma}\right)\left(1 - \frac{f}{f_c}\right)^{\beta},$$

$$\mathcal{RA}_3^*(f) = \exp\left(\frac{\alpha_1\left(f_c^{\gamma} - (f_c - f)^{\gamma}\right)}{\gamma} + \frac{\alpha_2\left(f_c^{2\gamma} - (f_c - f)^{2\gamma}\right)}{2\gamma} + \frac{\alpha_3\left(f_c^{3\gamma} - (f_c - f)^{3\gamma}\right)}{3\gamma}\right)$$
$$\times \left(1 - \frac{f}{f_c}\right)^{\beta}. \tag{1.129}$$

Consider separately the case when critical behavior occurs at infinity so that

$$\mu^{(p)}(f) = \sum_{k=1}^{p} b_k f^{\beta + (1-k)\gamma} = b_1 f^{\beta}(1 + \frac{b_2}{b_1} f^{-\gamma} + \frac{b_3}{b_1} f^{-2\gamma} + \ldots), \tag{1.130}$$

and $p \geq 1$. The critical index β could be positive or negative, and corrections defined through positive γ. The critical index β can be estimated from the following representation in terms of the derivative

$$\mathcal{B}_a(f) = \frac{d}{df}\log(\mu(f)) \simeq \frac{\beta}{f_0 + f}, \tag{1.131}$$

as $f \to -f_0$, thus defining critical index S as the residue in the corresponding single pole located at $-f_0$, with $f_0 > 0$. For all f, well outside of the immediate vicinity of the pole, a non-diagonal Padé approximant $P_{n,n+1}$ is assumed for the residue estimation, as described above in the chapter. The pole is located at negative f, in the non-physical region, so that relevant singularity only manifests at infinity. Let us define the following ansatz in the form of additive approximant, also valid for all f, which is specifically constructed to take into account confluent singularities at infinity,

$$\mathcal{B}_a(f) = \frac{d}{df}\log(\mu(f)) \simeq \frac{\beta}{f_0 + f} + \sum_{k=1}^{n}\alpha_k(f_0 + f)^{-1-\gamma k}, \tag{1.132}$$

and $n \geq 1$. Here the values of f_0 and α_k to be determined from asymptotic conditions. An explicit formula can be found for $\mu(f)$ from (1.127) and (1.132) in arbitrary order, since the integration can be performed. For convenience, let us consider

$$\mathcal{RA}_0^*(f) = \left(1 + \frac{f}{f_0}\right)^{\beta}.$$

This result corresponds to the single-pole approximation (1.131) per se. $\mathcal{RA}_0^*(f)$ enters as a factor to all higher-order approximations. As the result, for all $n \geq 1$ one can find the following recursion,

$$\mathcal{RA}_n^*(f) = \mathcal{RA}_{n-1}^*(f)\exp\left(\frac{\alpha_n\left(f_0^{-\gamma n} - (f_0 + f)^{-\gamma n}\right)}{\gamma n}\right). \tag{1.133}$$

In particular, in the low-orders it follows that

$$\mathcal{RA}_1^*(f) = e^{\frac{\alpha_1\left(f_0^{-\gamma} - (f_0+f)^{-\gamma}\right)}{\gamma}}\left(1 + \frac{f}{f_0}\right)^{\beta},$$

$$\mathcal{RA}_2^*(f) = \exp\left(\frac{\alpha_1\left(f_0^{-\gamma} - (f_0+f)^{-\gamma}\right)}{\gamma} + \frac{\alpha_2\left(f_0^{-2\gamma} - (f_0+f)^{-2\gamma}\right)}{2\gamma}\right)\left(1 + \frac{f}{f_0}\right)^{\beta},$$

$$\mathcal{RA}_3^*(f) = \exp\left(\frac{\alpha_1\left(f_0^{-\gamma} - (f_0+f)^{-\gamma}\right)}{\gamma} + \frac{\alpha_2\left(f_0^{-2\gamma} - (f_0+f)^{-2\gamma}\right)}{2\gamma} + \frac{\alpha_3\left(f_0^{-3\gamma} - (f_0+f)^{-3\gamma}\right)}{3\gamma}\right)$$

$$\times \left(1 + \frac{f}{f_0}\right)^{\beta}.$$

$$\tag{1.134}$$

Recursions (1.128), (1.133) are called $DLog$ additive recursive approximants, or simply $DLog$ additive approximants. Procedure of $DLog$ exponentiation allows to reduce computations to operations with additive quantities, instead of dealing with factors. Although the final forms are easily factorized and remain positive. The parameters of the approximants could be found from the asymptotic equivalence with the low-concentration expansions, which are usually easier to find analytically. When the position of the pole (threshold, critical point) is known from other considerations, the calculations are simplified immensely and reduced to linear systems. Sometimes, a few terms from the high-concentration expansions are available and could be incorporated into the parameters of the recursion. Further generalizations to a "multi-pole" approximations with corresponding confluent singularities considered around each pole, may be considered as well. The technique of $DLog$ additive recursive approximants will be heavily employed in the study of elastic problems.

Consider now a concrete example of the recursion (1.128). In practice, the composite often consists of a uniform background-host reinforced by a large number (high concentration) of unidirectional rod-or fiber-like inclusions. Consider the two-dimensional composite corresponding to the regular square lattice (square array) arrangement of an ideally conducting cylinders of radius r embedded into the matrix made of a conducting material (Gluzman et al., 2017). The volume fraction (concentration) of cylinders is equal to $f = \pi r^2$. We are interested primarily in the case of a high contrast regular composites, when the conductivity of the inclusions is much larger than the conductivity of the host. So, the highly conducting inclusions are replaced by the ideally conducting inclusions with infinite conductivity. The transverse effective conductivity $\sigma_e = \sigma_e(f)$ of regular composites is presented in the form of a rather long power series. The coefficients c_n in the expansion of $\sigma_e(f) = 1 + \sum_{n=1}^{\infty} c_n f^n$, are obtained from the exact formula, see page 164 in Gluzman et al. (2017). Below, only the starting terms in the series are presented,

$$\sigma_e(f) = 1 + 2f + 2f^2 + 2f^3 + 2f^4 + 2.6116556664543236 f^5$$
$$+ 3.2233113329086476 f^6 + 3.8349669993629707 f^7 + \ldots,$$
$$(1.135)$$

while the terms up to $O(f^{27})$ are available. The effective conductivity is expected to tend to infinity as a power-law, as the concentration f tends to the maximal value $f_c = \frac{\pi}{4}$ for the square array (Gluzman et al., 2017),

$$\sigma_e \simeq \frac{1}{2} \pi^{\frac{3}{2}} (f_c - f)^{-1/2} - 1.85. \qquad (1.136)$$

In this case, the critical index $\beta = -\frac{1}{2}$, and the correction index $\gamma = \frac{1}{2}$. And, explicitly, the following approximants are found, after amplitudes α_i were calculated from asymptotic equivalence with the series (1.135), (1.136),

$$\sigma_2^*(f) = \frac{1.59694 e^{0.707739 f - 0.664472\sqrt{0.785398 - f}}}{\sqrt{0.785398 - f}},$$

$$\sigma_3^*(f) = \frac{2.48208 e^{-0.633592(0.785398 - f)^{3/2} - 0.664472\sqrt{0.785398 - f} + 0.146232 f}}{\sqrt{0.785398 - f}},$$

$$\sigma_4^*(f) = \frac{0.699942 \exp\left(f\left(-1.18506\sqrt{0.785398 - f} - 1.02606 f + 2.56384\right) + 0.26627\sqrt{0.785398 - f}\right)}{\sqrt{0.785398 - f}}.$$

Let us use the method of corrected approximants, when all 26 terms from the weak-coupling limit and two terms from the strong-coupling limit are utilized, so that

$$\sigma^*(f) = \sigma_4^*(f)\, \mathbf{Cor}(f). \tag{1.137}$$

The correction term is sought as the diagonal Padé approximant $P_{13,13}$, given as follows:

$$\mathbf{Cor}(f) = \frac{P(f)}{Q(f)},$$

$$\begin{aligned}
P(f) = {} & 1 + 0.404305\,f + 0.393235\,f^2 + 0.0285778\,f^3 - 1.77771\,f^4 + 1.54073\,f^5 \\
& - 0.616216\,f^6 + 0.243378\,f^7 - 0.0561403\,f^8 - 2.26205\,f^9 \\
& + 0.738228\,f^{10} - 0.428836\,f^{11} - 0.231561\,f^{12} - 2.26654\,f^{13},
\end{aligned}$$

$$\begin{aligned}
Q(f) = {} & 1 + 0.404305\,f + 0.393235\,f^2 - 0.111694\,f^3 - 1.39465\,f^4 + 1.07257\,f^5 \\
& - 0.245803\,f^6 - 0.0494292\,f^7 - 0.418391\,f^8 - 1.16378\,f^9 \\
& - 0.266856\,f^{10} + 0.749836\,f^{11} - 0.803724\,f^{12} - 2.36854\,f^{13}.
\end{aligned}$$

$$\tag{1.138}$$

The formula (1.137) appears to be very accurate, with maximal error of negligible 0.0039%, when compared with the numerical data of (Perrins et al., 1979). This result is significantly better than our previous best formula (6.7.81) (Gluzman et al., 2017), which brings the maximal error of only 0.0193%.

Algorithms, computations and structural information

2

2.1 Computing in modern computational sciences

An interdisciplinary research often leads to a new symbolic (non-numerical) algorithms. Emerging new computational objects require special techniques and methods for calculating their values. Some frequently asked question relates to a possibility of generalizing given algorithm developed for a specific problem, to solve another problem. Computational science based on analytical methods, requires new symbolic computation algorithms, as well as more traditional algorithms arising in the modeling process.

As stated in von zur Gathen and Gerhard (2013, page 1), *computer algebra* is rapidly expanding area of computer science, where mathematical tools and computer software are developed for the exact solution of equations. One can also characterize this field as "development, implementation, and application of algorithms that manipulate and analyze mathematical expressions" (Cohen, 2002). Computer algebra has many names (Buchberger et al., 1985), e.g., it is called algebraic computations, analytic computations, computer analysis. This kind of computing is most often called *symbolic computations*. It is worth noting that rapid advancement of computer technology has brought a renaissance of algorithmic mathematics, which gave rise to the creation of new disciplines, like symbolic computations (Paule et al., 2009). We can, therefore, consider symbolic computations algorithms as a new research field stemming from the traditional algorithms.

Another aspect of symbolic computations is the design of computer representations of analytical objects. It often happens that a certain procedure, solution or model is constructive from a mathematical point of view, but it is not computable using a given programming language. Therefore, it is important to develop techniques enabling the implementation of analytical concepts or objects by creating their *computational representations*. All symbolic algorithms discussed below, are presented using the *Mathematical Pseudo-Language* (MPL) (see Appendix A.5), described in details by Cohen (2002), covering basic operations common to popular Computer Algebra Systems (CAS).

2.2 Symbolic representations of coefficients for the effective conductivity

In this section we provide a self-contained mathematical statement of the problems that will be tackled in two subsequent sections. This section and the following two

Applied Analysis of Composite Media. https://doi.org/10.1016/B978-0-08-102670-0.00011-1

sections can be understood even without preliminary knowledge of the background material. Following Susbsection 1.2.3 of the present book and Gluzman et al. (2017), the effective conductivity of macroscopically isotropic 2D composites is expressed as the series (1.10)

$$\sigma_e = 1 + 2\varrho f \sum_{q=0}^{\infty} A_q f^q. \tag{2.1}$$

Coefficients A_q reflecting (1.11), can be formally introduced by the following

Definition 2.1 (A_q coefficient). Let a_m $(m = 1, 2, \ldots, N)$ be complex numbers. The A_q coefficients are defined as

$$A_0 = 1, \quad A_q = \sum_{m=1}^{N} f_m \psi_m^{(q)}(a_m), \quad q = 1, 2, 3 \ldots, \tag{2.2}$$

where the functions $\psi_m^{(q)}(z)$ are found by recurrent formulas (Berlyand and Mityushev, 2005, Theorem 2.1)

$$\psi_m^{(0)}(z) = 1,$$

$$\psi_m^{(q+1)}(z) = \varrho \sum_{k=1}^{N} \left[f_k E_2(z - a_k)\overline{\psi_{0,k}^{(q)}} + f_k^2 E_3(z - a_k)\overline{\psi_{1,k}^{(q-1)}} + \ldots \right.$$
$$\left. + f_k^{q+1} E_{q+2}(z - a_k)\overline{\psi_{q,k}^{(0)}} \right], \quad q = 0, 1, 2, \ldots \quad . \tag{2.3}$$

Here, $\psi_{l,k}^{(q)}$ is the l-th coefficient of the Taylor expansion of $\psi_k^{(q)}(z)$. The function $\psi_k(z)$ is represented in the form of Taylor series convergent in the disk $|z - a_k| \leq r_k$. $E_k(z)$ $(k = 2, 3 \ldots)$ are Eisenstein functions (see Appendix A.2).

Both computational problems we are facing here are of a symbolic nature. First, we need to express A_q in terms of structural sums in order to develop abstract computations of $\psi_k^{(q)}(z)$ functions. The second task is to simplify symbolically dependent structural sums in the expression for A_q.

2.2.1 Algorithm 1: Express A_q in terms of structural sums

In this section, we describe an algorithm that tackles the first problem, namely expressing A_q in terms of structural sums. First, we introduce the coefficients A_q corresponding to (2.2), and write (2.3) in more convenient form

$$
\begin{aligned}
A_q &= \sum_{m=1}^{N} f_m \psi_q(a_m), \quad q = 1, 2, 3 \ldots, \\
\psi_0(z) &= 1, \\
\psi_q(z) &= \varrho \sum_{k=1}^{N} \sum_{j=1}^{q} E_{j+1}(z - a_k) \overline{\psi_{q-j,j-1}(a_k)} f_k^j, \quad q = 1, 2, 3, \ldots,
\end{aligned}
\tag{2.4}
$$

where

$$
\psi_{q,n}(z) = \frac{1}{n!} \frac{d^n}{dz^n} \psi_q(z).
\tag{2.5}
$$

Here, $\psi_q^{(n)}(z)$ stands for nth derivative of $\psi_q(z)$.

We now proceed to outline a straightforward symbolic algorithm for computing A_q. In order to compute the coefficient A_q, we have to calculate functions ψ_q. Moreover, every function ψ_k ($k = 1, 2, 3, \ldots, q - 1$) is called during computations 2^{q-k-1} times in order to perform derivative operation, hence additional memory for memoization is required. Thereby, we cannot derive A_q directly from A_{q-1}, i.e., the first order recurrence relation for coefficients A_q does not exist. Finally, to transform results into the structural sum notation, one can construct some pattern–matching algorithm for *recognizing* structural sums in expressions. Moreover, construction of such an algorithm is non-trivial, hence the structural sum representation of the coefficient A_q cannot be achieved without a suitable auxiliary transformation.

The next section contains formulation of an efficient algorithm for computing coefficients A_q ($q = 1, 2, \ldots$) in the form of recurrence relation of the first order, based solely on structural sums. In turn, this leads to effective symbolic computation of A_q.

2.2.1.1   Algorithm

First, we verify the following

Lemma 2.1. *The expression of $\psi_q^{(n+1)}(z)$ can be symbolically derived from $\psi_q^{(n)}(z)$ by substituting every term, according to the following rule*

$$
E_{p_1}(z - a) \longmapsto -\frac{p_1}{n+1} E_{p_1+1}(z - a)
\tag{2.6}
$$

for arbitrary a.

Proof. In view of (A.31) we have

$$
\psi_q^{(n)}(z) = (-1)^n \varrho \sum_{k=1}^{N} \sum_{j=1}^{q} f_k^j \frac{(j+n)!}{j!} E_{j+n+1}(z - a_k) \overline{\psi_{q-j,j-1}(a_k)},
$$

which implies

$$\psi_q^{(n+1)}(z) = (-1)^{n+1} \varrho \sum_{k=1}^{N} \sum_{j=1}^{q} f_k^j \frac{(j+n)!}{j!} (j+n+1) E_{j+n+2}(z-a_k) \overline{\psi_{q-j,j-1}(a_k)},$$

since (A.30) holds. Hence and by (2.5), in order to derive symbolically $\psi_q^{(n+1)}(z)$ from $\psi_q^{(n)}(z)$, one can transform every term of the sum by applying the rule (2.6).
   The lemma is proved.   □

   Now let us formulate and prove the theorem that tackles the first problem.

Theorem 2.1. *The algorithm given by (2.2) and (2.3) takes the form of a recurrence relation of the first order*

$$A_1 = \pi^{-1} \varrho e_2,$$
$$A_2 = \pi^{-2} \varrho^2 e_{22},$$
$$A_q = \pi^{-1} \beta A_{q-1}, \quad q = 3, 4, 5...,$$

where β is the substitution operator modifying every structural sum in A_{q-1} according the transformation rule

$$e_{p_1, p_2, \dots, p_n} \longmapsto \varrho e_{2, p_1, p_2, \dots, p_n} - \frac{p_2}{p_1 - 1} e_{p_1 + 1, p_2 + 1, p_3, \dots, p_n}. \tag{2.7}$$

Proof. Constant factor π^{-1} is omitted for shortness. Introduce the values

$$\Psi_1 = \varrho \sum_{m,k}^{N} f_k E_2(a_m - a_k) \overline{\psi_{q-1,0}(a_k)}$$

and

$$\Psi_2 = \varrho \sum_{m,k}^{N} \sum_{j=2}^{q} f_k^j E_{j+1}(a_m - a_k) \overline{\psi_{q-j,j-1}(a_k)}$$

in such a way that

$$A_q = \Psi_1 + \Psi_2.$$

Using (2.5) we have

$$\Psi_1 = \varrho \sum_{m,k}^{N} E_2(a_m - a_k) \overline{f_k \psi_{q-1}(a_k)}. \tag{2.8}$$

Since $f_k \psi_{q-1}(a_k)$ is the kth term of A_{q-1}, hence A_{q-1} is transformed into Ψ_1 by application of the following rule to every structural sum

$$e_{p_1, p_2, \dots, p_n} \longmapsto \varrho e_{2, p_1, p_2, \dots, p_n}. \tag{2.9}$$

Here, conjugation applies automatically due to index shifting in structural sum.
One can rewrite Ψ_2 as

$$\varrho \sum_{m,k}^{N} \sum_{j=1}^{q-1} f_k^{j+1} E_{j+2}(a_m - a_k) \overline{\psi_{q-j-1,j}(a_k)},$$

which has similar form to

$$A_{q-1} = \varrho \sum_{m,k}^{N} \sum_{j=1}^{q-1} f_k^{j} E_{j+1}(a_m - a_k) \overline{\psi_{q-j-1,j-1}(a_k)}. \tag{2.10}$$

Hence, it remains to find the proper transformation rules to derive symbolically Ψ_2 from A_{q-1}. Applying the following rule

$$E_{p_2}(a_k - a) \longmapsto -\frac{p_2}{j} E_{p_2+1}(a_k - a), \tag{2.11}$$

the expression $\psi_{q-j-1,j-1}(a_k)$ is transformed into $\psi_{q-j-1,j}(a_k)$ by Lemma 2.1. Note that the value of the denominator in (2.11) is related to the subscript of E_{j+1} in (2.10), hence one can rewrite rule (2.11) as follows

$$E_{p_2}(a_k - a) \longmapsto -\frac{p_2}{p_1 - 1} E_{p_2+1}(a_k - a). \tag{2.12}$$

The rest of the expression is transformed as follows

$$f_k^{j} E_{p_1}(a_m - a_k) \longmapsto f_k^{j+1} E_{p_1+1}(a_m - a_k). \tag{2.13}$$

Rules (2.12) and (2.13), taken together give the following rule expressed in structural sum notation

$$e_{p_1,p_2,\ldots,p_n} \longmapsto -\frac{p_2}{p_1 - 1} e_{p_1+1,p_2+1,p_3,\ldots,p_n}, \tag{2.14}$$

since (1.16) holds. Both (2.9) and (2.14) give (2.7).
The theorem is proved. $\qquad\qquad\square$

Consecutive construction rules for the structural sums can be illustrated by means of the tree displayed in Fig. 2.1. The tree graphically demonstrates how far have we advanced in the theoretical study and computer simulations of the micro-structure of composites. In the constructive homogenization (Gluzman et al., 2017, Chapter 3), expansions of two types are applied: the cluster expansion in f and the contrast expansion in ϱ. The effective conductivity up to $O(f^p)$ includes $A_1, A_2, \ldots, A_{p-1}$. Therefore, we go from left to right following arrows within the tree to involve higher powers of f in the cluster expansion.

The contrast parameter expansion, within the accuracy $O(\varrho^q)$ for the effective conductivity, contains the structural sums of order $q-1$. For instance, any formula for the

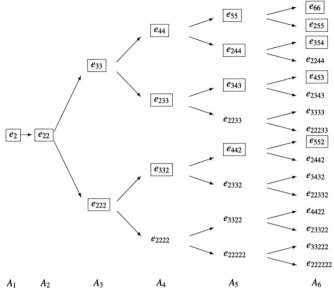

A_1 $\quad$ A_2 $\qquad$ A_3 $\qquad$ A_4 $\qquad$ A_5 $\qquad$ A_6

Figure 2.1 Graph of structural sums. One can see that number of terms of A_q doubles with each subsequent q. The subtree for the structural sums e_2, e_{pp} and $e_{p_1 p_2 p_3}$ is displayed in frames.

effective conductivity having the accuracy $O(\varrho^2)$, contains only the structural sum e_2. Any formula within the accuracy $O(\varrho^3)$ contains the structural sums e_2 and the infinite number of e_{pp} ($p = 2, 3, \ldots$), located on top of the graph in Fig. 2.1. It is worth noting that the Torquato–Milton parameter is based on accuracy $O(\varrho^3)$ and was exactly written in terms of e_2 and e_{pp} in Gluzman et al. (2017, Chapter 4, Sec. 2.3). Any formula within the accuracy $O(\varrho^4)$ contains the structural sums e_2, e_{pp} and $e_{p_1 p_2 p_3}$, etc. The fragment of corresponding infinite subtree of structural sums is displayed in frames in Fig. 2.1. Therefore, we go from left-top to right-bottom by the same arrows within the tree to involve higher powers of ϱ in the contrast expansion.

2.2.1.2 Pseudocode

Now we implement the algorithm given by Theorem 2.1 in MPL. The main recurrence procedure is presented in Algorithm 2.1. The procedure *next* is defined by the sequence of transformation rules shown in Algorithm 2.2. Note that the construction of algorithm provides an expanded form of A_q. For example, let us track calculations of the coefficient A_5:

$$A(5) = next(A(4)) = next(next(A(3))) = next(next(next(A(2))))$$
$$= next(next(next(\varrho^2 e_{22}))) = next(next(\varrho^3 e_{222} - 2\varrho^2 e_{33}))$$
$$= next(next(\varrho^3 e_{222}) + next(-2\varrho^2 e_{33}))$$
$$= next(\varrho^4 e_{2222} - 2\varrho^3 e_{332} - 2\varrho^3 e_{233} + 3\varrho^2 e_{44})$$

Algorithm 2.1 Recurrence procedure for computing coefficients A_q.

Require: q
Ensure: $A(q)$ – symbolic representation of the coefficient A_q
  1: **if** $q = 1$ **then**
  2:       **return** ϱe_2
  3: **if** $q = 2$ **then**
  4:       **return** $\varrho^2 e_{22}$
  5: **else**
  6:       **return** $next(A(q-1))$

Algorithm 2.2 Sequence of transformation rules defining the *next* procedure; c is an arbitrary constant.

Require: expression u
Ensure: expression u transformed by operator β (see Theorem 2.1)
  1: **NEXT_1.**   **if** $Kind(u) = $ " $+$ "   **then**

$$next(u) \longmapsto Map(next, u)$$

  2: **NEXT_2.**   **if** $u = c\varrho^k e_{p_1, p_2, \ldots, p_n}$ **then**

$$next(u) \longmapsto c\varrho^{k+1} e_{2, p_1, p_2, \ldots, p_n} - c\varrho^k \frac{p_2}{p_1 - 1} e_{p_1 + 1, p_2 + 1, p_3, \ldots, p_n}$$

  3: **NEXT_3.** $next(u) \longmapsto \text{``}next\text{''}(u),$

$$= next(\varrho^4 e_{2222}) + next(-2\varrho^3 e_{332}) + next(-2\varrho^3 e_{233}) + next(3\varrho^2 e_{44})$$
$$= \varrho^5 e_{22222} - 2\varrho^4 e_{3322} - 2\varrho^4 e_{2332} + 3\varrho^3 e_{442} - 2\varrho^4 e_{2233} + 6\varrho^3 e_{343}$$
$$+ 3\varrho^3 e_{244} - 4\varrho^2 e_{55}.$$

2.2.2   Algorithm 2: Reduction of dependent structural sums in A_q

In this section, we describe an algorithm that tackles the second problem. Namely, how to reduce dependent structural sums in a given expression of A_q. But first, we need to develop some theory.

2.2.2.1   Reduction of mirror terms in A_q

The structural sum representation is very convenient for the reduction of A_q. The following theorem describes a correspondence between the so-called *mirror sums*.

Theorem 2.2. *Let* $\alpha = \sum_{j=1}^{N} p_j$. *Then*

$$e_{p_1, p_2, p_3, \ldots, p_n}^{f_0, f_1, f_2, \ldots, f_n} = (-1)^{\alpha} \mathbf{C}^{n+1} e_{p_n, p_{n-1}, p_{n-2}, \ldots, p_1}^{f_n, f_{n-1}, f_{n-2}, \ldots, f_1}.$$

Proof. In view of (1.15) we have

$$
e_{p_1,p_2,p_3,\dots,p_n}^{f_0,f_1,f_2,\dots,f_n} = \frac{(-1)^{\alpha}}{\eta^{1+\frac{1}{2}(p_1+\cdots+p_n)}} \mathbf{C}^{n+1} \sum_{k_0,k_1,\dots,k_n} f_{k_n}^{t_n} f_{k_{n-1}}^{t_{n-1}} f_{k_{n-2}}^{t_{n-2}} \cdots f_{k_1}^{t_1} E_{p_n}(a_{k_n}-a_{k_{n-1}})
$$
$$
\times E_{p_{n-1}}(a_{k_{n-1}}-a_{k_{n-2}}) E_{p_{n-2}}(a_{k_{n-2}}-a_{k_{n-3}})\cdots \mathbf{C}^{n+1} E_{p_1}(a_{k_1}-a_{k_0}),
$$
$$(2.15)$$

since $E_p(z) = (-1)^p E_p(-z)$ holds (Weil, 1999). The definition of A_q (2.2) implies $t_n = 1$. Thus, one can rewrite the recurrence relation (1.17) as

$$
\begin{aligned}
&t_n = 1, \\
&t_s = p_{s+1} - t_{s+1}, \quad s = 0,1,2,\dots,n-1.
\end{aligned}
\qquad (2.16)
$$

Then, the right-hand side of (2.15) mirrors it left-hand side and equals $(-1)^{\alpha}\mathbf{C}^{n+1} e_{p_n,p_{n-1},p_{n-2},\dots,p_1}^{f_n,f_{n-1},f_{n-2},\dots,f_1}$.

The theorem is proved. $\square$

Note that α is odd for every structural sum in A_q. Hence, every pair of terms including mirror structural sums, reduces to either $2c\,\mathrm{Re}\,e_{p_1,p_2,p_3,\dots,p_n}$ or $2ce_{p_1,p_2,p_3,\dots,p_n}$, where Re stands for the real part of complex number and c is a constant factor. For example, the coefficient

$$
A_4 = \varrho^4 e_{2222} - 2\varrho^3 e_{233} - 2\varrho^3 e_{332} + 3\varrho^2 e_{44}
$$

can be written in simplified form

$$
A_4 = \varrho^4 e_{2222} - 4\varrho^3 e_{233} + 3\varrho^2 e_{44}.
$$

Computations performed at the end of the Section A.6.2 demonstrate that about half of terms vanishes (see Table 2.1), due to the Theorem 2.2.

One can implement the reduction of A_q by augmenting the set of simplification rules by yet another rule

$$
c_1 e_{p_1,p_2,p_3,\dots,p_n} + c_2 e_{p_n,p_{n-1},p_{n-2},\dots,p_1} \longmapsto c_1 e_{p_1,p_2,p_3,\dots,p_n} + c_2 \mathbf{C}^{n+1} e_{p_1,p_2,p_3,\dots,p_n},
$$

where c_1 and c_2 are arbitrary constants, and leave all the work to automatic simplification process of given CAS. However, we want such simplifications to be applied to A_q only, and the automatic simplification process may violate the final result by erroneously simplifying intermediate expressions (i.e. A_k ($k < q$)).

Constructing a procedure searching for pairs of mirror operands, seems to be a better option. One can traverse all terms of the particular coefficient A_q to find corresponding mirror structural sum among remaining operands, and reduce them. It requires asymptotically $\mathcal{O}(Length(A_q)^2)$ operations. However, a small modification of the algorithm given by Algorithms 2.1–2.2, provides for sorted operands in A_q. Therefore, the binary search can be applied to find and reduce all pairs of mirror terms. First, introduce an ordering relation in the set of operands as follows. We will say that

Table 2.1 Number of structural sums of order q ($q = 3, 4, \ldots, 18$) before and after reduction of mirror terms.

Number of terms	A_3	A_4	A_5	A_6	A_7	A_8	A_9	A_{10}
Before reduction	2	4	8	16	32	64	128	256
After reduction	2	3	6	10	20	36	72	136

Number of terms	A_{11}	A_{12}	A_{13}	A_{14}	A_{15}	A_{16}	A_{17}	A_{18}
Before reduction	512	1024	2048	4096	8192	16384	32768	65536
After reduction	272	528	1056	2080	4160	8256	16512	32896

expression $c_1 e_{p_1, p_2, \ldots, p_n}$ precedes expression $c_2 e_{w_1, w_2, \ldots, w_s}$ if vector $(p_1, p_2, \ldots, p_n)$ precedes $(w_1, w_2, \ldots, w_s)$ in lexicographic order. Since all multi-indexes of structural sums of given A_q differ, such a relation is sufficient to perform an exact search. Let us prove the following

Lemma 2.2. *Let A_q^1 and A_q^2 be expressions derived from A_{q-1} by application of rules (2.9) and (2.14), respectively. And let A_q be computed as $A_q^1 + A_q^2$. Such an arrangement of computations provides lexicographic order of operands in the output expression of A_q.*

Proof. The lemma is true for $q = 1, 2$. Suppose that it holds for $q = k$, where $k > 2$. Then operands of both A_{k+1}^1 and A_{k+1}^2 are sorted. One can see that all operands of A_q^1 precede operands of A_q^2. Hence, by mathematical induction, the lemma is proved. □

2.2.2.2 Pseudocode

Since modification of the Algorithms 2.1–2.2 seems to be an easy task, let us present pseudocodes of procedures for both reducing and comparing operands. For reducing operator see *Mirror_reduce*. Assume that we have implemented *Binary_search(x, L, j, k)* procedure, that searches for the expression x among the elements $L[m]$ ($j \leq m \leq k$) of the list L, and returns the index corresponding to x if it is found. Otherwise, it returns -1. We also omit describing implementation of *Mirror(u)* returning the expression u transformed according the rule

$$e_{p_1, p_2, \ldots, p_n} \longmapsto e_{p_n, p_{n-1}, \ldots, p_1}.$$

Here, n stands for the number of indexes of structural sum included in term $L[k]$. *Binary_search* uses the procedure comparing two different operands. For implementation in MPL of such a procedure, see *Less* operator in Algorithm 2.4.

Algorithm 2.3 *Mirror_reduce* operator; *n* stands for the number of indexes of structural sum included in *k*th term $L[k]$.

Require: expression *u* representing coefficient A_q
Ensure: *Mirror_reduce(u)* – expression *u* with reduced mirror sums
  1: $L := Operand\_list(u)$
  2: $k := 1$
  3: **while** $k < Length(L)$ **do**
  4:     $esum := Mirror(L[k])$
  5:     **if** $esum \neq L[k]$ **then**
  6:        $i := Binary\_search(esum, L, k+1, Length(L))$
  7:        $L := Delete(L[i], L)$
  8:        $L[k] := L[k] + C^{n+1}L[k]$
  9:     $k := k+1$
10: **return** $Construct(Sum, L)$

Algorithm 2.4 $Less(c_1 e_{p_1,p_2,...,p_q}, c_2 e_{w_1,w_2,...,w_s})$ operator; c_1, c_2 are arbitrary constants.

Require: two terms $c_1 e_{p_1,p_2,...,p_q}, c_2 e_{w_1,w_2,...,w_s}$ of A_q
Ensure: True if the first operand precedes the second one, otherwise – False
  1: $imax := Min(\{q, s\})$
  2: $i := 1$
  3: **while** $p_i = w_i$ **do**
  4:     **if** $i = imax$ **then**
  5:        **return** $q < s$
  6:     $i := i+1$
  7: **return** $p_i < w_i$

For example, let us trace several first iterations of the loop in the Algorithm 2.3.

$$
\begin{aligned}
A(6) =\ & \varrho^6 e_{222222} - 2\varrho^5 e_{22233} - 2\varrho^5 e_{22332} + 3\varrho^4 e_{2244} - 2\varrho^5 e_{23322} \\
& + 6\varrho^4 e_{2343} + 3\varrho^4 e_{2442} - 4\varrho^3 e_{255} - 2\varrho^5 e_{33222} + 4\varrho^4 e_{3333} \\
& + 6\varrho^4 e_{3432} - 12\varrho^3 e_{354} + 3\varrho^4 e_{4422} - 12\varrho^3 e_{453} - 4\varrho^3 e_{552} + 5\varrho^2 e_{66}
\end{aligned}
$$

$k = 1$ (condition in line 5 is false; the algorithm skips to the next iteration)

$$
\begin{aligned}
[\ &\boxed{\varrho^6 e_{222222}}, -2\varrho^5 e_{22233}, -2\varrho^5 e_{22332}, 3\varrho^4 e_{2244}, -2\varrho^5 e_{23322}, \\
& 6\varrho^4 e_{2343}, 3\varrho^4 e_{2442}, -4\varrho^3 e_{255}, -2\varrho^5 e_{33222}, 4\varrho^4 e_{3333}, \\
& 6\varrho^4 e_{3432}, -12\varrho^3 e_{354}, 3\varrho^4 e_{4422}, -12\varrho^3 e_{453}, -4\varrho^3 e_{552}, 5\varrho^2 e_{66}\ ]
\end{aligned}
$$

$k = 2$ (mirror terms are found; actual element is modified (line 8); the corresponding
       one is deleted (line 7) and, therefore, it is not taken into account in further

iterations)

$$[\quad \varrho^6 e_{222222}, \boxed{-2\varrho^5 e_{22233}}, -2\varrho^5 e_{22332}, 3\varrho^4 e_{2244}, -2\varrho^5 e_{23322},$$
$$6\varrho^4 e_{2343}, 3\varrho^4 e_{2442}, -4\varrho^3 e_{255}, \boxed{-2\varrho^5 e_{33222}}, 4\varrho^4 e_{3333},$$
$$6\varrho^4 e_{3432}, -12\varrho^3 e_{354}, 3\varrho^4 e_{4422}, -12\varrho^3 e_{453} - 4\varrho^3 e_{552}, 5\varrho^2 e_{66}]$$

$k = 3$ (mirror terms are found)

$$[\quad \varrho^6 e_{222222}, -4\varrho^5 e_{22233}, \boxed{-2\varrho^5 e_{22332}}, 3\varrho^4 e_{2244}, \boxed{-2\varrho^5 e_{23322}},$$
$$6\varrho^4 e_{2343}, 3\varrho^4 e_{2442}, -4\varrho^3 e_{255}, 4\varrho^4 e_{3333},$$
$$6\varrho^4 e_{3432}, -12\varrho^3 e_{354}, 3\varrho^4 e_{4422}, -12\varrho^3 e_{453}, -4\varrho^3 e_{552}, 5\varrho^2 e_{66}]$$

When the loop terminates, the list of operands of $A(6)$ shrinks to

$$[\quad \varrho^6 e_{222222}, -4\varrho^5 e_{22233}, -4\varrho^5 e_{22332}, 6\varrho^4 \mathrm{Re}\, e_{2244}, 12\varrho^4 \mathrm{Re}\, e_{2343},$$
$$3\varrho^4 e_{2442}, -8\varrho^3 e_{255}, 4\varrho^4 e_{3333}, -24\varrho^3 e_{354}, 5\varrho^2 e_{66}]$$

2.3   Algorithms for computing structural sums

The main goal of this section is to present a detailed investigation of algorithms and methods for structural sums computation. All algorithms presented can be relatively easy implemented by means of modern scientific computing packages, maintaining both efficiency of calculations and high level of abstraction.

In the previous section, we presented algorithms for computing symbolic representations of the coefficients A_q, in terms of structural sums only. Now, we focus on efficient algorithms for computing structural sums. In Section 2.3.1 we present some simplified formulas for e_{pp} ($p = 2, 3, \ldots$) and e_{2222}. Such structural sums are a special case of multidimensional discrete convolutions of functions introduced in Section 2.3.2. In the same section, we prove the theorem that leads to efficient calculation of the values of the convolutions. As we will see, an exponential running time of a naive approach to calculation of multidimensional discrete convolution of functions can be reduced to a linear one. This result allows the development of algorithms and methods for computing structural sums in Section 2.3.3. Another application is introduced in Section 2.3.4, where the recurrence algorithm for computing sets of structural sums of order q ($q = 1, 2, 3, \ldots$) is provided.

2.3.1   Simplified formulas for e_{pp} and e_{2222}

According to (A.32), e_m becomes the classical Eisenstein–Rayleigh sum S_m in the case $N = 1$.

The number of sums in formula (1.12) is equal to $(q + 1)$ and can be reduced as follows. Introduce the function

$$F_p(z) = \frac{1}{N} \sum_{k=1}^{N} E_p(z - a_k). \tag{2.17}$$

Then the sum e_{pp} can be written in the form

$$e_{pp} = \frac{1}{N^p} \sum_{k_0, k_1} E_p(a_{k_0} - a_{k_1}) \overline{F_p(a_{k_1})} = \frac{(-1)^p}{N^{p-1}} \sum_{k_1} F_p(a_{k_1}) \overline{F_p(a_{k_1})}$$

$$= \frac{(-1)^p}{N^{p+1}} \sum_{m=1}^{N} \left| \sum_{k=1}^{N} E_p(a_m - a_k) \right|^2. \tag{2.18}$$

Here, it is used that $E_p(-z) = (-1)^p E_p(z)$.

In order to transform e_{2222} introduce the function

$$G_2(z) = \frac{1}{N^2} \sum_{k_1, k_2} E_2(z - a_{k_1}) \overline{E_2(a_{k_1} - a_{k_2})}. \tag{2.19}$$

Then, the sum e_{2222} can be transformed similarly to (2.18)

$$e_{2222} = \frac{1}{N} \sum_{k=1}^{N} |G_2(a_k)|^2. \tag{2.20}$$

Formulas (2.18) and (2.20) contain one summation on N less than the definition (1.12). Such improvement is important for computations.

2.3.2 Calculation of multidimensional discrete chain convolutions of functions

In this subsection, we define multidimensional discrete chain convolution and estimate running time of the naive approach to computations. Furthermore, we prove a theorem leading to an efficient calculation.

Definition 2.2 (Multidimensional discrete chain convolution). Let $A = \{a_1, a_2, a_3, \ldots, a_N\}$ be a set of arbitrary complex numbers. The following sum

$$\sum_{k_0, k_1, k_2, \ldots, k_n} F_1(a_{k_0}, a_{k_1}) F_2(a_{k_1}, a_{k_2}) F_3(a_{k_2}, a_{k_3}) \cdots F_n(a_{k_{n-1}}, a_{k_n}), \quad n \in \mathbb{N}, \tag{2.21}$$

is called *multidimensional discrete chain convolution of functions* F_j $(j = 1, 2, 3, \ldots, n)$.

One can omit the summation over k_0 and k_n in (2.21) and obtain the function of two variables k_0 and k_n. The rest of the sums, over k_j, form a chain expression in a_{k_j}.

Example 2.1. The structural sum (1.12) is a multidimensional discrete chain convolution (2.21).

Remark 2.1. In the theory of signals and image analysis, the multidimensional discrete convolution is introduced as the following function of $s_0, s_1, \ldots, s_n \in \mathbb{Z}$

$$\sum_{k_0, k_1, k_2, \ldots, k_n \in \mathbb{Z}} f(k_0, k_1, \ldots, k_n) \, g(s_0 - k_0, s_1 - k_1, \ldots, s_n - k_n). \qquad (2.22)$$

Convolutions (2.21) and (2.22) are different mathematical objects, though formally (2.21) is the particular case of (2.22).

The sum (2.21) involves N^{n+1} terms, since the summation runs over all $(n + 1)$-tuples of N-element set. Each term is a product of n factors. In a naively implemented algorithm, we perform N^{n+1} additions and $(n-1)N^{n+1}$ multiplications. In case n is fixed, we have polynomial running time $O(N^n)$. Unfortunately, the varying n leads to an exponential running time $O(nN^n)$ as N is fixed. In the case of structural sums, the number of functions in convolution grows, while A is fixed. Hence, our main goal is to reduce the exponential complexity of the naive approach. Let us introduce the following matrix corresponding to function F_j:

$$C_{F_j} = \begin{pmatrix} F_j(a_1, a_1) & F_j(a_2, a_1) & \cdots & F_j(a_N, a_1) \\ F_j(a_1, a_2) & F_j(a_2, a_2) & \cdots & F_j(a_N, a_2) \\ \vdots & \vdots & \ddots & \vdots \\ F_j(a_1, a_N) & F_j(a_2, a_N) & \cdots & F_j(a_N, a_N) \end{pmatrix}. \qquad (2.23)$$

Let

$$C_j = \begin{cases} (1, 1, 1, \ldots, 1)^{\mathbf{T}} \in \mathbb{R}^N & \text{if } j = 0 \\ C_{F_j} \cdot C_{j-1} & \text{otherwise} \end{cases}, \qquad (2.24)$$

where $\mathbf{T}$ denotes the operation of the matrix transposition, and the dot "$\cdot$" stands for the matrix multiplication. Now let us formulate the theorem that tackles our problem and prove it by induction.

Theorem 2.3. *The sum of all coordinates of the vector[1] C_n equals the value of convolution (2.21).*

Proof. Let

$$\mathcal{F}_n(z) = \sum_{k_0, k_1, k_2, \ldots, k_{n-1}} F_1(a_{k_0}, a_{k_1}) F_2(a_{k_1}, a_{k_2}) F_3(a_{k_2}, a_{k_3}) \cdots F_n(a_{k_{n-1}}, z), n \in \mathbb{N}.$$

Since the value of convolution (2.21) is equal to $\displaystyle\sum_{k=1}^{N} \mathcal{F}_n(a_k)$, it suffices to show that

$$C_n = (\mathcal{F}_n(a_1), \mathcal{F}_n(a_2), \mathcal{F}_n(a_3), \ldots, \mathcal{F}_n(a_N))^{\mathbf{T}}. \qquad (2.25)$$

---

[1] We treat one-column matrix as a vector.

Let $n = 1$. Then (2.24) yields

$$C_1 = \left(\sum_{k=1}^{N} F_1(a_k, a_1), \sum_{k=1}^{N} F_1(a_k, a_2), \sum_{k=1}^{N} F_1(a_k, a_3), \ldots, \sum_{k=1}^{N} F_1(a_k, a_N) \right)^{\mathbf{T}}$$

$$= (\mathcal{F}_1(a_1), \mathcal{F}_1(a_2), \mathcal{F}_1(a_3), \ldots, \mathcal{F}_1(a_N))^{\mathbf{T}}.$$

Assume that (2.25) holds for $n = m$. Thus

$$C_{m+1} = C_{F_{m+1}} \cdot C_{F_m} = \left(\sum_{k=1}^{N} \mathcal{F}_m(a_k) F_{m+1}(a_k, a_1), \sum_{k=1}^{N} \mathcal{F}_m(a_k) F_{m+1}(a_k, a_2), \right.$$

$$\left. \sum_{k=1}^{N} \mathcal{F}_m(a_k) F_{m+1}(a_k, a_3), \ldots, \sum_{k=1}^{N} \mathcal{F}_m(a_k) F_{m+1}(a_k, a_N), \right)^{\mathbf{T}}.$$

Hence, by mathematical induction, the theorem is proved.   □

We call the vector C_n the *vector representation* of the convolution (2.21). One can use the alternative *matrix representation*, given as the following

Remark 2.2. One can similarly prove, setting

$$\mathcal{F}_n(z_1, z_2) = \sum_{k_0, k_1, k_2, \ldots, k_{n-2}} F_1(a_{k_0}, a_{k_1}) \cdots F_{n-1}(a_{k_{n-2}}, z_1) F_n(z_1, z_2), n \in \mathbb{N},$$

that the sum of all entries of the product

$$C_{F_n} \cdot C_{F_{n-1}} \cdot \ldots \cdot C_{F_2} \cdot C_{F_1} \tag{2.26}$$

equals the value of convolution (2.21).

The computation using vector representation (2.24) involves n matrix-vector multiplications of order N, hence it takes $O(nN^2)$ time. On the other hand, the use of a matrix representation (2.26) takes $O(nN^3)$ time. Hence, the running time of the naive approach has been significantly reduced, e.g., it is linear as n grows and N is fixed. This result is very important for the constructive theory of composites. It applies to the fixed system of disks, where the number of functions involved in structural sums grows, as the number of leading terms in approximation of the effective conductivity $\sigma_e(f)$ increases.

2.3.3 Efficient computation of structural sums

This subsection covers the application of Subsection 2.3.2 to structural sums computation. It is also concerned with implementation of algorithms in modern scientific computing software, using algebraic operations on multidimensional arrays. The subsection also provides pseudocodes of algorithms and demonstration of their effectiveness.

Let us show that structural sums form convolutions (2.21). Let k, j run over 1 to N. One can rewrite (1.15) in the following form

$$
\begin{aligned}
e^{f_0,f_1,f_2,\dots,f_n}_{p_1,p_2,p_3,\dots,p_n} = \frac{1}{\eta^{\delta+1}} \sum_{k_0,k_1,\dots,k_n} & f^{t_0}_{k_0} E_{p_1}(a_{k_0} - a_{k_1}) f^{t_1}_{k_1} \overline{E_{p_2}(a_{k_1} - a_{k_2})} \\
& \times f^{t_2}_{k_2} E_{p_3}(a_{k_2} - a_{k_3}) \cdots f^{t_{n-1}}_{k_{n-1}} \mathbf{C}^{n+1} E_{p_n}(a_{k_{n-1}} - a_{k_n}) f^{t_n}_{k_n}.
\end{aligned}
\tag{2.27}
$$

Hence, to get the form of convolution, it suffices to set

$$
F_m(a_k, a_j) = \begin{cases} f^{t_0}_k E_{p_1}(a_k - a_j) f^{t_1}_j & \text{if } m = 1 \\ \mathbf{C}^{m+1} E_{p_m}(a_k - a_j) f^{t_m}_j & \text{if } m > 1 \end{cases}.
\tag{2.28}
$$

Moreover, in case of identical disks, where $f_j = 1$ $(j = 1, 2, 3, \dots, N)$, the above formula reduces to

$$
F_m(a_k, a_j) = \mathbf{C}^{m+1} E_{p_m}(a_k - a_j).
\tag{2.29}
$$

2.3.3.1 Implementation notes

The complexity analysis in Section 2.3.2 does not take into account the process time spent on the computation of functions F_j $(j = 1, 2, 3, \dots, n)$. Notice that in (2.21) every function is called N^{n-1} times. On the other hand, the new approach allows to store all values as entries of matrices and to reuse them during computations. Moreover, usage of either vector or matrix representation of the convolution implies efficiency and simplicity of the computation implementation, due to highly optimized design of matrix operations in modern scientific packages. For example, Python module *NumPy* (van der Walt et al., 2011) is equipped with vectorized operations on data, stored in contiguous chunks of memory. Furthermore, the operands of algebraic operators can be either scalars or multidimensional arrays, ensuring efficient calculations, while maintaining a high level of abstraction. Hence, one can easily use operations on vectors and matrices such as Hadamard product of matrices, scalar multiplication, matrix addition, etc.

During the implementation process, one can take advantage of algebraic dependencies among functions involved in convolution. As an example, let us discuss dependencies occurring in structural sums. The Eisenstein functions $E_n(z)$ $(n = 2, 3, \dots)$ and the Weierstrass function $\wp(z)$ are related by the identities (A.26). It follows from the elliptic function theory (Akhiezer, 1990) that

$$
\wp''(z) = 6\wp(z)^2 - 30S_4.
\tag{2.30}
$$

Thus, each function $E_n(z)$ is an algebraic combination of $\wp(z)$ and $\wp'(z)$. One can implement both functions $\wp(z)$ and $\wp'(z)$ via the series expansion (Akhiezer, 1990, Table X, p. 204). Dependencies (A.26) and (2.30) can be implemented in any CAS, in order to calculate the symbolic representations of $E_n(z)$. Hence, to determine matrices (2.23) for Eisenstein functions combined in a given structural sum, it is sufficient to perform a set of algebraic operations on the *base* matrices $C_\wp$ and $C'_\wp$. In addition, it will be convenient to set $\wp(0) := 0$.

Following the discussion, let us sketch the scheme of calculations for a single structural sum, with the use of the vector representation of convolution. Hereafter, the following convention on operators is adopted. Let a be a scalar and $\mathbf{b} = (b_1, b_2, b_3, \ldots, b_N)$ be a vector. We define the following operations for matrices A and B

$$
\begin{aligned}
(A + B)_{i,j} &:= A_{i,j} + B_{i,j} & A \cdot \mathbf{b} &:= A \cdot \mathbf{b}^\mathsf{T} \\
(a + B)_{i,j} &:= a + B_{i,j} & \left(\overline{A}\right)_{i,j} &:= \overline{A_{i,j}} \\
(A * B)_{i,j} &:= A_{i,j} B_{i,j} & (A^a)_{i,j} &:= A^a_{i,j} \\
(a * B)_{i,j} &:= a B_{i,j} & (\mathbf{b}^a)_i &:= \mathbf{b}^a_i \\
(\mathbf{b} * B)_{i,j} &:= v_i B_{i,j} &
\end{aligned}
$$

For illustration, let us consider the sum e_{277}. By (A.26) we have

$$
E_7(z) = \frac{1}{2}(3 S_4 - \wp(z)^2)\wp'(z). \tag{2.31}
$$

Assume the entries of matrices $C_\wp$ and $C'_\wp$ are already computed, therefore it follows from the definition of the value of E_n at the origin (see (A.32)), that

$$
C_{E_2} = C_\wp + S_2
$$

and

$$
C_{E_7} = \left[\frac{1}{2}(3 S_4 - C_\wp{}^2) * C_{\wp'}\right] * J + S_7 * I,
$$

where I denotes the identity matrix, and J is the all-ones matrix with the main diagonal being zeroed. Let

$$
\mathbf{v} = (f_1, f_2, f_3, \ldots, f_N)
$$

and assume the precedence of the matrix multiplication "$\cdot$" before the operation "$*$". Thus, from (2.28), (1.16) and (2.24), the sum of all elements of the array

$$
\mathbf{v}^1 * C_{E_2} \cdot (\mathbf{v}^6 * \overline{C_{E_7}} \cdot (\mathbf{v}^1 * C_{E_7} \cdot \mathbf{v}^1))
$$

equals $(\sum_{j=1}^{N} f_j)^9 e_{277}$.

This scheme can be easily implemented. For example, the code snippet on Algorithm 2.5 presents how operators of the above expressions appear in NumPy. The above process of determining the value of e_{277} can be generalized in the form of Algorithm 2.6. Here, lines 1 and 4 correspond to dependencies (1.16). In case of monodisperse disks, the procedure simplifies to the Algorithm 2.7.

Algorithm 2.5 Example of implementation in Python.

```
import numpy as np

# Cp, Cpp represents the base matrix for Weierstrass function $\wp$, $\wp'$
# S2, S4, S6, S7 represent constants S_2, S_4, S_6, S_7

I = np.eye(len(A), dtype=int)
J = np.ones((len(A), len(A)), dtype=np.int) - I

# required matrices
C2  = Cp + S2
C7  = 0.5 * (3*S4 - Cp**2) * Cpp * J + S7 * I

# vector representation
repr = v * C[2].dot(v**6 * np.conjugate(C[7]).dot(v * C[7].dot(v)))

# the value of e277
delta = (2 + 7 + 7)/2
e277 = repr.sum() / sum(v)**(delta + 1)
```

Algorithm 2.6 Calculate structural sum $e_{p_1,p_2,p_3,\ldots,p_n}$ for polydisperse disks.

Require: multi-indexes $p_1, p_2, p_3, \ldots, p_n$ of structural sum
Ensure: value of structural sum $e_{p_1,p_2,p_3,\ldots,p_n}$
1: $t = p_1 - 1$
2: $temp = \mathbf{v}^{p_1-1} * C_{E_{p_1}} \cdot \mathbf{v}$
3: **for** k in $(2,3,4,\ldots,n)$
4: $\quad t = p_k - t$
5: $\quad temp = \mathbf{v}^t * \mathbf{C}^{k+1} C_{E_{p_k}} \cdot temp$
6: $\delta = \frac{1}{2}\sum_{j=1}^{n} p_j$
7: $\eta = \sum_{j=1}^{N} f_j$
8: **return** $\left(\sum_{j=1}^{N} temp_j\right)/\eta^{\delta+1}$

Algorithm 2.7 Calculate structural sum $e_{p_1,p_2,p_3,\ldots,p_n}$ for monodisperse disks.

Require: multi-indexes $p_1, p_2, p_3, \ldots, p_n$ of structural sum
Ensure: value of structural sum $e_{p_1,p_2,p_3,\ldots,p_n}$
1: $temp = (1,1,1,\ldots,1) \in \mathbb{R}^N$
2: **for** k in $(1,2,3,\ldots,n)$
3: $\quad temp = \mathbf{C}^{k+1} C_{E_{p_k}} \cdot temp$
4: $\delta = \frac{1}{2}\sum_{j=1}^{n} p_j$
5: **return** $\left(\sum_{j=1}^{N} temp_j\right)/N^{\delta+1}$

Algorithm 2.8 Calculate structural sum e_{p_1,p_2} via naive approach.

Require: multi-indexes p_1, p_2 of structural sum
Ensure: value of structural sum e_{p_1,p_2}
1: $s = 0$
2: **for** k_0 **in** $(1, 2, 3, \ldots, N)$
3:     **for** k_1 **in** $(1, 2, 3, \ldots, N)$
4:         **for** k_2 **in** $(1, 2, 3, \ldots, N)$
5:             $s = s + E_{p_1}(a_{k_0} - a_{k_1}) E_{p_2}(a_{k_1} - a_{k_2})$
6: $\delta = \frac{1}{2}(p_1 + p_2)$
7: **return** $s/N^{\delta+1}$

2.3.3.2 Computational effectiveness

Let us investigate the effectiveness of proposed techniques implemented with Python and NumPy. The naive approach algorithm was implemented as multiple nested `for` loops procedure. For example, the pseudo-code of the procedure computing structural sum e_{p_1,p_2} for monodisperse disks takes the form of Algorithm 2.8. Since the native to Python `for` loop is a bit slow, all functions of naive approach were compiled with the use of Cython (Bradshaw et al., 2011). On the other hand, procedures Algorithm 2.6 and Algorithm 2.7 were implemented without any optimization, hence the high level of abstraction has been preserved. The average processing time was measured using `timeit` module of Python. The values obtained are presented on Table 2.2. Note that these results do not include the processing time spent on the calculation of either the values of functions $E_n(z)$, or the entries of required matrices C_{F_n}. Computations involving the vector representation remain efficient for higher values of n (see Table 2.3). All calculations were performed on a Laptop PC with I7-4712MQ CPU and 16 GB RAM under Debian operating system.

2.3.4 Computing structural sums of a given order

In this subsection, we investigate the problem of determining values of all structural sums of a given order. As we will see, the vector representation allows to construct recurrence algorithms for computing structural sums directly, using the representations of lower order sums.

Note that the coefficients A_q involve mirror structural sums. We will apply Theorem 2.2 in order to consider independent sums only. Let M_q be the set of all structural sums of order q. By the virtue of Theorem 2.1, consecutive sets M_q can be generated iteratively as follows:

$$
\begin{aligned}
&M_1 = \{e_2\}, \quad M_2 = \{e_{22}\}, \quad M_3 = \{e_{222}, e_{33}\} \\
&M_4 = \{e_{2222}, e_{233}, e_{332}, e_{44}\} \\
&M_5 = \{e_{22222}, e_{2233}, e_{2332}, e_{244}, e_{3322}, e_{343}, e_{442}, e_{55}\}. \\
&M_6 = \{e_{222222}, e_{22233}, e_{22332}, e_{2244}, e_{23322}, e_{2343}, e_{2442}, e_{255}, \\
&\qquad\quad e_{33222}, e_{3333}, e_{3432}, e_{354}, e_{4422}, e_{453}, e_{552}, e_{66}\}.
\end{aligned}
\tag{2.32}
$$

Table 2.2 Impact on performance in terms of average processing time in seconds. Data are for: $e_{p_1,\ldots,p_n}$, where $p_j = 2$ $(j = 1, 2, \ldots, n)$ and $n = 2, 3, \ldots, 8$. Note that some values were omitted due to exponential increase of running time. Calculations were performed for random samples of N complex numbers.

n	Naive approach (sec)	Vector representation (sec)	Matrix representation (sec)
$(N = 10)$			
2	0.0000472	0.0000096	0.0000087
3	0.0006007	0.0000124	0.0000109
4	0.0081279	0.0000125	0.0000143
5	0.1031724	0.0000131	0.0000174
6	1.2209936	0.0000150	0.0000205
7	14.3845958	0.0000161	0.0000227
8	165.9913234	0.0000180	0.0000275
$(N = 100)$			
2	0.0414119	0.0000738	0.0007620
3	6.0027449	0.0000812	0.0014421
4	807.3853987	0.0001074	0.0021484
5	–	0.0001266	0.0028418
6	–	0.0001479	0.0035734
7	–	0.0001635	0.0042517
8	–	0.0001895	0.0049600
$(N = 1000)$			
2	40.0561751	0.0057021	0.6622977
3	–	0.0068790	1.3288249
4	–	0.0109440	2.0072163
5	–	0.0132987	2.7095607
6	–	0.0161475	3.3095384
7	–	0.0174779	4.0035864
8	–	0.0213148	4.6563258

Let G_q denote subset of M_q, containing independent sums only. For instance, the set G_6 takes the following form:

$$G_6 = \{e_{222222}, e_{23322}, e_{2442}, e_{33222}, e_{3333}, e_{3432}, e_{4422}, e_{453}, e_{552}, e_{66}\}.$$

Whether we consider structural sums as components of the effective conductivity formula (2.1), or as the set of components of the structural features vector defined in Section 2.5, the key problem is to determine values of structural sums in several initial orders. For example, the approximation of (2.1) as the polynomial of degree $q + 1$, requires to compute all elements of $G'_q = \bigcup_{j=1}^{q} G_j$.

Table 2.3 Impact on performance in terms of average processing time in seconds for fixed $N = 1000$. Data are for: $e_{p_1, \ldots, p_n}$, where $p_j = 2$ ($j = 1, 2, \ldots, n$) and $n = 9, \ldots, 20$.

n	Vector representation CPU (sec)	Matrix representation CPU (sec)	n	Vector representation CPU (sec)	Matrix representation CPU (sec)
9	0.028	5.225	15	0.044	9.181
10	0.029	5.801	16	0.047	9.858
11	0.032	6.443	17	0.049	10.475
12	0.037	7.290	18	0.051	11.224
13	0.039	7.857	19	0.053	11.681
14	0.042	8.555	20	0.057	12.371

In the naive approach, all sums are calculated separately with the use of the vector representation of convolution. Alternatively, we can store the vector representations of structural sums during calculations. In such a case, the vector representation of $e_{p_1, \ldots, p_n}$ can be determined by multiplying the matrix $C_{E_{p_n}}$ by vector representation of $e_{p_1, \ldots, p_{n-1}}$. For instance, representations of sums e_{222222}, e_{23322}, e_{2442} from G_6 above, can be simply produced from representations of e_{22222}, e_{3322}, e_{442} from G_5 by only one matrix-vector multiplication per sum.

2.3.4.1   Recurrence algorithm

The above considerations suggest an algorithm, where structural sums of consecutive orders are calculated using representations of lower order sums via matrix-vector multiplications. Let *res* be a dictionary structure,[2] with keys being tuples corresponding to multi-indexes $(p_1, \ldots, p_n)$ of structural sums. Each tuple is associated with one-dimensional array being vector representation denoted by $e(p_1, \ldots, p_n)$. In the first step we compute and store matrices C_{E_k} ($k = 2, 3, \ldots, n$) (see Subsection 2.3.3). Then we proceed with computations of structural sums of consecutive orders, starting from order 2. The calculations are performed via recursive procedure equipped with a memoization technique[3]: once computed representations are stored in the dictionary *res*, in order to reuse them in calculations of higher orders sums. The pseudocode of the procedure returning $e(p_1, \ldots, p_n)$ in case of (2.29), where centers of monodisperse disks are considered, has the form of Algorithm 2.9. Here, the precedence of the conjugation operator **C** before the matrix multiplication is assumed. The recursive call and storing results are performed in line 6, whereas lines 1–2 and 3–4 contain the base cases.

When it comes to (2.28), where disks of different radii are considered, the code requires only minor changes. The modified pseudocode takes the form of Algorithm 2.10. Here, lines 6–8 correspond to the dependencies (1.16), and $\mathbf{v} = (f_1, f_2, f_3, \ldots, f_N)$ is the vector of the constants (1.14) describing polydispersity.

---

[2] For example, `dictionary` object in Python.

[3] In Python one can apply `lru_cache` decorator from `functools` module.

Algorithm 2.9 Recurrence procedure for computing representations of structural sums from G_q' for monodisperse disks.

Require: multi-indexes $p_1, p_2, p_3, \ldots, p_n$ of structural sum
Ensure: vector representation $e(p_1, p_2, p_3, \ldots, p_n)$
  1: **if** $(p_1, \ldots, p_n) \in res$ **then**
  2:      **return** $res[p_1, \ldots, p_n]$
  3: **if** $n = 1$ **then**
  4:      $res[p_1, \ldots, p_n] = C_{E_{p_1}} \cdot (1, 1, 1, \ldots, 1)^{\mathbf{T}}$
  5: **else**
  6:      $res[p_1, \ldots, p_n] = \mathbf{C}^{n+1} C_{E_{p_n}} \cdot e(p_1, \ldots, p_{n-1})$
  7: **return** $res[p_1, \ldots, p_n]$

Algorithm 2.10 Recurrence procedure for computing representations of structural sums from G_q' for polydisperse disks.

Require: multi-indexes $p_1, p_2, p_3, \ldots, p_n$ of structural sum
Ensure: vector representation $e(p_1, p_2, p_3, \ldots, p_n)$
  1: **if** $(p_1, \ldots, p_n) \in res$ **then**
  2:      **return** $res[p_1, \ldots, p_n]$
  3: **if** $n = 1$ **then**
  4:      $res[p_1, \ldots, p_n] = \mathbf{v}^{p_1 - 1} * C_{E_{p_1}} \cdot \mathbf{v}$
  5: **else**
  6:      $t_n = 1$
  7:      **for** p **in** $(p_1, \ldots, p_n)$
  8:          $t_n = p - t_n$
  9:          $res[p_1, \ldots, p_n] = \mathbf{v}^{t_n} * \mathbf{C}^{n+1} C_{E_{p_n}} \cdot e(p_1, \ldots, p_{n-1})$
 10: **return** $res[p_1, \ldots, p_n]$

2.3.4.2 Effectiveness of algorithm

The above algorithms were implemented in Python. The simulation of computation of G_q' up to $q = 20$ was performed in order to observe and quantify the efficiency improvement. Presented approaches were compared with regard to the average number of matrix-vector products per sum in determining the set G_q'. In the naive approach every sum in G_q' is calculated separately, hence it requires $\frac{q+1}{2}$ multiplications. When the recurrence procedure is used, starting from a certain value of q, a single structural sum from G_q' consumes 1.8 operations in average. The results are presented in Table 2.4. Note that application of memoization technique requires a certain amount of memory resources for storage. The memory usage was estimated in terms of maximal number of representations stored during the calculations of G_q' (see Table 2.4). For randomly generated sample of disks the impact on performance is quantified in terms of average processing time spent on calculations of G_q'. It is presented in Table 2.5. Again, the processing time spent on computing of the underlying matrices and the values of Eisenstein functions has not been taken into account. Note that thousands of

Table 2.4 The average number of matrix-vector products per sum and the maximum number of vector representations stored in memory due to memoization. Data are for G'_q ($q = 1, 2, 3, \ldots, 20$).

q	Number of sums in $\bigcup_{j=1}^{q} M_j$	Number of sums in G'_q	Naive approach products per sum	Algorithms 2.9 and 2.10	
				Products per sum	max. storage
1	1	1	1.0	1.0000	0
2	2	2	1.5	1.0000	1
3	4	4	2.0	1.2500	1
4	8	7	2.4	1.2857	3
5	16	13	2.9	1.4615	6
6	32	23	3.4	1.4783	11
7	64	43	3.9	1.6047	20
8	128	79	4.4	1.6203	37
9	256	151	4.9	1.7020	72
10	512	287	5.4	1.7143	137
11	1024	559	6.0	1.7621	272
12	2048	1087	6.5	1.7700	529
13	4096	2143	7.0	1.7961	1056
14	8192	4223	7.5	1.8006	2081
15	16384	8383	8.0	1.8143	4160
16	32768	16639	8.5	1.8167	8257
17	65536	33151	9.0	1.8237	16512
18	131072	66047	9.5	1.8249	32897
19	262144	131839	10.0	1.8285	65792
20	524288	263167	10.5	1.8291	131329

samples in different configurations are used in applications. In such cases the memory consumption seems less important, in light of benefits gained from faster computing.

2.3.5 Convolutions of functions in case of elastic composites

In the next chapters, in order to study the elastic composites, we will need yet another type of structural sums (1.18), denoted by $e_{\mathbf{n}}^{(\mathbf{j})}$, and shown below. One can prove the following property of $e_{\mathbf{n}}^{(\mathbf{j})}$ similar to the mirror structural sums from Subsection 2.2.2.

Lemma 2.3 (Drygaś, 2016a). *Let* $\alpha = \sum_{s=1}^{p} (n_s - j_s)$. *Then, the following relations hold*

$$e_{n_1,\ldots,n_p}^{(j_1,\ldots,j_p)(l_1,\ldots,l_p)} = (-1)^{\alpha} e_{n_p,\ldots,n_1}^{(j_p,\ldots,j_1)(l_p,\ldots,l_1)}$$

and

$$e_{n_1,\ldots,n_p}^{(j_1,\ldots,j_p)(l_1+1,\ldots,l_p+1)} = C e_{n_1,\ldots,n_p}^{(j_1,\ldots,j_p)(l_1,\ldots,l_p)} = \overline{e_{n_1,\ldots,n_p}^{(j_1,\ldots,j_p)(l_1,\ldots,l_p)}}.$$

Table 2.5 Impact on performance in terms of average processing time in seconds for fixed $N = 100$. Data are for $n = 2, 3, 4, \ldots, 20$. In naive approach, every sum is computed separately via Algorithm 2.7 and Algorithm 2.6.

n	Monodispersed disks (2.29)		Polydispersed disks (2.28)	
	Naive approach	Algorithm 2.9	Naive approach	Algorithm 2.10
2	0.000276	0.000127	0.004734	0.000133
3	0.000254	0.000157	0.000417	0.000287
4	0.000478	0.000271	0.000841	0.000508
5	0.000988	0.000537	0.001920	0.001037
6	0.001987	0.000942	0.003823	0.002029
7	0.004161	0.001951	0.007924	0.003722
8	0.008373	0.003398	0.016268	0.006780
9	0.017690	0.006711	0.034786	0.009845
10	0.037012	0.012927	0.044799	0.014101
11	0.077664	0.032321	0.086187	0.027468
12	0.158632	0.047103	0.182847	0.053663
13	0.335148	0.093760	0.371405	0.107147
14	0.700936	0.186772	0.806131	0.211722
15	1.477388	0.375465	1.650290	0.428172
16	3.061254	0.748324	3.587924	0.859925
17	6.405137	1.503426	7.596746	1.704431
18	13.618074	3.063568	15.757380	3.501950
19	29.939919	6.517275	32.971304	6.988692
20	61.547285	12.119296	69.307875	14.073094

Subsection 2.3.2 gives the general framework for calculation of the multidimensional convolutions of functions. This scheme was realized with additional optimizations in Subsections 2.3.3 and 2.3.4, in case of structural sums arisen in the effective conductivity functional $\sigma_e(f)$. However, the analogous sums are encountered in the problem of elastic composites with circular inclusions presented in Chapter 4. Indeed, we can rewrite the right-hand side of generalized structural sums (1.18) in a more readable form:

$$\sum_{k_0, k_1, \ldots, k_N} f_0(a_{k_0}) \mathbf{C}^{l_1} E_{p_1}^{(j_1)}(a_{k_0} - a_{k_1}) f_1(a_{k_1}) \mathbf{C}^{l_2} E_{p_2}^{(j_2)}(a_{k_2} - a_{k_2}) \ldots$$

$$\times f_{N-1}(a_{k_{N-1}}) \mathbf{C}^{l_N} E_{p_N}^{(j_N)}(a_{k_{N-1}} - a_{k_N}) f_N(a_{k_N}), \quad n \in \mathbb{N}, \tag{2.33}$$

which reflects the pattern (2.21) with

$$F_m(a_{k_1}, a_{k_2}) = \begin{cases} f_0(a_{k_1}) \mathbf{C}^{l_m} E_{p_1}^{(j_1)}(a_{k_1} - a_{k_2}) f_1(a_{k_2}) & \text{if } m = 1, \\ \mathbf{C}^{l_m} E_{p_m}^{(j_m)}(a_{k_1} - a_{k_2}) f_m(a_{k_2}) & \text{if } m > 1, \end{cases} \tag{2.34}$$

where k_1, k_2 run over 1 to n, j equals 0 or 1, and f as function of variable k is really a function of variables R_k, ϱ_{1k}, ϱ_{2k} and ϱ_{3k}.

2.4 Symbolic calculations for systems of functional equations

Let us present a general model of symbolic computations, designed for the solution of the system of functional equations (6.40), in the form of analytic approximation (6.57). The main difficulty in programming of such a procedure consists in handling of *nested* summations. In fact, they are *indefinite*, i.e., we do not know in advance the number of functions u_k. Hence, we need a representation of the sum that maintains the summation index in a symbolic form.

2.4.1 Indefinite symbolic sums

Let $A_1, A_2, A_3, \ldots, A_n$ be arbitrary sets, and let $f(\mathbf{a})$ be an arbitrary function of $\mathbf{a} = [a_1, a_2, a_3, \ldots, a_n] \in A_1 \times A_2 \times A_3 \times \ldots \times A_n$. Let us define the following sum:

$$\sum_{\mathbf{a}} f(\mathbf{a}) := \sum_{a_1 \in A_1, \ldots, a_n \in A_n} f(a_1, \ldots, a_n) = \sum_{a_1 \in A_1} \sum_{a_2 \in A_2} \ldots \sum_{a_n \in A_n} f(a_1, \ldots, a_n). \qquad (2.35)$$

We do not discuss the existence of (2.35), and we assume that the information about the form of A_k does not matter. Hereafter, all formulas are considered as *symbolic expressions*. Let us introduce the following *indefinite symbolic sum* as a representation of (2.35):

$$sum(f(a_1, \ldots, a_n), [a_1, \ldots, a_n]), \qquad (2.36)$$

where the operands are, correspondingly, a summand and a sequence of symbols over which the sum is taken. For example, using this representation, the sum $\sum_{m,n=1}^{N} |a_m - a_m|$ is expressed as follows:

$$sum(|a_1 - a_2|, [a_1, a_2]).$$

The key point of the presented approach to computations using *sum*, is to prevent the expression undergoing symbolic manipulations from generating nested *sum* subexpressions. This leads to the following rules in the *automatic simplification* algorithms (Cohen, 2002):

Rule 1. $sum(f(\mathbf{a}) + g(\mathbf{a}), \mathbf{a}) \rightarrow sum(f(\mathbf{a}), \mathbf{a}) + sum(g(\mathbf{a}), \mathbf{a})$

Rule 2. $f(\mathbf{a}) sum(g(\mathbf{b}), \mathbf{b}) \rightarrow sum(f(\mathbf{a})g(\mathbf{b}), \mathbf{b})$

Rule 3. If $\mathbf{a} \cap \mathbf{b} = \emptyset$, then

$$sum(sum(f(\mathbf{a}, \mathbf{b}), \mathbf{b}), \mathbf{a}) \rightarrow sum(f(\mathbf{a}, \mathbf{b}), [\mathbf{a}, \mathbf{b}]),$$

where $[\mathbf{a}, \mathbf{b}] = [a_1, \ldots, a_n, b_1, \ldots, b_n,]$.

Rule 4. $sum(0, \mathbf{a}) \rightarrow 0$

2.4.2 Symbolic representation of successive approximations

The following system expresses the generalized form of (6.40):

$$u_k(\mathbf{x}) = c_k + f(\mathbf{x}) + \sum_{m \neq k} g(\mathbf{x}, a_m)(u_m \circ s)(\mathbf{x}, a_m), \quad k = 1, 2, \ldots, n, \qquad (2.37)$$

where $u_k(\mathbf{x})$ $(k = 1, 2, \ldots, n)$, c_k are unknown and $\circ$ denotes function composition. The generalized form of the system can be applied, for instance, to the effective conductivity of 2D composite materials (Gluzman et al., 2017, eq. (2.3.85)). The system (2.37) can be rewritten in a symbolic form:

$$u_k(\mathbf{x}) = c_k + f(\mathbf{x}) + sum \left(g(\mathbf{x}, a_m)(u_m \circ s)(\mathbf{x}, a_m), [a_m] \right), \quad k = 1, 2, \ldots, n,$$

assuming that summation excludes a_k. The successive approximations yield following relations:

$$u_{k,0}(\mathbf{x}) = c_{k,0} + f_0(\mathbf{x})$$
$$u_{k,q}(\mathbf{x}) = c_{k,q} + f_q(\mathbf{x}) + sum(g(\mathbf{x}, a_{m,q-1})(u_{m,q-1} \circ s)(\mathbf{x}, a_{m,q-1}), [a_{q-1}])$$
$$k = 1, 2, \ldots, n,$$

where $u_{k,q}$ denotes the approximation of u_k in qth iteration. Moreover, the remaining expressions indexed by q, correspond to expressions introduced to the solution by the qth iteration. For the sake of implementation, one can omit indexes k and m. This yields following relations:

$$u_0(\mathbf{x}) = c_0 + f_0(\mathbf{x}),$$
$$u_q(\mathbf{x}) = c_q + f_q(\mathbf{x}) \qquad (2.38)$$
$$+ sum(Algebraic\_expand(g(\mathbf{x}, a_{q-1})(u_{q-1} \circ s)(\mathbf{x}, a_{q-1})), [a_{q-1}]).$$

It is easy to prove by induction and applying rules 1–4, that each approximation $u_q(\mathbf{x})$ can be rewritten as a sum of non-nested *sum* expressions and of the term $f(\mathbf{x}) + c_q$. Note that the *Algebraic_expand* procedure in (2.38) keeps the summand expanded, and lets the rules work properly. The relation (2.38), as well as the rules 1–4, can be implemented in CAS. One can revert the symbolic expression $u_q(\mathbf{x})$ to an analytic formula, through the two simple steps:

- symbols c_q, a_q correspond to c_k, a_k in (2.37);
- symbols c_m, a_m for $m < q$ transform to the summation, e.g.

$$sum(h(c_3, c_4, a_3, a_4), [a_3, a_4]) \text{ corresponds to } \sum_{\substack{m \neq k \\ s \neq m}} h(c_s, c_m, a_s, a_m).$$

2.4.3 Application to the 3D effective conductivity

Let us sketch an explicit algorithm for computing the analytic approximation (6.57) of functions $u_k(\mathbf{x})$, up to $O(r_0^{q+1})$, using presented model. The procedure follows

Section 6.3, and is expressed in MPL (see Appendix A.5). The construction of the procedure is based on the following *transformation rules*, describing symbolic operations of substituting terms in a given form:

Rule 5. $sum(f(\mathbf{a}), \mathbf{a}) \rightarrow sum(Algebraic\_expand(f(\mathbf{a})), \mathbf{a})$

Rule 6. $sum(f(\mathbf{a}), \mathbf{a}) \rightarrow sum(Series(f(\mathbf{a}), r_0, 0, q), \mathbf{a})$

The *Series* operator in the rule 6 produces a power series expansion for $f(\mathbf{a})$ with respect to r_0, about the point 0 to order q. Some CAS have the capability of implementation of the transformation rules through the *rule-based programming*. Hence, the Algorithm 2.11 can be directly realized. In lines 8–11 constants c_m are found by calculating z_q explicitly up to $O(r_0^{q+1})$ by successive approximations. Then, c_m are substituted to the solution in lines 12–13. The $Reindex(expr, q, m)$ operator substitutes all symbols of the form z_m, a_m into $expr$ by z_{2m-q}, a_{2m-q}, respectively, in order to prevent indices from duplicating in the resulting expression. Note that negative indices may appear. However, it does not matter in the model used in computations. More details on the procedure $u(q)$, as well as an example of its implementation, will be published elsewhere.

Algorithm 2.11 Procedure for computing approximations $u_q(\mathbf{x})$.

Require: q
Ensure: $u(q)$ – symbolic representation of the approximation $u_q(\mathbf{x})$
1: $expr := u_q(\mathbf{x})$ // computed via (2.38)
2: $expr := Substitute(expr, \text{Rule 6})$
3: $zexpr := expr$
4: $zexpr := Substitute(zexpr, \mathbf{x} \rightarrow \mathbf{a}_q)$
5: **for each** $m \leq q$:
6:       $zexpr := Substitute(zexpr, c_m \rightarrow a_{m_j} - z(m))$
7: $zexpr := Solve(zexpr = 0, z_q)$ // based on (6.53)
8: **while** *not* $Free\_of(zexpr, z)$:
9:       **for each** $m < q$:
10:           $zexpr := Substitute(zexpr, z_m \rightarrow Reindex(zexpr, q, m))$
11:           $zexpr := Sequential\_substitute(zexpr, [\text{Rule 5, Rule 6}])$
12: **for each** m:
13:       $expr := Substitute(expr, c_m \rightarrow a_{m_j} - Reindex(zexpr, q, m))$
14: $expr := Sequential\_substitute(expr, [\text{Rule 5, Rule 6}])$
15: $Return(expr)$

2.5 Structural information and structural features vector

In this section we present a general approach to extracting new information processing tools, based on the experience gained from computational science. Then, we realize a concrete scheme, based on the so-called *structural features vector*, for the data represented by configurations of circles or points on the plane. Such systems may, for example, represent objects extracted from digital images. The section ends with a discussion related to extracting of an analogous vector for the three-dimensional data.

2.5.1 Pattern recognition and feature vectors

The main goal of computer science is *information processing* and construction of related new tools. Data processing is crucial in most cases of *pattern recognition*. Pattern recognition consists of signal processing, image processing and artificial intelligence (Daintith and Wright, 2008). The main problem consists in an automatic detection of regularities in data. Such detection can facilitate the decision making (Bishop, 2006). For instance, allow for classification of samples, based on existing knowledge or on statistical information extracted from data, as well as from its *representations* (Murty and Devi, 2011).

The important part of applications is the *pre-processing* stage, or transforming data into a form amenable to an algorithm. It often results in construction of a data representation in the form of a vector of numbers, called *feature vector*. The key element of this process is *feature extraction*, meaning isolating and detecting desired features for identifying or interpreting meaningful information gathered from the data. This is an extremely important activity in machine learning, where the transformation of the data into a new *feature space* precedes attending to classification or regression problems. In the space of feature vectors the solution of the problem may be more feasible.

In order to map the data to the feature space a specific tool for information processing is needed. We present below a generalized scheme illustrating how such tools are derived from the available solutions to various problems of computational sciences. The realization of this scheme will also be presented based on the structural feature vector.

2.5.2 From computational science to information processing

Computational sciences can suggest new tools for information processing. Assume that we want to extract a feature vector $\mathbf{X}$ related to the information $\mathcal{I}$. Let us suggest the following generalized scheme of finding $\mathbf{X}$:

1. Formulate a problem (e.g. physical problem) including information $\mathcal{I}$ in *implicit form*; there should be a premise that the solution $\mathcal{S}(\mathcal{I})$ of the problem is potentially related to $\mathcal{I}$.
2. Create a mathematical model $\mathcal{M}(\mathcal{I})$ based on physical or engineering principles.
3. Develop a constructive analytical computational model $\mathcal{C}(\mathcal{I})$, enabling you to get $\mathcal{S}(\mathcal{I})$ in an analytical form depending on the given information $\mathcal{I}$.
4. Adapt the model $\mathcal{C}(\mathcal{I})$ to implementation in a programming language – this can be accomplished by creating new algorithms and computational representations of analytic concepts; the derived tools should automatically generate a solution $\mathcal{S}(\mathcal{I})$ in a symbolic form.
5. Identify and extract components $\{x_i\}$ of the solution $\mathcal{S}(\mathcal{I})$, describing the information $\mathcal{I}$, and formalize them in the form of self-contained objects. In this way, we

receive the representation of the information in *explicit form*. The fact of describing the information $\mathcal{I}$ should be justified in the context of the original problem or its model $\mathcal{M}(\mathcal{I})$. One can illustrate the fact of describing $\mathcal{I}$ by $\{x_i\}$ through empirical experiments. Moreover, a *complete* construction of $\{x_i\}$ carrying the *whole* information $\mathcal{I}$ is desirable.

6. Design a vector **X** representing the information $\mathcal{I}$ on the basis of $\{x_i\}$. Assume that $\{x_i\}$ is an infinite set. In such a case, it is necessary to define finite approximations of **X** and determine to what extent they describe the information $\mathcal{I}$.

Even when the vector **X** is obtained in explicit form, one may not fully understand its meaning and be forced to consider it as a *black box*. In such case we need to study the meaning of the vector. Such research is then a new kind of scientific activity.

2.5.3   Structural information

The simplest representation of a two-dimensional image is in the form of a vector of pixel colors. Usually, transforming raw input into the form that more directly captures the relevant information, can greatly improve the final result. For example, in the male-female classification case, the use of *geometric information* such as the distance between the eyes or the width of the mouth, would increase the precision of classifiers (Barber, 2012). In fact, the elements of the face forms *a system* or *a configuration of objects*. Our goal is to use geometric features of such a system in feature vector construction.

One can consider even more complex systems, frequently studied in biological and medical imaging, for instance the data derived from the observation of the bacterial reproduction process. The bacteria form a system of objects, but how to describe its geometry? How to compare configurations created by different strains of bacteria? Similar questions appear in computational materials science, where multi-phase composite materials are studied. Each of the phases is characterized by a certain geometrical information, that one would like to use in analysis or classification.

In general, consider a problem of construction of a feature vector of system of objects arranged on a given subset of the plane. Assume that a certain configuration of objects has been extracted from the image. Many questions arise then. For example, how do we show that configurations differ, whether models fall into the same class, i.e., do objects have the same properties of their plane distributions? Can we model a certain distribution of objects? If so, how accurate is the model? Assuming that we obtained a series of data representing perturbations of objects, how can this process be visualized?

Configurations of objects in considered examples are characterized by their *structure*. Hence, our goal is to create a feature space describing *structural information* encoded in the data represented by configurations of objects on the plane. Let us present the process of creation such a vector according to the scheme formulated above.

Point 1. Consider a class of problem of determining *effective properties* of deterministic and random composite materials. Let us limit ourselves to the case of composites with non-overlapping circular inclusions (see, Fig. 1.3). Real composites can be seen as exemplary random heterogeneous materials, composed of different phases (component materials), distinguishable on the microscopic scale. Each phase has a conductivity property (either thermal or electrical). Heterogeneous material can be viewed as a continuum on the macroscopic scale (i.e. homogeneous material) with the effective conductivity (EC) determined as the overall macroscopic property of the material. Note that the EC of the composite containing non-overlapping inclusions depends on physical parameters (i.e. conductivity of the host, conductivity of inclusions, contrast parameter, concentrations of phases) as well as on geometric characteristics (i.e. centers of inclusions and their radii). If the configuration of inclusion changes, we can expect a change in the EC. This constitutes a credible premise that the EC formula, being an analytical solution of the problem, contains in implicit form the structural information. Note that the EC problem is sufficient for our purposes, since a non-circular object can be approximated by cluster of disks.

Points 2 and 3. The mathematical model of the problem is the two-dimensional Laplace equation with boundary conditions. The formulation of the model as well as the analytical solution to the problem are presented in (Gluzman et al., 2017) and is summarized in the formula (2.1). Analyticity of the solution is crucial, since we have to keep the dependence of the EC on the structural information. Such an approach is not possible in the case where numerical methods (e.g. finite element method) are applied to certain configuration of objects. It yields only very specific solution, without possibility of obtaining general dependence on considered information. Although such dependence can be investigated by Monte Carlo methods, it requires a tedious, time-consuming computer simulations. It is rather impossible at the present time to get an accurate results. For more details on different levels of *exactness* of solutions, see (Gluzman et al., 2017, Chapter 1).

Point 4. Section 2.2 presents a computational representation of the analytical model as well as the algorithm allowing for straightforward implementation in any programming language.

Point 5. The stage of extracting components of the solution is possible due to observation by Mityushev (2006), that the EC formula can be rewritten in the form of structural sums (1.15). Structiral sums are the only objects in the EC function, which depend on the centers of the circles locations and their radii and do not contain any physical parameters. Moreover, structural sums perform the same role in description of microstructure as the n-point correlation functions. Therefore, the sums are good candidates for the role of structural information carriers. The algorithm for generating the infinite set of all structural sums included in the EC formula is derived in Section 2.2. It allows for the final extraction of structural information from the EC problem. Hence, we obtained the complete set of features in an explicit form, representing the structure encoded in configurations of objects on the plane.

Point 6. The set of all sums describing a whole structural information is infinite, hence the straightforward application (e.g. in machine learning algorithms) needs further approximations. Section 2.5.4 below is devoted to the construction of the vector **X**.

The area of potential applications of structural sums as tools for data analysis involves, for instance, plane objects frequently studied in biological and medical images. One can consider a set of disks centers as a point process pattern, hence the structural sums can be applied to characterization of point fields, in important area of statistics. Summarizing, the presented approach can be applied to other types of data, as long as the data can be represented by distributions of points or non-overlapping disks on the plane.

It is worth noting already existing applications of structural sums for analysis of different kind of data. In the paper by (Mityushev and Nawalaniec, 2015) the structural sums are used for the systematic investigation of dynamically changing microstructures. The paper is the first work using the sums solely for the analysis of the geometry. Other applications cover, for instance, the analysis of collective behavior of bacteria (Czapla and Mityushev, 2017), the description of non-overlapping walks of disks on the plane (Nawalaniec, 2015), and parametrized comparison of composite materials obtained in different technological processes (Kurtyka and Rylko, 2017). All results clearly show that structural sums carry useful geometrical information. However, they are based on the observation of differences of particular sums of low orders for identical disks. In general, the sums are used as a black box, without examining their geometrical meaning. We present an approach to the study of such new tools with machine learning, as presented in Section 2.6.

2.5.4   Construction of structural features vector

In order to define finite approximations to the vector $\mathbf{X}$, one can apply results of Section 2.2, where the algorithm for generating consecutive approximations of the EC formula in the form of first order recurrence relation was developed. In theory, the EC has the form of an infinite series (2.1). Each subsequent iteration of the relation increases the order of the approximating polynomial. Let the properties of the composite be fixed. Then, the fundamental problem of composites consists in construction of a homogenization operator $\mathcal{H} : G \to M$, where G stands for microstructure (geometry) and M for the macroscopic physical constants. The key point is a precise and convenient description of the geometrical set G. Based on the reasoning of the preceding section, we suggest to choose the geometric parameters as the following set of structural sums:

$$G = \{e_{\mathbf{m}},\ \mathbf{m} \in \mathcal{M}_e\},$$

where the set $\mathcal{M}_e$ is introduced by the recursive rules based on Theorem 2.1:

1. $(2), (2, 2) \in \mathcal{M}_e$;
2.   If $(m_1, \ldots, m_q) \in \mathcal{M}_e$ and $q > 1$, then
     $(2, m_1, \ldots, m_q), (m_1 + 1, m_2 + 1, m_3, \ldots, m_q) \in \mathcal{M}_e$.

It is difficult to explicitly characterize the geometric meaning of structural sums. The elements $e_{\mathbf{m}}$ can be treated as weighted moments (integrals over the cell) of the

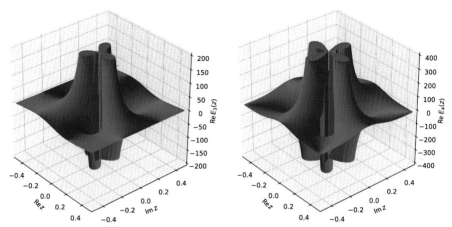

Figure 2.2 Plots of the real part of Eisenstein functions E_n for $n = 3, 4$ for the square lattice, where $\omega_1 = 1$ and $\omega_2 = i$.

correlation functions (Gluzman et al., 2017). Use of $e_{\mathbf{m}}$ allows to avoid heavy computations of the correlation functions, and compute implicitly their weighted moments of high orders.

One can also consider structural sums as *generalized directed distances*. Indeed, each term of the sum (1.15) can be intuitively interpreted as a chain of reciprocals of powers of directed distances (measured in complex numbers). Hence, a particular sum is some kind of a summary characteristic of the network formed by disks. The higher the degree of function E_n is, the more complicated form it has.

For instance, the function E_n is doubly periodic and has a pole of order n at $z = 0$. The magnitude of $E_n(z)$ grows as the order increases, thereby the function shows different levels of sensitivity to the distance between points. Moreover, one can observe that the frequency of oscillations of values along the curve $|z| = \text{const}$, is also related to the degree of E_n. See, for example, plots of real values of $E_3(z)$ and $E_4(z)$ presented in Fig. 2.2.

One can consider the set G as the base of the general form of the feature vector of random composites modeled by non-overlapping disks on the plane. The set $\mathcal{M}_e$ is infinite, therefore applications require finite approximations. Let us consider sets M_q and G_q defined in Section 2.3.4. Note that $G = \bigcup_{j=1}^{\infty} M_j$. Hence, we define *structural features vector of order q* as follows:

$$X_q = \{G_1, G_2, G_3, \ldots, G_q\}, \qquad (2.39)$$

and consider it as an approximation of $\mathbf{X}$. For example, the feature vector of order 4 has the following form:

$$X_4 = \{e_2, e_{2,2}, e_{2,2,2}, e_{3,3}, e_{2,2,2,2}, e_{2,3,3}, e_{4,4}\}.$$

Since the elements of X_q are complex numbers, one can use different vectors derived from X_q, for instance:

- $|X_q|$ – absolute values of elements of X_q;
- $\text{Re}X_q$ – real parts of elements of X_q;
- $\text{Im}X_q$ – imaginary parts of elements of X_q;
- $\text{Arg}X_q$ – arguments (angles) of elements of X_q;
- $\left[\text{Re}X_q, \text{Im}X_q\right]$ – both real and imaginary parts of elements of X_q.

Note that we described the construction of the general form of the feature vector. In applications, one can select a certain subset of structural sums that gives satisfactory results. For instance, the following set:

$$X'_q = \{e_{p,p} : 2 \le p \le q\}$$

is the smallest subset involving Eisenstein functions E_n $(n = 2, 3, \ldots, q)$. Some particular datasets may reveal another correlations between sums, allowing further elimination of dependent parameters. For example, considering concrete models of macroscopically isotropic composites, one can apply so-called Keller's identity (for more details, see Mityushev and Rylko (2012)).

The vector X_q incorporates all structural sums included in the polynomial of order $q + 1$ approximating (2.1). One can formulate the following hypothesis. Since by increasing of the order of approximating polynomial, the accuracy of the EC grows, hence the vector X_q of higher order should supply more information. As a justification one can use the example from Gluzman et al. (2017, Example 4, p. 88). In the example the Keller's identity was applied in formulating the isotropy criterion of composite materials. It was observed that increasing the order of sums resulted in more precise criterion. The hypothesis is finally justified experimentally in Section 2.6.

2.5.5 Structural information in three-dimensional space

Considered structural sums operate on two-dimensional data. The natural question is whether it is possible to construct such a tool for processing data in the three-dimensional space[4]? It occurs that the realization of the scheme of extracting new tools requires an approach different than in the two-dimensional case. Similarly, one can consider the problem of the effective conductivity of random composite materials with spherical disjoint inclusions, see Fig. 2.3. In Chapter 6 we apply the method of functional equations using the so-called Eisenstein summation to obtain analytical approximation of the EC formula for random three-dimensional composites, with the accuracy $O(f^{\frac{10}{3}})$, where f denotes the concentration of spherical inclusions, see formulas (6.87) and (6.88). The obtained formula demonstrates faithfully the dependence of the properties on geometrical configuration of spheres. In fact, we succeed in extracting four new computational objects dependent only on the structure. Such objects constitute three-dimensional structural sums, for instance (6.71) and (6.86)).

---

[4] Generalization to arbitrary dimensionality is set aside for future studies.

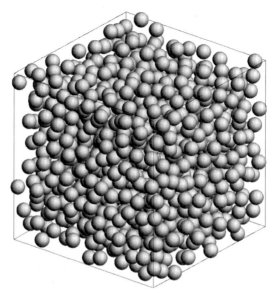

Figure 2.3 Example configuration of disjoint spherical objects.

2.6   Classification and analysis of random structures

In order to explore structural sums as an information black box, we discuss below two examples of classification of simulated composites, based on the general structural sums vector's approximations. The examples serve to illustrate two important issues. First, we want to answer the question, whether the higher-order sums are worth analyzing, or is all useful information encoded in the lower-order parameters. Second, it will be shown that different kind of data may require a specific modification of the vector X_q. The examples discuss composites with inclusions modeled by distributions of disks and distributions of shapes formed by disks (clusters).

All examples fit the same mold, known in machine learning as *classification problem*, consisting in taking input vectors and deciding which of c classes they belong to, based on *training (learning)* from exemplars of each class (Marsland, 2014). Let C_j ($j = 1, 2, 3, \ldots, c$) denote considered classes of distributions. For each C_j we generate a set of 100 samples drawn from a given distribution. In order to create a *training set*, we randomly pick k samples from each class. The training set is used to build a data model. In order to decide how well the algorithm has learned using a given approximation of feature vector, we predict the outputs on a *test set* consisting of $(100 - k)c$ remaining samples (i.e. the portion of data not seen in the training set). Note that in our examples we use $k \leq 40$, hence we are dealing with a relatively large test set (60%–96% of the original dataset). For the feature vector, we use different modifications of the structural sums vector (2.39) of varying order q ($q = 1, 2, \ldots, 10$).

As the classifier we apply a simple Naive Bayes model with Gaussian Naive Bayes classification algorithm, being capable of handling multiple classes directly, imple-

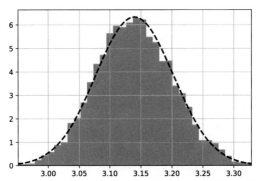

Figure 2.4 Histogram of values of sum e_2 computed for 5000 samples of $N = 100$ disks of concentration $f = 0.5$ generated via random walk in the square cell. Black dashed line is for best-fit normal distribution with mean $\mu = 3.14$ and standard deviation $\sigma = 0.063$.

mented in Scikit-Learn Python module (Pedregosa et al., 2011). The Naive Bayes model also assumes that features are independent of each other. In general, this assumption may not be true, however it yields computational simplification. Moreover, the likelihood of features is assumed to be Gaussian among a given class, which seems to be justified by observation that values of structural sums tend to be normally distributed (see Fig. 2.4). For more details on the Naive Bayes classifier, see the Scikit-Learn module documentation or Marsland (2014).

Let us measure the accuracy of prediction as the ratio of correctly classified samples to all the samples. In each case we build 10 random splits of the set, as described above, and the final score is counted as the arithmetic mean of the resulting accuracy values. Our goal is to observe how the accuracy changes depending on the type of modification of the feature vector, the order q, and the size kc of a training set.

2.6.1 Circular inclusions

In the first experiment we generate various distributions of disks on the plane, and verify whether they are distinguishable by the structural sums feature vector. In order to generate sample data, we use a specific random walk algorithm with varying parameters. Let us describe the general Markov-chain (MC) protocol for Monte Carlo simulations. Assume that initially all centers are placed in the cell, as a regular array of disks or employing the Random Sequential Adsorption (RSA) protocol, where consecutive objects are placed randomly in the cell, rejecting those that overlap previously adsorbed ones. For each a_k ($k = 1, 2, \ldots, N$) we execute the following procedure at every step. Let ϕ be chosen as a realization of the random variable uniformly distributed in the interval $(0, \pi)$. First, a straight line passing through a_k and inclined on the angle ϕ to the real axes is constructed. Next, the segment with endpoints $a_k + d_{\max}e^{i\phi}$ and $a_k - d_{\min}e^{i\phi}$ is taken on this line. Furthermore, the non-negative values $d_{\max}$ and $d_{\min}$ are computed in such a way that the disk can move along the directions ϕ or $\phi + \pi$, respectively, without collisions with other disks (see Fig. 2.5). If collisions do not occur, a_k can move up to the boundary of the cell Q.

Otherwise, the point a_k moves along the direction ϕ (or $\phi + \pi$), with the distance d considered to be a realization of a random variable. The variable is distributed in $(d_{\min}, d_{\max})$, and the distance

$$d = d_{\min} + (d_{\max} - d_{\min})Z,$$

where Z is a random variable distributed in the interval $[0, 1]$. Negative values of d correspond to the direction $\phi + \pi$. Hence, every center obtains a new complex coordinate a'_k after the performed step.

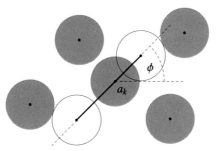

Figure 2.5 Illustration of the random move of disk a_k in a direction ϕ. The solid segment represents all possible values of distance d of movement. Endpoints of the segment correspond to the extremal values $d_{\max}$ and $d_{\min}$ of d.

This move is repeated with new coordinates for each $k = 1, 2, \dots, N$. We say that a *cycle* is performed if k runs from 1 to N. The number of cycles, denoted by t, corresponds to the time of random walks of disks. A location of points a_k at time t forms a probabilistic event described by distribution $\mathcal{U}_t(f)$. After a sufficient number of walks t, the obtained location of the centers can be considered a statistical realization of the distribution $\mathcal{U}_\infty(f)$.

One can obtain various classes of distributions of disks on the plane via two parameters: random variable Z and the distribution of radii of disks. In our experiments we consider following three examples in place of variable Z. Let Z_1 be the random variable uniformly distributed in the interval $[0, 1]$ and let

$$Z_2 = \frac{Z_{\mathcal{N}}}{6} + \frac{1}{2},$$
$$Z_3 = frac\left(\frac{Z_{\mathcal{N}}}{6} + 1\right),$$

where $Z_{\mathcal{N}}$ is the random variable with the standard normal distribution $\mathcal{N}(\mu = 0, \sigma^2 = 1)$ truncated to the range $[-3, 3]$ and $frac(x)$ is the fractional part of x. The PDFs of considered random variables are shown on top of Fig. 2.6. Together with Z_j ($j = 1, 2, 3$), we look into different distributions of radii of disks (i.e. identical disks, uniformly distributed radii and normally distributed radii). Hence, we end up with 9 classes being combinations of those two parameters. Our simulations are based on distributions of 256 disks of concentration $f = 0.5$ generated by the MC protocol described above. Fig. 2.6 presents sample configurations from each class.

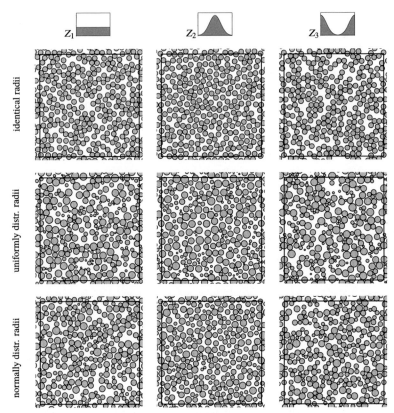

Figure 2.6 Sample configurations from each class of distributions of disks used in classification problems. Rows correspond to considered radii distributions; columns specify distributions Z_k ($k = 1, 2, 3$) related to the distance d of the circle's displacement.

The results of classification, dependent on the size of training data and the order q of the modified feature vector X_q, are presented in Table 2.6. One can see that very good results are obtained for a relatively small training set (7%–19% of the dataset) for vectors $|X_q|$ and $\text{Re}X_q$ of orders 8–10. Note that a purely random classifier has an accuracy of 0.11.

Both a high accuracy of $|X_q|$ and a poor performance of $\text{Arg}X_q$ seem to be justifiable. Differences between presented classes of distributions appear to be strongly related to distances between inclusions. On the other hand, since the samples show the uniformity in all orientations (i.e. isotropy), the vector $\text{Arg}X_q$ seems to be less relevant. It was also observed that in case of such isotropic configurations, the imaginary part for a large number of sums vanishes. Hence, the different accuracy of $\text{Re}X_q$ and $\text{Im}X_q$ is explained.

Besides calculating the general accuracy of the model, one can also analyze the *confusion matrix* that contains all the classes in both the horizontal and vertical directions. The element of the matrix at (i, j) tells us how many samples belonging to

Table 2.6 Accuracy of classification of distributions of disks for $|X_q|$, $\mathrm{Re}X_q$, $\mathrm{Im}X_q$, and $\mathrm{Arg}X_q$.

Train size	Order of feature vector									
	1	**2**	**3**	**4**	**5**	**6**	**7**	**8**	**9**	**10**
$|X_q|$										
4%	0.120	0.349	0.427	0.500	0.537	0.556	0.690	0.773	0.840	0.858
7%	0.129	0.340	0.427	0.544	0.593	0.637	0.772	0.861	0.902	0.918
10%	0.135	0.350	0.457	0.559	0.618	0.663	0.800	0.883	0.927	0.941
13%	0.135	0.343	0.463	0.570	0.626	0.670	0.806	0.895	0.925	0.936
16%	0.140	0.348	0.472	0.585	0.650	0.700	0.841	0.913	0.944	0.954
19%	0.141	0.354	0.476	0.600	0.658	0.709	0.843	0.915	0.948	0.957
22%	0.142	0.354	0.479	0.599	0.666	0.718	0.850	0.924	0.953	0.961
25%	0.140	0.338	0.483	0.612	0.682	0.734	0.864	0.935	0.958	0.965
28%	0.142	0.345	0.489	0.610	0.676	0.727	0.864	0.934	0.958	0.964
31%	0.147	0.339	0.492	0.609	0.679	0.731	0.862	0.934	0.960	0.965
34%	0.136	0.344	0.490	0.614	0.676	0.730	0.867	0.932	0.959	0.963
37%	0.140	0.338	0.494	0.606	0.675	0.726	0.864	0.932	0.956	0.961
40%	0.138	0.334	0.491	0.611	0.686	0.740	0.878	0.937	0.961	0.966
$\mathrm{Re}X_q$										
4%	0.120	0.350	0.428	0.501	0.538	0.561	0.689	0.777	0.833	0.851
7%	0.129	0.340	0.427	0.544	0.591	0.632	0.772	0.858	0.898	0.914
10%	0.136	0.351	0.457	0.560	0.617	0.664	0.802	0.881	0.925	0.941
13%	0.135	0.343	0.463	0.570	0.625	0.670	0.814	0.896	0.926	0.939
16%	0.140	0.348	0.472	0.584	0.644	0.695	0.838	0.913	0.941	0.954
19%	0.141	0.356	0.476	0.600	0.662	0.711	0.846	0.917	0.949	0.960
22%	0.141	0.354	0.479	0.599	0.667	0.715	0.853	0.922	0.953	0.962
25%	0.140	0.338	0.483	0.612	0.680	0.729	0.863	0.936	0.959	0.967
28%	0.142	0.345	0.489	0.610	0.679	0.729	0.865	0.936	0.959	0.964
31%	0.147	0.339	0.492	0.609	0.679	0.733	0.863	0.933	0.959	0.964
34%	0.137	0.345	0.490	0.614	0.673	0.729	0.869	0.934	0.958	0.962
37%	0.138	0.338	0.494	0.606	0.673	0.728	0.862	0.930	0.956	0.961
40%	0.138	0.335	0.490	0.610	0.687	0.738	0.876	0.938	0.960	0.966
$\mathrm{Im}X_q$										
16%	0.139	0.139	0.147	0.181	0.223	0.256	0.312	0.378	0.470	0.554
28%	0.146	0.146	0.151	0.184	0.233	0.271	0.341	0.426	0.542	0.637
40%	0.140	0.140	0.151	0.188	0.241	0.274	0.346	0.441	0.551	0.654
$\mathrm{Arg}X_q$										
16%	0.139	0.139	0.141	0.138	0.144	0.143	0.158	0.186	0.219	0.277
28%	0.145	0.145	0.144	0.141	0.151	0.152	0.169	0.202	0.250	0.310
40%	0.141	0.141	0.135	0.138	0.145	0.152	0.172	0.214	0.256	0.336

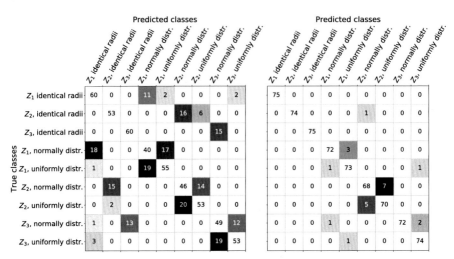

Figure 2.7 Confusion matrices for models built for the vector $|X_q|$ for $q = 5$ (left) and $q = 10$ (right).

class C_i have been assigned to class C_j by the algorithm. Hence, the leading diagonal contains all the correct predictions. All presented confusion matrices were generated using training:testing ratio 25:75 and the 3-fold cross-validation procedure (Bishop, 2006).

Let us analyze the confusion matrix for the model built on the vector $|X_q|$ for $q = 5$ (see Fig. 2.7 (left)). First of all, algorithm almost always assigns proper variable Z_j ($j = 1, 2, 3$). Secondly, sample belonging to the class with identical radii is never confused with another identical radii class (note the white 3×3 square in the top-left corner of the matrix), rather with a polydisperse one with normally distributed radii and the same Z_k (first three rows of the matrix). Finally, another source of error is to confuse uniformly distributed radii with normally distributed ones among the same Z_k (see black spots around the leading diagonal of the 6×6 square in the bottom-right corner of the matrix). As the order q grows, the accuracy of the model increases, see the confusion matrix for $q = 10$ (Fig. 2.7 (right)). One can see that there is still a small problem with confusing radii distribution among the Z_2 class. Fig. 2.8 shows the confusion matrix for the vector $\mathrm{Arg}X_q$ for $q = 10$. The matrix reflects sources of the poor performance of $\mathrm{Arg}X_q$.

2.6.2 Non-circular inclusions

Inclusions of non-circular shape can be approximated by clusters of disks. Consider similar shapes shown in Fig. 2.9, consisting of 21 identical disks each. In order to generate samples of distributions, we apply the RSA protocol for each shape. Since the chosen protocol is the same for each shape, one can assume that the potential differences between classes will be caused by the shape of inclusions, rather than their planar distribution. In the present section we analyze the shape-sensitivity of structural sums. Simulated samples are configurations of concentration $f = 0.3$ (see examples in

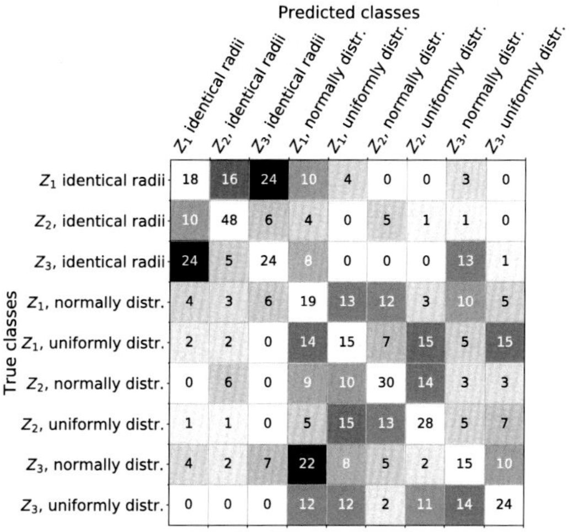

Figure 2.8 Confusion matrix for model built on the vector $\text{Arg}X_q$ for $q = 10$.

Fig. 2.9). We generated a set of 100 samples for each class of distributions and adopted the same classification scheme as described on page 89. The results, in varying both the size of training data and the order of modified feature vector X_q, are presented in Table 2.7.

One can see that application of vectors $|X_q|$ and $\text{Re}X_q$, which performed well in the preceding example, yields poor results. This clearly shows that the methods based only on distances between centers of disks may be insufficient in some cases. In contrast to the preceding experiment, very good results are obtained for a relatively small training set (4%–16% of the dataset) for vectors $\text{Im}X_q$ and $\text{Arg}X_q$ of orders 6–10. It is worth noting that the analysis of distances is equivalent to analysis based on the autocorrelation (2-point correlation) function. Also note that a purely random classifier has the expected accuracy of 0.1.

Let us find out what are the sources of the poor performance of $|X_q|$. In order to do that, we analyze the confusion matrix for $q = 3$ (Fig. 2.10). Black squares around the leading diagonal tell us that its main difficulty is to distinguish between mirrored shapes. Otherwise, it performs surprisingly well, considering its small size.

2.6.3 Irregularity of random structures and data visualization

It is known from the theory (Mityushev and Rylko, 2012) that structural sums attain their extrema for optimally distributed disks. For example, in the case of identical disks the regular hexagonal array (see Fig. 2.11) allows to achieve the maximal concentration. All distributions displayed in Fig. 2.6 have the same concentration $f = 0.5$, however each of them shows different level of *irregularity*. For example, the sample in the second column and the first row (Z_2, identical radii) seems to be the most regular among the others. On the other hand, the pattern in the third column and the second row (Z_3, normally distributed radii) looks very irregular. Let us apply the structural

Table 2.7 Accuracy of classification of distributions of shapes for $|X_q|$, $\mathrm{Re}X_q$, $\mathrm{Im}X_q$, and $\mathrm{Arg}X_q$.

Train size	Order of feature vector									
	1	2	3	4	5	6	7	8	9	10
$|X_q|$										
16%	0.135	0.279	0.505	0.504	0.503	0.505	0.500	0.507	0.505	0.502
28%	0.133	0.277	0.502	0.499	0.493	0.499	0.497	0.499	0.497	0.498
40%	0.136	0.278	0.513	0.511	0.495	0.502	0.499	0.499	0.496	0.501
$\mathrm{Re}X_q$										
16%	0.137	0.278	0.505	0.503	0.501	0.502	0.502	0.508	0.505	0.498
28%	0.133	0.278	0.501	0.499	0.495	0.498	0.500	0.501	0.496	0.498
40%	0.135	0.278	0.515	0.510	0.501	0.503	0.505	0.499	0.502	0.502
$\mathrm{Im}X_q$										
4%	0.137	0.137	0.131	0.142	0.426	0.621	0.748	0.770	0.871	0.924
7%	0.143	0.143	0.138	0.145	0.468	0.700	0.852	0.890	0.958	0.976
10%	0.142	0.142	0.140	0.150	0.491	0.728	0.886	0.928	0.981	0.988
13%	0.142	0.142	0.143	0.157	0.497	0.752	0.909	0.943	0.989	0.994
16%	0.140	0.140	0.144	0.154	0.500	0.750	0.910	0.943	0.987	0.990
19%	0.141	0.141	0.144	0.160	0.513	0.777	0.919	0.947	0.991	0.995
22%	0.141	0.141	0.146	0.159	0.510	0.776	0.918	0.945	0.990	0.994
25%	0.139	0.139	0.147	0.155	0.517	0.780	0.931	0.961	0.994	0.995
28%	0.142	0.142	0.146	0.159	0.519	0.782	0.935	0.966	0.997	0.998
31%	0.139	0.139	0.143	0.160	0.526	0.791	0.927	0.956	0.995	0.997
34%	0.147	0.147	0.151	0.167	0.525	0.803	0.932	0.962	0.996	0.996
37%	0.146	0.146	0.150	0.159	0.530	0.806	0.947	0.974	0.998	0.998
40%	0.142	0.142	0.150	0.162	0.534	0.804	0.936	0.963	0.995	0.996
$\mathrm{Arg}X_q$										
4%	0.139	0.139	0.123	0.113	0.288	0.457	0.469	0.413	0.309	0.305
7%	0.141	0.141	0.131	0.130	0.370	0.609	0.721	0.673	0.718	0.693
10%	0.141	0.141	0.139	0.137	0.419	0.675	0.814	0.799	0.817	0.789
13%	0.144	0.144	0.143	0.139	0.426	0.688	0.824	0.857	0.871	0.821
16%	0.139	0.139	0.143	0.140	0.433	0.712	0.830	0.880	0.878	0.840
19%	0.140	0.140	0.138	0.138	0.433	0.710	0.839	0.895	0.907	0.881
22%	0.142	0.142	0.141	0.139	0.423	0.709	0.838	0.898	0.906	0.881
25%	0.139	0.139	0.139	0.141	0.441	0.737	0.858	0.914	0.920	0.891
28%	0.141	0.141	0.145	0.143	0.447	0.736	0.850	0.917	0.915	0.890
31%	0.139	0.139	0.140	0.137	0.448	0.730	0.844	0.930	0.932	0.899
34%	0.144	0.144	0.155	0.144	0.454	0.740	0.857	0.938	0.941	0.924
37%	0.147	0.147	0.141	0.138	0.443	0.743	0.860	0.937	0.935	0.923
40%	0.143	0.143	0.147	0.133	0.453	0.740	0.853	0.931	0.941	0.933

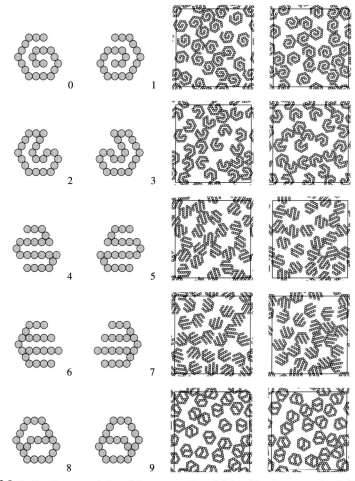

Figure 2.9 Similar shapes consisting of the same number of disks of fixed concentration (left) and corresponding samples of distributions generated via the RSA protocol (right).

sums in order to express this property quantitatively. For this purpose, we will present our data in two-dimensional space. To this end, let us apply the following simple *exhaustive* feature selection scheme. We train Naive Bayes model for only two features, considering all pair combinations of sums from $|X_{10}|$. Moreover, we consider only structural sums of the form $e_{p,p}$ ($2 \leq p \leq 10$). Experiments are performed using training:testing ratio 25:75 and the 3-fold cross-validation procedure. The average of resulting accuracy scores is assigned to each pair of features. Table 2.8 shows pairs with the accuracy score greater than 0.95. Let us select sums of lowest orders for further considerations, namely $e_{3,3}$ and $e_{8,8}$.

Consider the hexagonal array of disks as the most regular one. In such a case both minimum of $e_{8,8}$ and the maximum of $e_{3,3}$ are equal to zero. Intuitively, the more

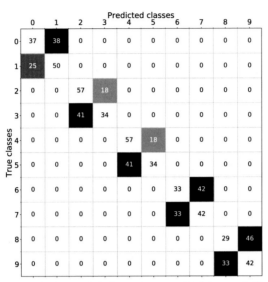

Figure 2.10 Confusion matrix for model built on the vector $|X_q|$ for $q = 3$. Numbers of classes are consistent with the numbering on Fig. 2.9.

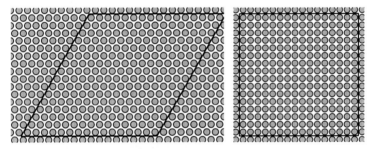

Figure 2.11 Regular arrays of identical disks: hexagonal array (left) and square array (right).

Table 2.8 Mean accuracy of classification of distributions of disks. The accuracy is computed as the average of the values computed via 3-fold cross-validation procedure using Gaussian Naive Bayes classification algorithm.

Accuracy	Features		Accuracy	Features	
0.987	$e_{4,4}$	$e_{10,10}$	0.973	$e_{5,5}$	$e_{8,8}$
0.987	$e_{3,3}$	$e_{10,10}$	0.972	$e_{5,5}$	$e_{9,9}$
0.977	$e_{5,5}$	$e_{10,10}$	0.969	$e_{3,3}$	$e_{8,8}$
0.977	$e_{4,4}$	$e_{9,9}$	0.968	$e_{3,3}$	$e_{9,9}$
0.973	$e_{4,4}$	$e_{8,8}$	0.951	$e_{6,6}$	$e_{10,10}$

regular system of disks is, the closer structural sums are to their extrema. Fig. 2.12, showing the plot of $-e_{3,3}$ against $e_{8,8}$, confirms this intuitive remark. One can also

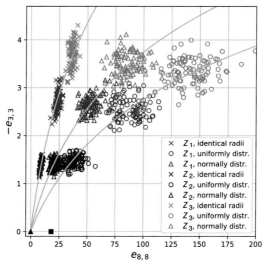

Figure 2.12 Values of $-e_{3,3}$ against $e_{8,8}$ for samples from considered distributions. The fitted curves are $3.118\log(0.061x+1)$ (identical radii, crosses), $2.526\log(0.034x+1)$ (normally distributed radii, triangles), $1.987\log(0.028x+1)$ (uniformly distributed radii, disks). Black triangle and square are for hexagonal and square regular arrays, respectively.

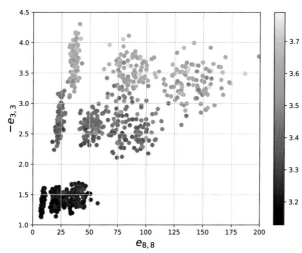

Figure 2.13 Values of $-e_{3,3}$ against $e_{8,8}$ for samples from considered distributions. The colors represent the values of the effective conductivity of the composite modeled by a given sample.

observe the convergence of the regularity of distributions to the hexagonal case. It occurs that the data can be fitted to a simple logarithmic model $a\log(bx+1)$ passing through the origin (see Fig. 2.12). The curves illustrate paths of convergence of the sums and also represent each distribution Z_j ($j=1,2,3$).

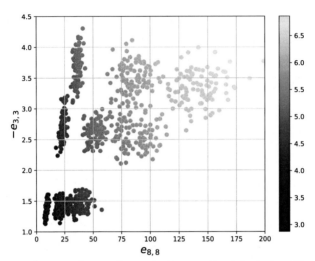

Figure 2.14 Values of $-e_{3,3}$ against $e_{8,8}$ for samples from considered distributions. The colors represent the values of the irregularity measure for a given sample.

Let us now attempt to describe geometric meaning of the selected sums. One can observe that $e_{8,8}$ is related to the heterogeneity of disks in a pattern. Moreover, $e_{3,3}$ seems to be reflecting *clustering* of disks. This can be explained on the basis of the effective conductivity of composites modeled by considered samples. The feature vector X_q provides components for an approximation of the effective conductivity (2.1). The values of σ_e are presented in Fig. 2.13. Since all samples have the same concentration of inclusions and same material parameters, hence the only possible source of differences in σ_e lies in the geometry of a system. One can see that the higher values of σ_e are common in classes, where distribution Z_3 was applied. It seems to be natural, since the more clusters appear in a configuration, the larger long-range connectivity is possible, bringing the sample closer to the so-called *percolation threshold*. Fig. 2.13 also demonstrates that the way inclusions are distributed (rather than the diversity of their radii) has much more impact on σ_e.

Both selected sums can be combined into single, entropy-like *irregularity measure* for sample s:

$$\mu(s) := \log\left[\left(1 - e_{3,3}(s)\right)\left(1 + e_{8,8}(s)\right)\right]. \tag{2.40}$$

Note that irregularity of the hexagonal array of identical disks is equal 0. The larger the value of $\mu(s)$ is, the more irregular model we expect (see Fig. 2.14). Table 2.9 shows the mean values of the irregularity measure for each class of distributions. Note that results are consistent with our initial guess.

Table 2.9 Mean values of $\mu(s)$ computed for each considered class of distributions of disks.

Distribution	$\langle\mu(s)\rangle$	Distribution	$\langle\mu(s)\rangle$
hexagonal array	0.000	Z_3, identical radii	5.163
square array	2.950	Z_1, normally distr. radii	5.271
Z_2, identical radii	3.164	Z_1, uniformly distr. radii	5.743
Z_2, normally distr. radii	4.003	Z_3, normally distr. radii	5.957
Z_1, identical radii	4.505	Z_3, uniformly distr. radii	6.429
Z_2, uniformly distr. radii	4.565		

2.7 The Python package for computing structural sums and the effective conductivity

Basicsums is a Python 3 package for computing structural sums and the effective conductivity of random composites (Nawalaniec, 2019a). The package forms a set of tools that can be divided into the following submodules presented in a bottom-up order (cf. Fig. 2.15).

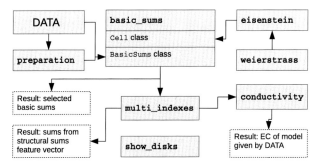

Figure 2.15 Modules dependency and data flow in the basicsums package.

- basicsums.weierstrass –incorporates Weierstrass class modeling Weierstrass $\wp$ elliptic function and its derivative $\wp'$. Note that structural sums operate on a subset of complex plane being a unit cell (parallelogram). Instances are created based on given half-periods of considered cell. Can be used as a standalone unit for calculations of $\wp(z)$ and $\wp'(z)$.
- basicsums.eisenstein – module for deriving Eisenstein–Rayleigh lattice sums S_k ($k = 2, 3, \ldots$) and Eisenstein functions E_k ($k = 2, 3, \ldots$) being algebraic combinations of $\wp(z)$ and $\wp'(z)$. Can be used as a standalone module for performing calculations with S_k and $E_k(z)$. The module is suitable for both numerical and symbolical calculations.
- basicsums.basic_sums – the key component of the basicsums package, providing the BasicSums class tackling calculations of structural sums being convolutions of Eisenstein functions. Moreover, the considered unit cell is modeled by the Cell

class, of which an instance determines and stores corresponding lattice sums and Eisenstein functions. For a particular cell, one can consider different configurations of points or disks. In order to compute structural sums for a given configuration, we create an instance of `BasicSums` class passing the data and the `Cell`'s object. At the time of creation, all required values of Eisenstein functions are computed and stored in a form of matrices, before any sum is called (for more details, see Section 2.3 and Nawalaniec (2017)).

- `basicsums.multi_indexes` – module incorporating procedures returning sets of multi-indexes of structural sums that are sufficient to achieve a given polynomial approximation of the effective conductivity or a certain approximation of the structural sums feature vector.

- `basicsums.conductivity` – designed for an original application of structural (basic) sums: computing both purely symbolic and symbolic-numeric formulas for the EC of random 2D composites with circular non-overlapping inclusions.

- `basicsums.preparation` – providing functions that help the user to prepare data for the application of structural sums. Moreover, the module contains a regularization function that transforms data to a particular system of disks of different radii in order to achieve more stable characteristics.

- `basicsums.show_disks` – this one-function module allows visualization of a system of disks in a given two-periodic cell.

Structural (basic) sums provide an object oriented architecture, strongly relying on recent rigorous results on algorithms and methods for calculation of structural sums. The idea is applied to the effective conductivity of random composites, as described in the present book and in Gluzman et al. (2017). Application of the specific vector-matrix representation of a convolution, significantly reduces computational complexity of calculations. It also allows to reuse intermediate results, cached during the computations (see Nawalaniec (2019b)). The software may help researchers from other fields, who may find it useful to experiment with structural sums as tools for data analysis.

It is worth noting that the package is independent of the commercial solutions, and allows for integration with Python scientific stack (Jones et al., 2001). Moreover, the software is available under the New BSD License. The project is supported by the documentation (Nawalaniec, 2019b), incorporating API description, the theory overview and the set of tutorials, that can be realized in an interactive notebooks ecosystem `Jupyter`. The software can be installed by the `pip` manager. The source code repository is hosted on Bitbucket (for more details, see Nawalaniec (2019b)).

Elasticity of hexagonal array of cylinders

3

3.1 Method of functional equations for elastic problems

Application of the method of functional equations to random and regular composites was presented in Gluzman et al. (2017, Chapter 8). The main attention was paid to 2D conductivity problems. Application to 3D conductivity problems was also outlined in Gluzman et al. (2017). In this section, we write the corresponding functional equations for 2D elastic composites. The equations were first derived in Mityushev (2000) and Mityushev and Rogosin (2000, Chapter 5), see also Drygaś and Mityushev (2016) and Gluzman et al. (2017, Chapter 10). Their applications are extended to high concentration of inclusions in the subsequent chapters of this book.

Different elastic constants are used in the 2D and 3D elastic problems. Some explanations are needed in order to precisely describe the designations used in the book. It is assumed that only isotropic elastic materials serve as matrix and inclusions. In fibrous composites with unidirectional fibers, corresponding elastic fields are considered as plane strain, distributed in the section perpendicular to fibers. Mind that stress state in thin flat plates is attributed to plane stress. Both plane strain and stress problems are reduced to mathematical 2D problems (Jasiuk et al., 1994). In the present book, only fibrous composites described with plane strain are considered. Below, a primed notation is used for 3D elastic constants and non-primed for 2D elastic constants in the main relations for plane strain elasticity, and it is known that:

- the 2D and 3D shear moduli coincide, hence, G is used for 2D and 3D shear moduli;
- for Poisson's ratios $v = \frac{v'}{1-v'}$;
- for Young's moduli $E = \frac{E'}{1-v'^2}$;
- for bulk moduli $k = k' + \frac{G}{3}$;
- for Muskhelishvili's constant $\kappa = 3 - 4v' = \frac{3-v}{1+v}$.

Some other useful relations between the elastic moduli from Eischen and Torquato (1993), Jasiuk et al. (1994), are shown below for completeness,

$$k = \frac{2G}{\kappa - 1} = \frac{G}{1 - 2v'} = \frac{E}{1 - v}, \quad k' = \frac{E'}{1 - 2v'}, \tag{3.1}$$

$$v = \frac{k - G}{k + G}, \quad v' = \frac{E'}{2G} - 1, \tag{3.2}$$

$$E' = 2G(1 + v'), \quad E = 2G(1 + v), \tag{3.3}$$

$$E = \frac{4kG}{k + G}. \tag{3.4}$$

Applied Analysis of Composite Media. https://doi.org/10.1016/B978-0-08-102670-0.00012-3

Besides separating 2D and 3D constants, we have to distinguish the elastic constants in inclusions and matrix. It is assumed that fibers and matrix are isotropic elastic materials described by the shear moduli G_1 and G, Young's moduli E_1 and E, where the subscript 1 denote the elastic constants for inclusions.

Consider the transverse effective moduli G_e, E_e and ν_e in the isotropy plane; and the longitudinal effective constants G^L, E^L and ν^L. The transverse effective constants are related by equation $E_e = 2G_e(1 + \nu_e)$ following from (3.3). Every transversely isotropic material is described by five independent elastic moduli (Lekhnitskii, 1981). However, the two-phase fibrous composite has only four independent entries G_1, G, E_1 and E. Hill (1964) had shown that transversely isotropic, two-phase fibrous composites are described by three independent elastic moduli since the two longitudinal effective moduli are expressed through other moduli. Namely,

$$E^L = E_1 f + E(1 - f) + 4 \left(\frac{\nu_1 - \nu}{\frac{1}{k_1} - \frac{1}{k}} \right)^2 \left(\frac{f}{k_1} + \frac{1 - f}{k} - \frac{1}{k_e} \right), \qquad (3.5)$$

$$\nu^L = \nu_1 f + \nu(1 - f) - \frac{\nu_1 - \nu}{\frac{1}{k_1} - \frac{1}{k}} \left(\frac{f}{k_1} + \frac{1 - f}{k} - \frac{1}{k_e} \right), \qquad (3.6)$$

where k_e denotes the effective transverse bulk modulus. The effective properties of elastic composites subject to thermal, electric and other fields can be tackled through combination of the methods developed in the present book and in (Gluzman et al., 2017). The corresponding homogenization equations can be found in Bardzokas et al. (2007), Wojnar (1995, 1997), Telega and Wojnar (1998, 2000, 2007), Wojnar et al. (1999).

In the present book, we deduce high order formulas in concentration for the constants G_e and k_e, correcting some numerical results from Supplement to (Drygaś and Mityushev, 2016). In this section, we consider a finite part of the hexagonal array with a finite number n of inclusions on the infinite plane. Approximate analytical formulas for the local fields are given in Subsection 3.1.1. In Subsection 3.1.2, we pass to the limit $n \to \infty$. Proper interpretations of the arising conditionally convergent sums are given in Appendix A.4.

Introduce the complex variable $z = x_1 + i x_2 \in \mathbb{C}$, where i denotes the imaginary unit. Consider non-overlapping domains D_k $(k = 1, 2, \cdots, n)$ in the complex plane $\mathbb{C}$ with smooth boundaries ∂D_k. Let $D := \mathbb{C} \cup \{\infty\} \setminus \left(\cup_{k=1}^{n} D_k \cup \partial D_k \right)$. Let domains D and D_k $(k = 1, 2, \cdots, n)$ to consist of the elastic materials with the constants G, ν and G_k, ν_k, respectively.

The following components of the stress tensor are determined by the Kolosov–Muskhelishvili formulas (Muskhelishvili, 1966)

$$\sigma_{11} + \sigma_{22} = \begin{cases} 4 \operatorname{Re} \varphi_k'(z), & z \in D_k, \\ 4 \operatorname{Re} \varphi_0'(z), & z \in D, \end{cases} \qquad (3.7)$$

$$\sigma_{11} - \sigma_{22} + 2i\sigma_{12} = \begin{cases} -2\left[\overline{z\varphi_k''(z)} + \overline{\psi_k'(z)} \right], & z \in D_k, \\ -2\left[\overline{z\varphi_0''(z)} + \overline{\psi_0'(z)} \right], & z \in D, \end{cases}$$

where Re denotes the real part, and the bar denotes the complex conjugation.

Let the stress tensor be applied at infinity

$$\sigma^\infty = \begin{pmatrix} \sigma_{11}^\infty & \sigma_{12}^\infty \\ \sigma_{12}^\infty & \sigma_{22}^\infty \end{pmatrix}. \tag{3.8}$$

Following (Muskhelishvili, 1966), let us introduce the constants

$$B_0 = \frac{\sigma_{11}^\infty + \sigma_{22}^\infty}{4}, \quad \Gamma_0 = \frac{\sigma_{22}^\infty - \sigma_{11}^\infty + 2i\sigma_{12}^\infty}{2}. \tag{3.9}$$

Then,

$$\varphi_0(z) = B_0 z + \varphi(z), \quad \psi_0(z) = \Gamma_0 z + \psi(z), \tag{3.10}$$

where $\varphi(z)$ and $\psi(z)$ are analytical in D and bounded at infinity. Functions $\varphi_k(z)$ and $\psi_k(z)$ are analytical in D_k and twice differentiable in the closures of the considered domains. Special attention will be paid to the independent elastic states, such as the uniform shear stress at infinity

$$\sigma_{11}^\infty = \sigma_{22}^\infty = 0, \ \sigma_{12}^\infty = \sigma_{21}^\infty = 1 \Leftrightarrow B_0 = 0, \ \Gamma_0 = i, \tag{3.11}$$

and the uniform simple tension at infinity

$$\sigma_{11}^\infty = \sigma_{22}^\infty = 2, \ \sigma_{12}^\infty = \sigma_{21}^\infty = 0 \Leftrightarrow B_0 = 1, \ \Gamma_0 = 0. \tag{3.12}$$

It is convenient to use such states (3.11) and (3.12), to estimate the effective shear and bulk moduli (Drygaś and Mityushev, 2016).

The strain tensor components ϵ_{11}, ϵ_{12}, ϵ_{22} are determined by Muskhelishvili (1966),

$$\epsilon_{11} + \epsilon_{22} = \begin{cases} \frac{\kappa_1 - 1}{G_1} \operatorname{Re} \varphi_k'(z), & z \in D_k, \\ \frac{\kappa - 1}{G} \operatorname{Re} \varphi_0'(z), & z \in D, \end{cases} \tag{3.13}$$

$$\epsilon_{11} - \epsilon_{22} + 2i\epsilon_{12} = \begin{cases} -\frac{1}{G_1}\left[\overline{z\varphi_k''(z)} + \overline{\psi_k'(z)} \right], & z \in D_k, \\ -\frac{1}{G}\left[\overline{z\varphi_0''(z)} + \overline{\psi_0'(z)} \right], & z \in D. \end{cases}$$

The condition of perfect bonding at the matrix-inclusion interface yields the following relations (Muskhelishvili, 1966):

$$\varphi_k(t) + t\overline{\varphi_k'(t)} + \overline{\psi_k(t)} = \varphi_0(t) + t\overline{\varphi_0'(t)} + \overline{\psi_0(t)}, \tag{3.14}$$

$$\frac{1}{G_1}\left[\kappa_1 \varphi_k(t) - t\overline{\varphi_k'(t)} - \overline{\psi_k(t)} \right] = \frac{1}{G}\left[\kappa \varphi_0(t) - t\overline{\varphi_0'(t)} - \overline{\psi_0(t)} \right]. \tag{3.15}$$

The force $\mathbf{F}$ acting on the arc AB can be represented in the complex form. For instance, the limit value on AB from the domain D (Muskhelishvili, 1966) is given by the formula

$$\mathbf{F} \equiv -i[\varphi_0(z) + z\overline{\varphi_0'(z)} + \overline{\psi_0(z)}]\Big|_B^A. \tag{3.16}$$

Then, Eq. (3.14) expresses the equality of normal forces from both sides of the boundary, up to an additive constant. The displacement vector could be expressed in the complex form, see Muskhelishvili (1966)

$$\mathbf{u}(x_1, x_2) \equiv \frac{1}{2G}[\kappa\varphi_0(z) - z\overline{\varphi_0'(z)} - \overline{\psi_0(z)}]. \tag{3.17}$$

Hence, (3.15) means the continuity of displacements on the boundary. The problem (3.14)–(3.15) is the classic boundary value problem of the plane elasticity. It was treated in many works and by various methods (Filshtinsky and Mityushev, 2014). Below, we concentrate on its analytical solution by the method of functional equations.

In the present book, we consider non–overlapping circular inclusions $D_k := \{z \in \mathbb{C} : |z - a_k| < r_k\}$ $(k = 1, 2, \cdots, n)$ where a_k denotes the complex coordinate of the k-th disk, r_k stands for its radius. In the present section, we concentrate our attention on the identical disks with $r = r_k$ for all k. The case of different radii will be considered separately.

Let $z_{(k)}^* = \frac{r^2}{\overline{z} - \overline{a_k}} + a_k$ denote the inversion with respect to the circle ∂D_k. Introduce the new unknown functions

$$\Phi_k(z) = \overline{z_{(k)}^*}\varphi_k'(z) + \psi_k(z), \quad |z - a_k| \le r, \tag{3.18}$$

analytic in D_k excepting at the point a_k, where its principal part has the form $r^2(z - a_k)^{-1}\varphi_k'(a_k)$.

Introduce the contrast parameters

$$\varrho_1 = \frac{\frac{G_1}{G} - 1}{\frac{G_1}{G} + \kappa_1}, \quad \varrho_2 = \frac{\kappa\frac{G_1}{G} - \kappa_1}{\kappa\frac{G_1}{G} + 1}, \quad \varrho_3 = \frac{\frac{G_1}{G} - 1}{\kappa\frac{G_1}{G} + 1}, \tag{3.19}$$

related by the identity

$$\varrho_1 = \frac{\varrho_3}{1 - \varrho_2 + \varrho_3} = \varrho_3\sum_{l=0}^{\infty}(\varrho_2 - \varrho_3)^l, \tag{3.20}$$

where the second equality is valid for $|\varrho_2 - \varrho_3| < 1$. The denominator of (3.20) is always positive and equals $\frac{\frac{G_1}{G} + \kappa_1}{\frac{G_1}{G}\kappa + 1}$.

The problem (3.14)–(3.15) was reduced to the system of functional equations first written by Mityushev (2000), by Mityushev and Rogosin (2000, Eqs. (5.6.11) and (5.6.16) in Chapter 5), see also Drygaś and Mityushev (2016), Gluzman et al. (2017).

We have the following functional–differential equations valid up to some additive constants. Such constants do not impact the stress-strain fields

$$\varphi_k(z) = \varrho_1 \sum_{m \neq k} \left[\overline{\Phi_m(z^*_{(m)})} - (z - a_m) \overline{\varphi'_m(a_m)} \right]$$

$$- \varrho_1 \overline{\varphi'_k(a_k)} (z - a_k) + \varrho_1 (1 + \varrho_3^{-1}) B_0 z + p_0, \qquad (3.21)$$

$$\Phi_k(z) = \varrho_2 \sum_{m \neq k} \overline{\varphi_m(z^*_{(m)})}$$

$$+ \varrho_3 \sum_{m \neq k} \left(\frac{r^2}{z - a_k} + \overline{a_k} - \frac{r^2}{z - a_m} - \overline{a_m} \right) \left[\left(\overline{\Phi_m(z^*_{(m)})} \right)' - \overline{\varphi'_m(a_m)} \right]$$

$$+ (1 + \varrho_3) \left[B_0 \left(\frac{r^2}{z - a_k} + \overline{a_k} \right) + \Gamma z \right] + \sum_{k=1}^{n} \frac{r^2 q_k}{z - a_k} + q_0,$$

$$|z - a_k| \leq r, \quad k = 1, 2, \ldots, n. \qquad (3.22)$$

The unknown functions $\varphi_k(z)$ and $\Phi_k(z) := \overline{z^*_{(k)}} \varphi'_k(z) + \psi_k(z)$ are analytic in $|z - a_k| < r$ and differentiable in $|z - a_k| \leq r$ $(k = 1, 2, \ldots, n)$, except for $z = a_k$, where $\Phi_k(z)$ has the principal part $\frac{r^2}{z - a_k} \varphi'_k(a_k)$.

The effective constants can be expressed in terms of functions $\varphi(z)$ and $\varphi_k(z)$ (Drygaś and Mityushev, 2016, Gluzman et al., 2017). Therefore, after solving the functional equations (3.21)–(3.22), we also need the function (Drygaś and Mityushev, 2016, Gluzman et al., 2017)

$$\varphi(z) = \frac{\varrho_3}{1 + \varrho_3} \sum_{m=1}^{n} \left[\overline{\Phi_m(z^*_{(m)})} - (z - a_m) \overline{\varphi'_k(a_k)} \right] + p'_0, \ z \in D, \qquad (3.23)$$

where p'_0 is a constant.

Following (Drygaś, 2016b), we introduce the Hardy–Sobolev space $\mathcal{H}^{(2,2)}(D_k)$ separately for each $k = 1, \ldots, n$. It is the space of functions analytic in D_k satisfying the conditions

$$\sup_{0 < r < r_k} \int_0^{2\pi} |f^{(q)} \left(re^{i\theta} + a_k \right)|^2 d\theta < \infty, \quad q = 0, 1, 2, \qquad (3.24)$$

where $f^{(q)}$ denotes the derivative $\frac{d^q f}{dz^q}$. The norm is introduced as follows

$$\|f\|^2_{\mathcal{H}^{(2,2)}(D_k)} = \|f\|^2_{\mathcal{H}^2(D_k)} + \|f'\|^2_{\mathcal{H}^2(D_k)} + \|f''\|^2_{\mathcal{H}^2(D_k)}, \qquad (3.25)$$

where the classic Hardy norm

$$\|f\|^2_{\mathcal{H}^2(D_k)} = \sup_{0 < r < r_k} \int_0^{2\pi} |f \left(re^{i\theta} + a_k \right)|^2 d\theta \qquad (3.26)$$

is used.

Using designations (3.24)–(3.26), we introduce the space $\mathcal{H}^{(2,2)}\left(\bigcup_{k=1}^{N} D_k\right)$ (denoted as $\mathcal{H}^{(2,2)}$) of functions f analytic in $\bigcup_{j=1}^{N} D_k$ endowed with the norm

$$\|f\|_{\mathcal{H}^{(2,2)}}^2 := \sum_{k=1}^{N} \|f_k\|_{\mathcal{H}^{(2,2)}(D_k)}^2,$$

where $f(z) = f_k(z)$ for $z \in D_k$ $(k = 1, \ldots, n)$.

The functional–differential equations (3.21)–(3.22) include the meromorphic functions $\Phi_k(z)$. Using (3.18) one can rewrite these equations shortly as operator equation

$$\phi = \mathcal{A}\phi + \eta, \tag{3.27}$$

on the vector-function $\phi(z) = (\varphi_k(z), \psi_k(z))^T$ in the space $\mathcal{H}^{(2,2)} \times \mathcal{H}^{(2,2)}$. The operator $\mathcal{A}$ and given vector-function η are determined by (3.21)–(3.22).

It was proved by Drygaś (2016b) that the operator $\mathcal{A}$ is compact in $\mathcal{H}^{(2,2)} \times \mathcal{H}^{(2,2)}$. The Fredholm properties of the operator equation (3.27), equivalent to the functional equations (3.21)–(3.22), justify application of the method of successive approximations for sufficiently small moduli of contrast parameters.

In order to study the structure of the effective elastic constants we introduce the scalar functions $\varphi(z) = \varphi_k(z)$ and $\Phi(z) = \Phi_k(z)$ (for $z \in D_k \cup \partial D_k$) determined in $\bigcup_{k=1}^{n}(D_k \cup \partial D_k)$. Then, the functional equations (3.21)–(3.22) can be written in the short form

$$\varphi(z) = \varrho_1(A_{11}\varphi)(z) + \varrho_1(A_{12}\Phi)(z) + \varrho_1(1 + \varrho_3^{-1})Bz + p_0, \tag{3.28}$$

$$\Phi(z) = \varrho_2(A_{21}\varphi)(z) + \varrho_3(A_{22}\varphi)(z) + \varrho_3(A_{23}\Phi)(z)$$
$$+ (1 + \varrho_3)\left[B\left(\frac{r^2}{z - a_k} + \overline{a_k} \right) + \Gamma z \right],$$
$$|z - a_k| \leq r, \ k = 1, 2, \ldots, n, \tag{3.29}$$

where the compact operators A_{kl} are determined by the right-hand side of Eqs. (3.21)–(3.22). It follows from Drygaś and Mityushev (2016) that every operator A_{kl} depends analytically on the parameter r^2.

Consider now the case of (3.11), when the free term in the right-hand side of (3.28)–(3.29) contains the elastic constants only in the form of multiplier $1 + \varrho_3$. Application of the successive approximations to Eqs. (3.28)–(3.29), and subsequent expansion of the obtained series on r^2, yield the following formulas for the complex potentials inside the inclusions

$$\varphi(z) = (1 + \varrho_3) \sum_{l_1,l_2,l_3=0}^{\infty} r^{2l_1} \varrho_2^{l_2} \varrho_3^{l_3} \varphi_{l_1 l_2 l_3}(z) + constant, \quad z \in \bigcup_{k=1}^{n} D_k, \tag{3.30}$$

where the functions $\varphi_{l_1 l_2 l_3}(z)$ can be calculated by means of the successive approximations by using (3.20).

The formula (3.30) can be considered as a decomposition of $\varphi(z)$ onto the linear combination of terms $\varphi_{l_1 l_2 l_3}(z)$ dependent only on locations of inclusions, while coefficients $\varrho_2^{l_2} \varrho_3^{l_3}$ depend only on the elastic constants multiplied by r^{2l_1}. The function $\Phi(z)$ has analogous structure. Then, the complex potential in the domain D can be calculated from (3.23) and (3.10) by using (3.30)

$$\varphi_0(z) = Bz + \sum_{l_1,l_2,l_3=0}^{\infty} r^{2l_1} \varrho_2^{l_2} \varrho_3^{l_3+1} \widetilde{\varphi}_{l_1 l_2 l_3}(z) + constant, \quad z \in D. \quad (3.31)$$

In the case of (3.12), the solution to (3.28)–(3.29) can be represented as a linear combination of two functions of two arguments z and ρ_j. After substitution into (3.23) one can see that the latter linear combination yields similar linear combination for $\varphi(z)$ in D, but with coefficients ρ_1 and ρ_3. Therefore, application of the successive approximations and of the corresponding expansions yield equivalent representations (3.30)–(3.31), but with different basic functions $\varphi_{l_1 l_2 l_3}(z)$ and $\widetilde{\varphi}_{l_1 l_2 l_3}(z)$.

The iterative method developed above, corresponds to the *contrast expansion* discussed in Mityushev (2015), Gluzman et al. (2017, Chapter 3), Torquato (2002, Chapter 20) for conductivity problems. The contrast expansion, for instance for the potential (3.30), should not include the dependence on radius. It can be formally written in the form

$$\varphi(z) = (1 + \varrho_3) \sum_{l_2,l_3=0}^{\infty} \varrho_2^{l_2} \varrho_3^{l_3} \left[\sum_{l_1=0}^{\infty} r^{2l_1} \varphi_{l_1 l_2 l_3}(z) \right] + constant. \quad (3.32)$$

The analytical dependence of the potentials on r^2 yields the *cluster expansion* (Gluzman et al., 2017, Chapter 3), (Torquato, 2002, Chapter 19). The corresponding series can be formally presented in the form

$$\varphi(z) = \sum_{l_1=0}^{\infty} r^{2l_1} \left[(1 + \varrho_3) \sum_{l_2,l_3=0}^{\infty} \varrho_2^{l_2} \varrho_3^{l_3} \varphi_{l_1 l_2 l_3}(z) \right] + constant. \quad (3.33)$$

The cluster expansion can be directly applied to computation of the local fields with a finite number of inclusions, located in the whole space or bounded domain.

We use the term concentration expansion for circular inclusions when an analytical dependence is in powers of f near $f = 0$. If inclusions have more complicated shape such a dependence may contain the power terms in $\sqrt{f}$ proportional to the radius r as in the formula (4.19) for the 2D conductivity problem from (Mityushev et al., 2019). The term *cluster expansion* is used here because of its constructive realization when double, triple etc interactions between inclusions serve as approximations of the local fields. Further application of this approach to the effective constants of random composites with infinite number of inclusions yields a conditionally convergent series or an integral.

To apply unthinkingly the scheme "double, triple, ... interactions" is methodologically wrong, see Gluzman et al. (2017, Chapter 3). The rigorous study of the cluster method requires a subtle mathematical investigation presented in Appendix A.4. Such advanced study is included to the book for the simple reason. Many scientists and engineers apply the cluster method in a wrong way. A brief rundown is given at the end of this section on page 119.

3.1.1 Local field. Analytical approximate formulas

In the present section, we describe the method of successive approximations applied to Eqs. (3.28)–(3.29) to derive approximate analytical formulas for the local elastic fields. We shall use the cluster and contrast approximations with subsequent expansion of the complex potentials into the Taylor series.

The proper application of the cluster method is based on the representation of the complex potentials φ_k and ψ_k in the form of power series in r^2

$$\varphi_k(z) = \sum_{s=1}^{\infty} \varphi_k^{(s)}(z) r^{2s}, \quad \psi_k(z) = \sum_{s=1}^{\infty} \psi_k^{(s)}(z) r^{2s}. \tag{3.34}$$

Selecting the terms with the same powers r^{2p}, we arrive at the iterative scheme for the solution of Eqs. (3.21)–(3.22). Straightforward computations give the approximate formulas

$$\varphi(z) = r^2 \varphi^{(1)}(z) + r^4 \varphi^{(2)}(z) + \ldots, \tag{3.35}$$

where

$$\varphi^{(1)}(z) = \overline{\Gamma_0} \varrho_3 \sum_{m=1}^{n} \frac{1}{z - a_m},$$

$$\varphi^{(2)}(z) = 2 \sum_{k=1}^{n} \left[B_0 \frac{\varrho_3(\varrho_3 - \varrho_2)}{2\varrho_3 - \varrho_2 + 1} \sum_{m \neq k} \frac{1}{(a_m - a_k)^2} + \overline{\Gamma_0} \varrho_3^2 \sum_{m \neq k} \frac{a_m - a_k}{(a_m - a_k)^3} \right]$$

$$\times \frac{1}{z - a_m}. \tag{3.36}$$

The third order approximation is written below in the case of (3.11)

$$\varphi^{(3)}(z) = -i \left[3\varrho_3^2 \sum_{l=1}^{n} \sum_{m \neq l} \frac{1}{(a_l - a_m)^4} (z - a_l)^{-2} \right.$$

$$+ \left(4\varrho_3^3 \sum_{l=1}^{n} \sum_{m \neq l} \sum_{m_1 \neq m} \frac{a_l - a_m}{(a_l - a_m)^3} \frac{\overline{a_m - a_{m_1}}}{(a_m - a_{m_1})^3} (z - a_l)^{-1} \right. \tag{3.37}$$

$$+ 6\varrho_3^2 \sum_{l=1}^{n} \sum_{m \neq l} \frac{1}{(a_l - a_m)^4} (z - a_l)^{-1}$$

$$+ \frac{\varrho_3^2(\varrho_3 - \varrho_2)}{2\varrho_3 - \varrho_2 + 1} \sum_{l=1}^{n} \sum_{m \neq l} \sum_{m_1 \neq m} \frac{1}{(a_l - a_m)^2} \frac{1}{(a_m - a_{m_1})^2} (z - a_l)^{-1}$$

$$- \frac{\varrho_3^2(\varrho_3 - \varrho_2)}{2\varrho_3 - \varrho_2 + 1} \sum_{l=1}^{n} \sum_{m \neq l} \sum_{m_1 \neq m} \frac{1}{(a_l - a_m)^2} \frac{1}{(a_m - a_{m_1})^2} (z - a_l)^{-1} \Bigg) \Bigg].$$

In the case of (3.12), corresponding approximations for $\varphi(z)$ are calculated by formulas

$$\varphi^{(1)}(z) = 0, \quad \varphi^{(2)}(z) = 2 \frac{\varrho_3(\varrho_3 - \varrho_2)}{2\varrho_3 - \varrho_2 + 1} \sum_{l=1}^{n} \sum_{m \neq l} \frac{1}{(a_m - a_l)^2} (z - a_l)^{-1}.$$

$$\varphi^{(3)}(z) = 2 \frac{\varrho_3(\varrho_3 - \varrho_2)}{2\varrho_3 - \varrho_2 + 1} \Bigg(2\varrho_3 \sum_{l=1}^{n} \sum_{m \neq l} \sum_{m_1 \neq m} \frac{a_l - a_m}{(a_l - a_m)^3} \frac{1}{(a_m - a_{m_1})^2} (z - a_l)^{-1}$$

$$- \sum_{l=1}^{n} \sum_{m \neq l} \frac{1}{(a_m - a_l)^3} (z - a_l)^{-2} \Bigg). \tag{3.38}$$

We apply now the method of successive approximations to (3.21)–(3.22), resulting in the approximations in powers of contrast parameters. Zero approximation is simply

$$\left(\varphi_k^{(0)} \right)'(z) = \frac{\varrho_1(1 + \varrho_3)}{\varrho_3(1 + \varrho_1)} B_0. \tag{3.39}$$

Suppose we know the approximation of the order $p - 1$. Then the p-th approximation is given by formulas

$$\varphi_k^{(p)}(z) = \varrho_1 \sum_{m \neq k} \left[\overline{\Phi_m^{(p-1)}(z_{(m)}^*)} - (z - a_m) \overline{\left(\varphi_m^{(p-1)} \right)'(a_m)} \right]$$

$$\varrho_1 \frac{\varrho_1(1 + \varrho_3)}{\varrho_3(1 + \varrho_1)} B_0 (z - a_k) + (1 + \varrho_3^{-1}) \varrho_1 B_0 z + p_0,$$

$$|z - a_k| \leq r_k, \ k = 1, 2, ..., N. \tag{3.40}$$

$$\omega_k^{(p)}(z) = \varrho_2 \sum_{m \neq k} \overline{\varphi_m^{(p-1)}(z_{(m)}^*)}$$

$$+ \varrho_3 \sum_{m \neq k} \overline{z_{(m)}^*} \left[\overline{\left(\Phi_m^{(p-1)}(z_{(m)}^*) \right)'} - \overline{\left(\varphi_m^{(p-1)} \right)'(a_m)} \right]$$

$$+ \sum_{m \neq k} \frac{r_m^2}{z - a_m} q_m + (1 + \varrho_3) \Gamma_0 z + q_0, \ |z - a_k| \leq r_k, \ k = 1, 2, ..., N,$$

$$\tag{3.41}$$

where $q_m = (\varrho_2 - \varrho_3)\varphi'_m(a_m) - \varrho_3\overline{\varphi'_m(a_m)}$, $m = 1, 2, ..., n$. Thus, we developed an explicit algorithm to solve functional equations (3.21)–(3.22) by the explicit iterative scheme.

An alternative, implicit iterative scheme for the system equations (3.21)–(3.22), is based on zeroth approximation of φ_k, given by (3.39), and the following zeroth approximation for ψ_k,

$$\psi_k^{(0)}(z) = (1+\kappa)\Gamma_0 z + q_0, \ |z - a_k| \le r_k, \ k = 1, 2, ..., N. \tag{3.42}$$

The p-th approximation for φ_k is given by (3.40), and for ψ_k is implicitly expressed by the relation

$$\psi_k^{(p)}(z) = \omega_k^{(p)}(z) + (\varrho_3 - \varrho_2)\overline{z_{(k)}^*}\left(\left(\varphi_k^{(p)}\right)'(z) - \left(\varphi_k^{(p)}\right)'(a_k)\right), \tag{3.43}$$

for $k = 1, 2, ..., N$, where $\omega_k^{(p)}$ is calculated from (3.41).

In order to extend the method to polydisperse problem with different radii of inclusions, we introduce the effective radius

$$r^2 = \frac{1}{n}\sum_{k=1}^{n} r_k^2.$$

Note that r^2 is proportional to the concentration f. Between r^2 and radii of each inclusion there is a relation, namely

$$r_k^2 = R_k r^2,$$

where R_k is the proportionality ratio.

Functions $\varphi_k^{(s)}$ and $\psi_k^{(s)}$ ($s = 0, 1, 2$) inside of each inclusion are presented by their Taylor series. It is sufficient to take only starting three terms

$$\varphi_k^{(s)}(z) = \alpha_{k,0}^{(s)} + \alpha_{k,1}^{(s)}(z - a_k) + \alpha_{k,2}^{(s)}(z - a_k)^2 + O((z - a_k)^3), \tag{3.44}$$

$$\psi_k^{(s)}(z) = \beta_{k,0}^{(s)} + \beta_{k,1}^{(s)}(z - a_k) + \beta_{k,2}^{(s)}(z - a_k)^2 + O((z - a_k)^3), \ \text{as } z \to a_k. \tag{3.45}$$

The reason for the precision $O((z - a_k)^3)$ is explained below, after Eqs. (3.46)–(3.48). For shortness let us introduce the auxiliary constants $\gamma_{m,l}^{(s)}$:

$$\gamma_{m,l}^{(s)} = (l + 2)r^2 R_m \alpha_{m,l+2}^{(s)} + \overline{a_m}(l + 1)\alpha_{m,l+1}^{(s)} + \beta_{m,l}^{(s)} \quad s = 0, 1; \ l = 1, 2. \tag{3.46}$$

Substitution of (3.34) into (3.21) and (3.22) yields

$$\sum_{p=0,1,2} r^{2p}\varphi_k^{(p)}(z) = \sum_{m \ne k}\sum_{l+s \le 2}\varrho_1 r^{2(s+l)}\overline{\gamma_{m,l}^{(s)}}(z - a_m)^{-l}$$

$$+ \left(\varrho_1 (1 + \varrho_3^{-1}) B_0 - \varrho_1 \sum_{s=0,1,2} r^{2s} \overline{\alpha_{k,1}^{(s)}} \right) (z - a_k) + p_1 + O(r^6), \quad (3.47)$$

and

$$\sum_{p=0,1,2} r^{2p} \left(\overline{a_k} \left(\varphi_k^{(p)}(z) \right)' + \psi_k^{(p)}(z) \right)$$

$$= - \sum_{s=0,1} R_k r^{2(s+1)} \left(\varphi_k^{(s)}(z) \right)' (z - a_k)^{-1}$$

$$+ \sum_{m \neq k} \sum_{l+s \leq 2} \varrho_2 R_m^l r^{2(l+s)} \overline{\alpha_{m,l}^{(s)}} (z - a_m)^{-l}$$

$$- \sum_{m \neq k} \varrho_3 R_k R_m r^4 \overline{\gamma_{m,1}^{(0)}} (z - a_k)^{-1} (z - a_m)^{-2}$$

$$+ \sum_{m \neq k} \sum_{l+s \leq 1} \varrho_3 l R_m^{l+1} r^{2(l+s+1)} \overline{\gamma_{m,l}^{(s)}} (z - a_m)^{-l-2}$$

$$- \sum_{m \neq k} \sum_{l+s \leq 2} \varrho_3 R_m^l l r^{2(l+s)} \overline{\gamma_{m,l}^{(s)}} (\overline{a_k - a_m})(z - a_m)^{-l-1}$$

$$+ (1 + \varrho_3) B_0 R_k r^2 (z - a_k)^{-1}$$

$$+ (1 + \varrho_3) \Gamma_0 (z - a_k) + \sum_{m \neq k} R_m r^2 q_m (z - a_m)^{-1} + q_0 + O(r^6), \quad (3.48)$$

where the sum $\sum_{l+s \leq 2}$ contains three terms with $l = 1, 2$ and $s = 0, 1$, satisfying the inequality $l + s \leq 2$. The sum $\sum_{m \neq k}$ means that m runs over $\{1, 2, ..., n\}$, except k. Other sums are defined analogously.

One can check that the higher order terms $(z - a_k)^l$ (for $l \geq 3$) in (3.44)–(3.45) produce only terms of order $O(r^6)$ in (3.47)–(3.48). Such a rule takes place in general case when the terms $(z - a_k)^l$ yield $O(r^{2l})$. Constants p_1, q_0, $\alpha_{m,0}^{(s)}$ and $\beta_{m,0}^{(s)}$ ($m = 1, .., n$, $s = 0, 1, 2$) in (3.47)–(3.48) are not relevant, since they determine parallel translations not influencing the stress and deformation fields.

3.1.2 Cluster approach to elasticity problem

Consider a fibrous composite. The axis x_3 is chosen to be parallel to the unidirectional fibers; the section perpendicular to x_3 forms the hexagonal array of inclusions on the plane of variables x_1 and x_2, called the isotropy plane. Let ω_1 and ω_2 be the fundamental pair of periods on the complex plane $\mathbb{C}$ introduced for the hexagonal array as follows

$$\omega_1 = \sqrt[4]{\frac{4}{3}}, \quad \omega_2 = \sqrt[4]{\frac{4}{3}} \left(\frac{1}{2} + i \frac{\sqrt{3}}{2} \right). \quad (3.49)$$

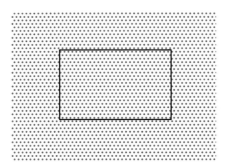

Figure 3.1 The rectangle ∂Q_n enclosed n inclusions of the hexagonal array.

Let $\mathbb{Z}$ denote the set of integer numbers. The points $m_1\omega_1 + m_2\omega_2$ ($m_1, m_2 \in \mathbb{Z}$) generate a doubly periodic hexagonal lattice $\mathcal{Q}$. Introduce also the zero-th cell

$$Q = Q_{(0,0)} = \{z = t_1\omega_1 + it_2\omega_2 \in \mathbb{C} : -\frac{1}{2} < t_1, t_2 < \frac{1}{2}\},$$

and the cells $Q_{(m_1,m_2)} = Q_{(0,0)} + m_1 + im_2$. Without loss of generality the area of Q is normalized to unity. Then, the concentration of disks f is simply

$$f = \pi r^2, \tag{3.50}$$

where r denotes the radius of inclusions.

In the present section, we modify the averaging computational scheme developed by Drygaś and Mityushev (2016). It is based on the averaging operators applied to complex potentials, not to the stress and strain fields as in (Drygaś and Mityushev, 2016). Such innovation essentially simplifies the symbolic-numerical computations and yields the high order formulas for the effective constants.

We calculate the averaged shear modulus $G_e^{(n)}$ of the considered finite composite. It is related to the effective shear modulus G_e for macroscopically isotropic composites through the limit $G_e = \lim_{n\to\infty} G_e^{(n)}$. In order to calculate $G_e^{(n)}$ it is sufficient to consider the uniform shear stress (3.11). Introduce the average over a sufficiently large domain Q_n containing the inclusions D_k

$$\langle w \rangle_n = \frac{1}{|Q_n|} \iint_{Q_n} w \, d^2\mathbf{x}, \tag{3.51}$$

where $d^2\mathbf{x} = dx_1 dx_2$. One can take ∂Q_n as a rectangle, symmetric with respect to coordinate axes as shown in Fig. 3.1.

The macroscopic shear modulus can be computed from (3.51)

$$G_e^{(n)} = \frac{\langle \sigma_{12} \rangle_n}{2\langle \epsilon_{12} \rangle_n}. \tag{3.52}$$

In what follows, ∂Q_n will tend to the infinite point, as $n \to \infty$, and we shall arrive at the macroscopic shear moduli $G_e = \lim_{n \to \infty} G_e^{(n)}$. The stress tensor components are calculated by (3.7), and the deformation tensor components by (3.13).

Instead of a direct computations in terms of the complex potentials, according to formula (3.52) from Drygaś and Mityushev (2016), it is simpler to compute another quantity

$$
P_n := \left\langle \frac{1}{2}(\sigma_{11} - \sigma_{22}) + i\sigma_{12} \right\rangle_n, \tag{3.53}
$$

and then just take its imaginary part to compute $\langle \sigma_{12} \rangle_n$. Using the definition of the average (3.51) and formulas (3.7), we obtain

$$
P_n = -\frac{1}{|Q_n|} \left\{ \iint_D \left[z\overline{\varphi_0''(z)} + \overline{\psi_0'(z)} \right] d^2\mathbf{x} + \sum_{k=1}^{n} \iint_{D_k} \left[z\overline{\varphi_k''(z)} + \overline{\psi_k'(z)} \right] d^2\mathbf{x} \right\}, \tag{3.54}
$$

where $z = x_1 + ix_2$. Green's formula in its complex form

$$
\iint_D \frac{\partial w(z)}{\partial \overline{z}} d^2\mathbf{x} = \frac{1}{2i} \int_{\partial D} w(t)\, dt, \tag{3.55}
$$

will be used below.

The boundary of D can be decomposed, i.e., $\partial D_n = \partial Q_n - \sum_{k=1}^{n} \partial D_k$, where ∂Q_n and ∂D_k are positively oriented. Application of (3.55) to (3.54) yields $P_n = P_n' + P_n''$, where

$$
P_n' = -\frac{1}{2i|Q_n|} \sum_{k=1}^{n} \int_{\partial D_k} \left[t\overline{\varphi_k'(t)} + \overline{\psi_k(t)} - t\overline{\varphi_0'(t)} - \overline{\psi_0(t)} \right] dt, \tag{3.56}
$$

$$
P_n'' = -\frac{1}{2i|Q_n|} \int_{\partial Q_n} \left[t\overline{\varphi_0'(t)} + \overline{\psi_0(t)} \right] dt.
$$

Green's formula (3.55) for $w = \overline{z}$ yields the area formula

$$
\frac{1}{2i} \int_{\partial Q_n} \overline{t}\, dt = |Q_n|. \tag{3.57}
$$

It follows from Drygaś and Mityushev (2016) that $\lim_{n \to \infty} P_n'' = i$. Using the boundary condition (3.14) we obtain

$$
P_n' = -\frac{1}{2i|Q_n|} \sum_{k=1}^{n} \int_{\partial D_k} [\varphi_0(t) - \varphi_k(t)]\, dt = -\frac{1}{2i|Q_n|} \sum_{k=1}^{n} \int_{\partial D_k} \varphi(t)\, dt. \tag{3.58}
$$

Here, we used equations $\int_{\partial D_k} \varphi_k(t)dt = 0$, following from the Cauchy integral theorem.

Therefore,

$$\lim_{n\to\infty} P_n = i + iA \;\Rightarrow\; \frac{1}{2}\langle\sigma_{12}\rangle = 1 + \mathrm{Re}\,A, \tag{3.59}$$

where

$$A = \lim_{n\to\infty} \frac{1}{2|Q_n|}\sum_{k=1}^{n}\int_{\partial D_k}\varphi(t)dt. \tag{3.60}$$

Similar manipulations can be performed with

$$R_n = R'_n + R''_n = \left\langle \frac{1}{2}(\epsilon_{11} - \epsilon_{22}) + i\epsilon_{12}\right\rangle_n, \tag{3.61}$$

where

$$R'_n = -\frac{1}{|Q_n|}\frac{1}{G}\int_D \left[\overline{z\varphi_0''(z)} + \overline{\psi_0'(z)}\right]d^2\mathbf{x}$$

$$= -\frac{1}{2i|Q_n|}\sum_{k=1}^{n}\int_{\partial D_k}\left[\frac{1}{G_1}\left(\kappa_1 t\overline{\varphi_k'(t)} + \overline{\psi_k(t)}\right) - \frac{1}{G}\left(\kappa t\overline{\varphi_0'(t)} + \overline{\psi_0(t)}\right)\right]dt, \tag{3.62}$$

and

$$R''_n = -\frac{1}{2i|Q_n|G}\int_{\partial Q_n}\left[t\overline{\varphi_0'(t)} + \overline{\psi_0(t)}\right]dt. \tag{3.63}$$

Using the boundary conditions (3.15) we get

$$\lim_{n\to\infty} R_n = \frac{1}{G}(i - i\kappa A) \;\Rightarrow\; \langle\epsilon_{12}\rangle = \frac{1}{G}(1 - \kappa\,\mathrm{Re}\,A). \tag{3.64}$$

Substituting (3.59), (3.64) into (3.52), and taking the limit as n tends to infinity, we obtain

$$\frac{G_e}{G} = \frac{1 + \mathrm{Re}\,A}{1 - \kappa\,\mathrm{Re}\,A}. \tag{3.65}$$

The integral (3.60) and other limits can be calculated explicitly by using approximations of the function $\varphi(z)$ obtained in Drygaś and Mityushev (2016) and described in the previous section. Thus, the first order approximation (Drygaś and Mityushev, 2016, formula (28)) has the form

$$\varphi^{(1)}(z) = -ir^2\varrho_3\sum_{m=1}^{n}\frac{1}{z - a_m}. \tag{3.66}$$

For any fixed k by the residue theorem we obtain

$$\int_{\partial D_k} \varphi^{(1)}(t)dt = 2\pi r^2 \varrho_3. \tag{3.67}$$

In the first order approximation

$$A = f\varrho_3 + O(f^2), \tag{3.68}$$

where the concentration $f = \lim_{n\to\infty} \frac{n\pi r^2}{|Q_n|}$. Substitution of (3.68) into (3.52) yields

$$\frac{G_e}{G} = \frac{1 + \varrho_3 f}{1 - \varrho_3 \kappa f} + O(f^2). \tag{3.69}$$

This expression is nothing else but the famous Hashin–Shtrikman lower bound for $G_1 \geq G$, $k_1 \geq k$, and the upper bound for $G_1 \leq G$, $k_1 \leq k$.

The averaged bulk modulus is calculated by the same method by formulas

$$k_e = \lim_{n\to\infty} k_e^{(n)} = \lim_{n\to\infty} \frac{\langle \sigma_{11} + \sigma_{22} \rangle_n}{2\langle \epsilon_{11} + \epsilon_{22} \rangle_n}. \tag{3.70}$$

It is sufficient to take the particular external stresses (3.12) to calculate the effective bulk modulus k_e.

Using (3.7) and (3.13) we obtain the following three integrals:

$$\frac{1}{|Q_n|} \iint_D d^2\mathbf{x}, \quad \frac{1}{|Q_n|} \iint_D \varphi'(z)d^2\mathbf{x}, \quad \frac{1}{|Q_n|} \iint_{D_l} \varphi'_l(z)d^2\mathbf{x}. \tag{3.71}$$

Green's formula (3.55) was used for the shear modulus. Now, it is convenient to apply Green's formula

$$\iint_D \frac{\partial w(z)}{\partial z} d^2\mathbf{x} = -\frac{1}{2i} \int_{\partial D} w(t)\, d\bar{t}. \tag{3.72}$$

The first integral tends to $(1 - f)$ as $n \to \infty$. The second integral becomes

$$\frac{1}{|Q_n|} \iint_D \varphi'(z)d^2\mathbf{x} = -\frac{1}{2i|Q_n|} \int_{\partial Q_n} \varphi'(t)d\bar{t} + \sum_{l=1}^n \frac{1}{2i|Q_n|} \int_{\partial D_l} \varphi'(t)d\bar{t}. \tag{3.73}$$

It follows from Drygaś and Mityushev (2016) that $\lim_{n\to\infty} \frac{1}{2i|Q_n|} \int_{\partial Q_n} \varphi'(t)d\bar{t} = 0$. The third integral of (3.71) and second integral of (3.73), are denoted respectively as

$$B = \lim_{n\to\infty} \frac{1}{2i|Q_n|} \sum_{l=1}^n \int_{\partial D_l} \varphi(t)d\bar{t}, \quad C = -\lim_{n\to\infty} \frac{1}{2i|Q_n|} \sum_{l=1}^n \int_{\partial D_l} \varphi_l(t)d\bar{t}. \tag{3.74}$$

Then, the limit of (3.70) as n tends to infinity, can be written in the form

$$\frac{k_e}{k} = \frac{1 - f + \operatorname{Re} B + \operatorname{Re} C}{1 - f + \operatorname{Re} B + \frac{k}{k_1} \operatorname{Re} C}. \tag{3.75}$$

Similar computations can be performed for the effective bulk modulus k_e. Using the relation $\frac{k}{k_1} = (1 - \varrho_2)(1 + \varrho_3)^{-1}$, the first-order approximation

$$\frac{k_e}{k} = \frac{1 + \frac{\varrho_1 - \varrho_3}{\varrho_3 + \varrho_1 \varrho_3} f}{1 + \frac{1 - \varrho_2}{1 + \varrho_3} \frac{\varrho_1 - \varrho_3}{\varrho_3 + \varrho_1 \varrho_3} f} + O(f^2) \tag{3.76}$$

gives the lower Hashin–Shtrikman bound for the bulk modulus when $G_1 \geq G, k_1 \geq k$, and with the upper bound when $G_1 \leq G, k_1 \leq k$.

3.1.3 Rayleigh's and Maxwell's approaches

The first order formulas (3.69) and (3.76) are derived in the previous section with rigorous justification of their validity up to $O(f^2)$. Though they were known all along as the Hashin–Shtrikman bounds, their precision is clearly determined for the first time. It is worth noting that during their derivation we do not make any physical assumption about particles interactions. We just dutifully calculated the effective constants up to $O(f^2)$. In the present section, we follow the same line and calculate the next order terms in f.

Beginning with the second order correction in f for a finite cluster, the terms computed in Drygaś and Mityushev (2016) contain the sums

$$S_2(n) = \sum_{l=1}^{n} \sum_{m \neq l} \frac{1}{(a_m - a_l)^2}, \quad S_3^{(1)}(n) = \sum_{l=1}^{n} \sum_{m \neq l} \frac{a_m - a_l}{(a_m - a_l)^3}. \tag{3.77}$$

In the limit $n \to \infty$, these sums become conditionally convergent series. For the hexagonal array they take the form

$$S_2 = \lim_{n \to \infty} S_2(n) = \sum_{(m_1, m_2) \in \mathbb{Z}^2 \setminus (0,0)} \frac{1}{(m_1 \omega_1 + m_2 \omega_2)^2}, \tag{3.78}$$

$$S_3^{(1)} = \lim_{n \to \infty} S_3^{(1)}(n) = \sum_{(m_1, m_2) \in \mathbb{Z}^2 \setminus (0,0)} \frac{m_1 \omega_1 + m_2 \omega_2}{(m_1 \omega_1 + m_2 \omega_2)^3}.$$

Here ω_j are given by (3.49). Conditional convergence of (3.78) implies that the result depends on the summation rule which ought to be determined by the form of ∂Q_n as its tending to infinity.

It is astonishing that different definitions of the limits of (3.78) are used dependent on their applications. This subtle convergence question is discussed in Appendix A.4 and the references therein. In this section, we present only final results relevant for the further numerical-symbolic computations.

Eisenstein summation is based on the iterated sum, when Q_n first is extended in the x_1-direction to infinity, and then, similarly, in the x_2-direction

$$\sum_{(m_1,m_2)\in\mathbb{Z}^2}^{(e)} = \lim_{N\to\infty} \sum_{m_2=-N}^{N} \lim_{M\to\infty} \sum_{m_1=-M}^{M}. \qquad (3.79)$$

The Eisenstein summation (3.79) leads to the values $S_2 = \pi$ (Rayleigh, 1892) and $S_3^{(1)} = \frac{\pi}{2}$ (Yakubovich et al., 2016).

But if we introduce the symmetric summation when ∂Q_n is a rhombus with the vertices $\pm N\omega_1 \pm N\omega_2$, so that

$$\sum_{(m_1,m_2)\in\mathbb{Z}^2}^{(sym)} = \lim_{N\to\infty} \sum_{m_1,m_2=-N}^{N}. \qquad (3.80)$$

One can see that in this case, by definition, $S_2 = S_3^{(1)} = 0$. Although the series S_2 in 2D conductivity problems can be determined in this way in principle, some induced dipole moment at infinity has to be introduced to equilibrate the total charge of the array.

The two variants were proposed by Rayleigh and Maxwell. First variant, or the Eisenstein–Rayleigh approach, is to define $S_2 = \pi$ to calculate the local fields and the effective conductivity. The second variant is based on the definition $S_2 = 0$ and on the Maxwell self-consistent approach for the effective conductivity. A rigorous explanation why we take two different values for the same object S_2 is given in Appendix A.4.

The same applies to the elasticity problem. The structure of the formula (3.65) corresponds to the Maxwell approach. Hence, the dipole moment will be taken into account if we take by the definition $S_2 = S_3^{(1)} = 0$, as it was done in (Mityushev, 2018, Drygaś and Mityushev, 2017).

Application of successive approximations to the functional equations (3.21)–(3.22) yields the local fields in symbolic form (see low order equations (3.35)–(3.38)). The integral (3.60) and the limit $n \to \infty$ are also computed symbolically in Mathematica. The development of main computational scheme follows from here (Drygaś and Mityushev, 2016), but instead of (3.79) the summation (3.80) is used.

The main theoretical result of the present section and Appendix A.4 can be summarized as follows. The different Eisenstein–Rayleigh and Maxwell approaches are based on the different definitions of the conditionally convergent series confronted in the course of constructive homogenization. It is demonstrated that determination of the local fields by an expansion method in the concentration or in the contrast parameters has to be based on the Eisenstein summation, with

$$S_2 = \pi, \quad S_3^{(1)} = \frac{\pi}{2}. \qquad (3.81)$$

At the same time, computation of the effective elastic constants has to be based on the symmetric summation, with

$$S_2 = 0, \quad S_3^{(1)} = 0. \qquad (3.82)$$

The two methods give different results formally, for the same series. Many respected authors ignored the difference, and derived various contradictory analytical formulas for the effective constants. In the present section and in Appendix A.4, we rigorously answer the question of conditional convergence. The explanation not only has a theoretical value, or shows the proper path of investigations, but engenders a constructive symbolic algorithm to derive high order formulas presented in the next sections.

3.2 General formula for the effective shear and bulk moduli

The effective shear modulus can be computed from the formula (3.65), after proper definition and subsequent calculation of the limit (3.60) in Section 3.1.3

$$\frac{G_e}{G} = \frac{1 + \operatorname{Re} A}{1 - \kappa \operatorname{Re} A}. \tag{3.83}$$

The value of Re A can be symbolically computed in the form of the expansion in f using the algorithm described in Chapter 2. In such a way we get the approximation

$$
\begin{aligned}
\operatorname{Re} A = {}& \varrho_3 f + \frac{48 \left(S_5^{(1)}\right)^2 \varrho_3^3}{\pi^4} f^5 - \frac{360 S_5^{(1)} S_6 \varrho_3^3}{\pi^5} f^6 + \frac{5 \varrho_3 S_6^2 \left(\varrho_1 \varrho_2 + 135 \varrho_3^2\right)}{\pi^6} f^7 \\
&+ \frac{2304 \left(S_5^{(1)}\right)^4 \varrho_3^5}{\pi^8} f^9 - \frac{34560 \left(S_5^{(1)}\right)^3 S_6 \varrho_3^5}{\pi^9} f^{10} \\
&+ \frac{60 \varrho_3^3 \left(15 \left(S_{11}^{(1)}\right)^2 + 8 \left(S_5^{(1)}\right)^2 S_6^2 \left(11 \varrho_1 \varrho_2 + 405 \varrho_3^2\right)\right)}{\pi^{10}} f^{11} \\
&- \frac{360 \varrho_3^3 \left(33 S_{11}^{(1)} S_{12} + 10 S_5^{(1)} S_6^3 \left(11 \varrho_1 \varrho_2 + 135 \varrho_3^2\right)\right)}{\pi^{11}} f^{12} \\
&+ \frac{\varrho_3 \left(110592 (S_5^{(1)})^6 \varrho_3^6 + 11 S_{12}^2 (\varrho_1 \varrho_2 + 3564 \varrho_3^2) + 25 S_6^4 (\varrho_1^2 \varrho_2^2 + 2970 \varrho_1 \varrho_2 \varrho_3^2 + 18225 \varrho_3^4)\right)}{\pi^{12}} f^{13} \\
&- \frac{2488320 \left(S_5^{(1)}\right)^5 S_6 \varrho_3^7}{\pi^{13}} f^{14} \\
&+ \frac{5760 \varrho_3^5 \left(2535 \left(S_{11}^{(1)}\right)^2 \left(S_5^{(1)}\right)^2 + 2 \left(S_5^{(1)}\right)^4 S_6^2 \left(43 \varrho_1 \varrho_2 + 2025 \varrho_3^2\right)\right)}{\pi^{14}} f^{15} + O(f^{16}). \tag{3.84}
\end{aligned}
$$

The lattice sums in the formula (3.84) are presented in a symbolic form. Their numerical values are $S_6 = 3.808150792274771$, $S_{12} = 2.5353168630599594$, $S_5^{(1)} = 4.242602158408933$, $S_{11}^{(1)} = 2.93753977651848$ for the considered hexagonal array. Moreover, the integer coefficients in the formula (3.65) are expressed through the binomial coefficients and written without rounding off the numbers. Therefore, the formulas (3.65) and (3.84) are exact expressions for the coefficients of the expansion

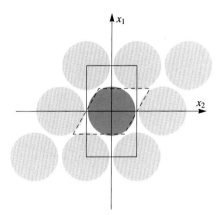

Figure 3.2 Hexagonal array in the section perpendicular to the x_3-axis. RVE (unit cell, fundamental domain) bounded by dashed line is used for analytical approach; double RVE enclosed by solid line is used for numerical computations in Section 3.6.2.

of Re A in f up to $O(f^{16})$. Furthermore, they are substituted into the ratio (3.65) for $\frac{G_e}{G}$.

The effective bulk modulus is calculated from (3.75)

$$\frac{k_e}{k} = \frac{1 - f + b + c}{1 - f + b + \frac{k}{k_1}c},$$
(3.85)

where

$$\frac{k}{k_1} = \frac{1 - \varrho_2}{1 + \varrho_3},$$

$$b = 10\frac{S_6^2}{\pi^6}\frac{\varrho_3 - \varrho_1}{1 + \varrho_1}f^7 + 12600\frac{S_{11}^{(1)} S_6^2}{\pi^{11}}\frac{\varrho_3(\varrho_3 - \varrho_1)}{1 + \varrho_1}f^{12},$$

and

$$c = \frac{\varrho_1(1 + \varrho_3)}{\varrho_3(1 + \varrho_1)}\left(f - \frac{10S_6^2}{\pi^6}\frac{\varrho_1 - \varrho_3}{1 + \varrho_1}f^7 - \frac{12600S_{11}^{(1)} S_6^2}{\pi^{11}}\frac{\varrho_3(\varrho_1 - \varrho_3)}{1 + \varrho_1}f^{12}\right).$$

As an example of direct computations, we discuss computations according to (3.65) and (3.84) substituting numerical values for the lattices sums. Restriction to the order $O(f^6)$ yields

$$\frac{G_e}{G} \simeq \frac{1 + \varrho_3 f + \varrho_3^3(8.86965\,f^5 - 19.0064\,f^6)}{1 - \kappa\left[\varrho_3 f + \varrho_3^3(8.86965\,f^5 - 19.0064\,f^6)\right]},$$
(3.86)

where ϱ_3 has the form (3.20). Below, the fraction (3.86) is expanded in f

$$\frac{G_e}{G} = 1 + f(\kappa + 1)\varrho_3 + f^2\kappa(\kappa + 1)\varrho_3^2 + f^3\kappa^2(\kappa + 1)\varrho_3^3$$

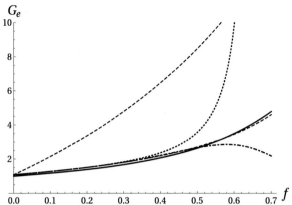

Figure 3.3 The effective shear modulus $\frac{G_e}{G}$ for glass fibers and resin matrix when $G_1 = 30.42$, $G = 1.087$, $v'_1 = 0.2$, $v' = 0.38$. The lower and upper Hashin–Shtrikman bounds are shown with dashed lines; the polynomial (3.87) with solid line; (3.86) with dot–dashed line; the truncated formula (3.86), i.e. $\frac{G_e}{G} = \frac{1+\varrho_3 f+\varrho_3^3 8.86965 f^5}{1-\kappa(\varrho_3 f+\varrho_3^3 8.86965 f^5)}$ with dotted line.

$$+ f^4\kappa^3(\kappa + 1)\varrho_3^4 + f^5\varrho_3^3(\kappa + 1)(\kappa^4\varrho_3^2 + 8.86965)$$
$$+ f^6\varrho_3^3(\kappa + 1)\left(\kappa^5\varrho_3^3 + 17.7393\kappa\varrho_3 - 19.0064\right) + O(f^7). \qquad (3.87)$$

Typical dependencies of $\frac{G_e}{G}$ on f calculated from (3.86), are presented in Fig. 3.3. First, one can see that G_e is close to the lower Hashin–Shtrikman bound for moderate f. This corroborates earlier Wall's observations (Wall, 1997), discussed in Introduction on page 3 where it is explained why accurate numerical results obtained for the hexagonal array by the self-consistent methods are close to the exact high order formulas.

The fraction (3.86) when expanded to $O(f^6)$ and up to $O(f^7)$ produces the higher and lower bounds, respectively, for the polynomial approximation (3.87). Higher order terms in f show similar unstable picture. Renormalization method will be applied below in the chapter to stabilize and extend the obtained fractions to the high concentration regime.

The next iterations are performed with the symbolic-numerical computations realized in *Mathematica*, see Chapter 2. Ultimately, we obtain an asymptotically equivalent polynomial form of (3.75)

$$\frac{k_e}{k} = \frac{1 + \frac{\varrho_1-\varrho_3}{\varrho_3(1+\varrho_1)} f + 0.1508\frac{(\varrho_1-\varrho_3)^2}{\varrho_3(1+\varrho_1)^2} f^7}{1 - 2\frac{\varrho_1}{1+\varrho_1} f - 0.3017\frac{\varrho_1(\varrho_1-\varrho_3)}{(1+\varrho_1)^2} f^7}, \qquad (3.88)$$

with the coeffcients given in terms of elastic constants.

3.3   Effective shear modulus for perfectly rigid inclusions embedded into matrix

Let us set $\frac{G_1}{G} \to \infty$. This is the "material" case corresponding to perfectly rigid (or super-rigid) inclusions. This yields $\varrho_1 = \varrho_2 = 1$ and $\varrho_3 = \frac{1}{\kappa}$. Then, the fraction (3.65) is expanded in f as follows

$$\frac{G_e}{G} = 1 + \sum_{i=1}^{10} c_i f^i + O(f^{11}), \tag{3.89}$$

where

$$c_1 = c_2 = c_3 = c_4 = 1 + \frac{1}{\kappa},$$

$$c_5 = \frac{\kappa + 1}{\kappa^3}\left(\kappa^2 + 8.86965\right), \quad c_6 = \frac{\kappa + 1}{\kappa^3}\left(\kappa^2 - 1.26709\right),$$

$$c_7 = \frac{\kappa + 1}{\kappa^3}\left(1.07542\kappa^2 - 1.22184\right), \quad c_8 = \frac{\kappa + 1}{\kappa^3}(1.15084\kappa^2 - 1.17663),$$

$$c_9 = \frac{\kappa + 1}{\kappa^5}\left(1.22627\kappa^4 - 1.13133\kappa^2 + 78.6706\right),$$

$$c_{10} = \frac{\kappa + 1}{\kappa^5}\left(1.30169\kappa^4 - 1.08608\kappa^2 - 101.148\right). \tag{3.90}$$

Setting the material parameter as $\frac{G_1}{G} \to \infty$, makes the coefficients appear to be independent on the Poisson's coefficient for inclusions. For this "material" case, the effective shear modulus depends only on the Poisson's coefficient of the matrix.

The 3D Poisson ratio is located in the interval $-1 \le \nu' \le 1/2$. Recall that the effective bulk modulus k (the inverse of compressibility) is expressed through the two other material properties by (3.2).

In principle, we are able to consider the effective properties in the limit of extremely compressible material, with the most negative Poisson coefficient $\nu' = -1$. The most highly compressible materials such as critical fluids are characterized by $\nu' \to -1$ (Lakes, 1987, Greaves et al., 2011). When $\nu' \to 1/2$, the material behaves like rubber and is referred to as incompressible.

It is possible also to study the effective properties in another particular case of compressible material with zero Poisson coefficient $\nu' = 0$. Corks and spongy materials correspond to such value. Interesting material with super-elastic performance like rubber and near-zero Poisson's ratio in all directions like cork was discovered by Wu et al. (2015).

The materials known as auxetics can have various negative Poisson ratio. They are of great practical interest because of their numerous potential applications, such as the design of fasteners, prostheses, piezo-composites, filters, earphones, seat cushions and superior dampers (Ho et al., 2014).

The effective shear modulus of the composite with perfectly rigid inclusions embedded into matrix characterized by its 3D Poisson ratio ν', is expected to diverge as a power-law

$$G_e(f, v') \simeq A\left(v'\right)(f_c - f)^{-S} + B, \tag{3.91}$$

as $f \to f_c = \frac{\pi}{\sqrt{12}} \approx 0.9069$. Here the super-elasticity index S is positive, A and B stand for the critical and sub-critical amplitudes. It is possible to evaluate the character of singularity as $f \to f_c$, from the series for small f with the coefficients (3.90). The corrective term, or confluent singularity, in (3.91) is usually assumed to be a constant (Flaherty and Keller, 1973). In what follows we will also consider higher-order corrections.

Formula (16.69) from Torquato (2002) was derived in the high-concentration limit, as $f \to f_c$

$$G_e\left(f, v'\right) = \frac{A\left(v'\right)}{\sqrt{f_c - f}}. \tag{3.92}$$

It implies a divergence of the critical amplitude $A(v')$ as $v' \to 1/2$,

$$A\left(v'\right) = \frac{0.647739}{0.5 - v'} + 2.59096. \tag{3.93}$$

This case should be considered separately, as it is also important from the physical standpoint, see Subsection 3.3.4.

3.3.1 Modified Padé approximations

Let us first extract the singularity and then construct the two-point Padé approximants, so that the final expressions satisfy both low- and high-concentration limits. The latter is given by (3.92). In the lower non-trivial order the expression for shear modulus can be expressed in a compact form,

$$Cor_1^*(f, v') = \frac{\frac{f(0.0569894 - 0.374485v')}{v' - 0.89529} + 0.952313}{\sqrt{0.9069 - f}\left(\frac{f(v'(1.34297 - 0.841908v') - 0.569973)}{v'(v' - 1.64529) + 0.671467} + 1\right)}. \tag{3.94}$$

In the higher-order it can still be expressed in the analytic form,

$$
\begin{aligned}
Cor_2^*(f, v') &= \frac{f^2 w_2(v') + f w_1(v') + w_0(v')}{f^2 w_4(v') + f w_3(v') + 1} \frac{1}{\sqrt{f_c - f}}, \\
w_0 &= 0.952313, \\
w_1(v') &= \frac{v'(v'(v'(3.7848 - 1.05622v') - 4.9428) + 2.86853) - 0.606921}{v'(v'(v'(v' - 4.12452) + 6.15224) - 4.06778) + 1.01382}, \\
w_2(v') &= \frac{v'(v'((0.14604v' - 0.186016)v' - 0.258805) + 0.520528) - 0.241963}{v'(v'(v'(v' - 4.12452) + 6.15224) - 4.06778) + 1.01382}, \\
w_3(v') &= \frac{v'(v'(4.90655 - 1.55778v') - 5.11436) + 1.90683}{v'(v'(v' - 3.37452) + 3.62135) - 1.35176}, \\
w_4(v') &= \frac{v'((0.555594v' - 1.55783)v' + 1.52841) - 0.549161}{v'(v'(v' - 3.37452) + 3.62135) - 1.35176}.
\end{aligned}
\tag{3.95}
$$

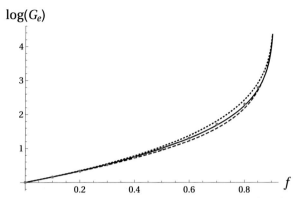

Figure 3.4 Results for the effective shear modulus for all concentrations, and with Poisson's coefficient for the matrix $\nu' = 0.3$, obtained with formula (3.94) (dotted), (3.95) (dashed) and (3.104) (solid). Numerical data from Eischen and Torquato (1993) are shown as well with dots.

Results for the effective shear modulus for all concentrations and for the Poisson's coefficient $\nu' = 0.3$, obtained with formulas (3.94), (3.95), (3.104), are shown in Fig. 3.4. Numerical data from Eischen and Torquato (1993) are shown as well. Strikingly, the expressions (3.94) and (3.95) give finite results for all ν', and all f, except strictly at the point $f = f_c$. Cancellation of the singular behavior at $\nu' = 1/2$ occurs spontaneously.

One can also attempt to construct some modified Padé approximants without invoking the high-concentration limit, and employ only the low-concentration expansion. But such approach leads to a gross underestimation of the effective modulus, even in the region of moderately-to-high concentration of inclusions.

3.3.2 Additive approximants

Let us first estimate the index $\mathcal{S}$ employing additive approximants introduced in Subsection 1.3.5, along the lines already pursued for the problem of effective conductivity (Gluzman et al., 2017). We start with finding a suitable starting approximation for the modulus and critical index. Mind that one can derive the expressions for modulus (and conductivity) from "left-to-right", i.e., extending the series from small f to large f. Alternatively, one can proceed from "right-to-left", i.e., extending the series from the large f (close to f_c) to small f (Gluzman et al., 2017).

Let us start with defining reasonable "right-to-left" zero-approximation, which is analogous to the form used in Gluzman et al. (2017) for the effective conductivity. The simplest way to proceed is to look for the solution in the whole region $[0, f_c)$ as the formal extension of the expansion in the vicinity of f_c. The extension leads to an additive self-similar approximant

$$\mathcal{A}_3^*(f, \nu') = \alpha_0(\nu')(f_c - f)^{-\mathcal{S}} + \alpha_1(\nu') + \alpha_2(\nu')(f_c - f)^{\mathcal{S}} + \alpha_3(\nu')(f_c - f)^{2\mathcal{S}}. \quad (3.96)$$

Consider first the case of $v' = v'_1 = 1/2$. All parameters in (3.96) will be obtained by matching it asymptotically with the truncated series $G_{e,4} = 1 + 2f + 2f^2 + 2f^3 + 2f^4$, with the following result,

$$A_3^*(f, 1/2) = \frac{4.69346}{(0.9069 - f)^{0.520766}} - 5.86967$$
$$+ 2.53246(0.9069 - f)^{0.520766} - 0.526588(0.9069 - f)^{1.04153}.$$
$$(3.97)$$

The value of $S = 0.520766$ for the critical index is quite good compared with the exact $1/2$. Moreover, the value does not depend on Poisson coefficient v', as can be seen from calculations with various v'.

Let us ensure the correct critical index and threshold already in the starting additive approximation, so that all parameters are obtained by matching it asymptotically with the truncated series in general form. Let us also have the limit (3.92) satisfied, so that

$$A_3^*(f, v') = f \left(0.324539 - \frac{0.0811347}{v' - 0.75} \right)$$
$$+ \frac{v'(3.93251 - 3.12943\sqrt{0.9069 - f}) + f(1.52486 - 1.52486v') + 3.37943\sqrt{0.9069 - f} - 3.93251}{(v' - 0.75)\sqrt{0.9069 - f}}.$$
$$(3.98)$$

We also bring below the lower-order additive approximants A_2^* and A_3^*

$$A_2^*(f, v') = \frac{1.94322 - 2.59096v'}{(0.5 - v')\sqrt{0.9069 - f}} + \frac{v'(4.1914 - 2.6276v') - 1.77889}{v'(v' - 1.25) + 0.375}$$
$$+ \frac{((0.952313v' - 1.42847)v' + 0.654715)\sqrt{0.9069 - f}}{v'(v' - 1.25) + 0.375},$$

$$A_3^*(f, v') = \frac{1.94322 - 2.59096v'}{\sqrt{0.9069 - f}(0.5 - v')} + \frac{v'(4.69799 - 2.96533v') - 2.1178}{v'(v' - 1.25) + 0.375}$$
$$+ \frac{0.372401(f - 0.9069)(v'(v' - 1.5) + 1.00349)}{v'(v' - 1.25) + 0.375} + \frac{(v'(1.6616v' - 2.4924) + 1.36647)\sqrt{0.9069 - f}}{v'(v' - 1.25) + 0.375},$$
$$(3.99)$$

obtained along similar lines.

The results for shear modulus for all concentrations and typical Poisson's coefficient $v' = 0.3$, obtained with formulas (3.98), (3.99), (3.104), are shown in Fig. 3.5. Numerical data from Eischen and Torquato (1993) are shown as well. Compared with the modified Padé approximants, additive formulas (3.98) and (3.99) provide much tighter bounds on the shear modulus. But they are not applicable as $v' = 1/2$. One can also attempt to construct additive approximants without invoking the high-concentration limit, and employ only the low-concentration expansion. Such approach leads to underestimation of the effective modulus, in the available region of moderately-to-high concentration of inclusions.

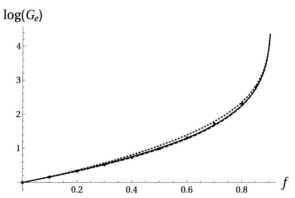

Figure 3.5 Results for the effective shear modulus for all concentrations and Poisson's coefficient of the matrix $v' = 0.3$, obtained with formula (3.98) (dotted), (3.99) (dashed) and (3.104) (solid). Numerical data from Eischen and Torquato (1993) are shown as well with dots.

3.3.3 Perfectly rigid inclusions. $DLog$ additive approximants

We would like to extend the region of applicability for our low-concentration approximations, without invoking the exact expressions for amplitudes from high-concentration limit. Consider an approach based on the high-concentration expansions written in general form. By analogy to the conductivity problem one can consider the following additive expansion for the effective modulus valid in the vicinity of f_c,

$$G_n(f, v') \simeq \sum_{k=0}^{n} A_k(v') (f_c - f)^{\frac{k-1}{2}}$$
$$= A_0 (v') (f_c - f)^{-1/2} \left(1 + \frac{A_1(v')}{A_0(v')}(f_c - f)^{1/2} + \frac{A_2(v')}{A_0(v')}(f_c - f) + \dots\right),$$

(3.100)

where $A_0 \equiv A$, $A_1 \equiv B$, and $n \geq 0$. Expression (3.100) is analogous to the expression for the effective conductivity (Gluzman et al., 2017). Generally speaking, such form describes the so-called confluent singularities, when in addition to the leading singularity there is a correction term, also called the confluent singularity, which tends to zero at the critical point, but can not be ignored in away from the critical point.

It turns out that such additive approximations work better when the following feature is added, general formulas in Subsection 1.3.5. It consists in modification of the standard $DLog$ technique for critical index calculation. Specifically, it incorporates the known value of the index and takes into account confluent additive corrections. Mind that the critical index S can be estimated also from standard representation for the derivative[1]

$$\mathcal{B}_a(f, v') = \frac{d}{df} \log(G_e(f, v')) \simeq \frac{S}{f_c - f},$$

(3.101)

as $f \to f_c$, thus defining critical index S as the residue in the corresponding single pole. For all f, well outside of the immediate vicinity of the critical point, a diagonal

---

[1] See also extensive discussion in Subsection 1.3.2

Padé approximant is assumed for the residue estimation (Baker and Graves-Moris, 1996). However, it is not able to take into account the confluent singularities (3.100).

Let us define a more general ansatz also valid for all f, but constructed to take into account confluent additive singularities, which can be called a modified "single-pole" approximation,

$$B_a(f, v') = \frac{d}{df} \log(G_e(f, v')) \simeq \frac{S}{f_c - f} + \sum_{k=1}^{n} \alpha_k (f_c - f)^{\frac{k-2}{2}}, \quad (3.102)$$

where the critical index is going to be fixed to the correct value $S = \frac{1}{2}$ and $n \geq 1$. In such formulation we can consider matrix with arbitrary v' and derive formulas satisfying both the low-concentration and high-concentrations expansions (3.100). Such formulas turn out to more accurate and more general than various low-concentration approximations based on considered above. Only properly incorporating interactions in the high-concentration limit allows to construct a consistent formula.

In principle, the formula for $G_e (f, v')$ can be found from (3.102) in arbitrary order, since the integration can be performed explicitly, as demonstrated in Subsection 1.3.5. For convenience let us introduce

$$G_0^* (f, v') \equiv \mathcal{R}\mathcal{A}_0^* (f, v') = \frac{1}{\sqrt{1 - \frac{f}{f_c}}}.$$

As the result, for all $k \geq 1$ one can find the recursion equivalent to the $DLog$ additive approximants (see Subsection 1.3.5),

$$\mathcal{R}\mathcal{A}_k^* (f, v') = \mathcal{R}\mathcal{A}_{k-1}^* (f, v') \exp \left(\frac{2\alpha_k \left(f_c^{k/2} - (f_c - f)^{k/2} \right)}{k} \right), \quad (3.103)$$

which follows directly from Eq. (1.128) on page 53.

It appears that for perfectly rigid inclusions only the low order approximants corresponding to $n = 1$ and $n = 3$, satisfy the Hashin–Shtrikman bounds. The approximants can be expressed analytically as follows:

$$\mathcal{R}\mathcal{A}_1^* (f, v') = \frac{0.952313 e^{-\frac{0.85455(v' - 1.3072)(\sqrt{0.9069 - f} - 0.952313)}{v' - 0.75}}}{\sqrt{0.9069 - f}}, \quad (3.104)$$

and

$$\mathcal{R}\mathcal{A}_3^* (f, v') = \sqrt{f_c}(f_c - f)^{-1/2}$$
$$\times \exp \left(\frac{2}{3} \sqrt{f_c - f}(\alpha_3(v')(f - f_c) - 3\alpha_1(v')) + \frac{2}{3} \sqrt{f_c}(3\alpha_1(v') + f_c\alpha_3(v')) + f\alpha_2(v') \right), \quad (3.105)$$

where

$$\alpha_1(v') = -\frac{24 f_c{}^3 c_1(v') c_2(v') - 8 f_c{}^3 c_1(v')^3 - 2 f_c{}^2 c_1(v')^2 + 4 f_c{}^2 c_2(v') - 24 f_c{}^3 c_3(v') + 3}{2\sqrt{f_c}},$$

$$\alpha_2(v') = -\frac{-48 f_c{}^3 c_1(v') c_2(v') + 16 f_c{}^3 c_1(v')^3 + 8 f_c{}^2 c_1(v')^2 - 2 f_c c_1(v') - 16 f_c{}^2 c_2(v') + 48 f_c{}^3 c_3(v') - 3}{2 f_c},$$

$$\alpha_3(v') = -\frac{24 f_c{}^3 c_1(v') c_2(v') - 8 f_c{}^3 c_1(v')^3 - 6 f_c{}^2 c_1(v')^2 + 12 f_c{}^2 c_2(v') - 24 f_c{}^3 c_3(v') + 1}{2 f_c{}^{3/2}}.$$

$$(3.106)$$

Formula (3.93) implies a divergence of the critical amplitude as $v' \to 1/2$, while our expressions for the critical amplitude $A_{0,1}(v')$, $A_{0,3}(v')$ derived from $\mathcal{RA}_1^*$ and $\mathcal{RA}_3^*$, respectively, remain finite in this limit. But they do demonstrate growth as this limit is approached,

$$A_{0,1}(v') = 2.14886 e^{-\frac{0.45345}{v'-0.75}},$$
$$A_{0,3}(v') = 147.211 \exp\left(\frac{v'(v'(8.72402 - 3.97811 v') - 6.37297) + 1.51965}{(v'-0.75)^3}\right). \quad (3.107)$$

Alternatively, we can construct the high-concentration approximants satisfying also the known value of critical amplitude. In the low-orders the following expressions hold,

$$\mathcal{RA}_1^*(f, v') = \frac{A(v')\left(\frac{A(v')}{\sqrt{f_c}}\right)^{-\frac{\sqrt{f_c-f}}{\sqrt{f_c}}}}{\sqrt{f_c-f}} = \frac{(2.59096 v' - 1.94322)\left(2.7207 - \frac{0.680175}{v'-0.5}\right)^{-1.05008\sqrt{0.9069-f}}}{(v'-0.5)\sqrt{0.9069-f}},$$

$$\mathcal{RA}_2^*(f, v') = \frac{A(v')\left(\frac{A(v')}{\sqrt{f_c}}\right)^{1-\frac{2\sqrt{f_c}\sqrt{f_c-f}+f}{f_c}} e^{\frac{(\sqrt{f_c}\sqrt{f_c-f}+f-f_c)(2 f_c c_1(v')-1)}{f_c}}}{\sqrt{f_c-f}}$$
$$= (v'(7.04921 v' - 10.5738) + 3.96518)$$
$$\times \frac{e^{1.10266(0.952313\sqrt{0.9069-f}+f-0.9069)\left(0.813799+\frac{1.8138}{3-4v'}\right)}\left(2.7207-\frac{0.680175}{v'-0.5}\right)^{-2.10015\sqrt{0.9069-f}-1.10266 f}}{\sqrt{0.9069-f}(0.5-v')^2}.$$

$$(3.108)$$

Let us construct for sake of comparison the simplest root approximant satisfying the same conditions as $\mathcal{RA}_1^*(f, v')$ from (3.108),

$$\mathcal{R}(f, v') = \sqrt{\frac{f A(v')^2}{f_c(f_c - f)} + 1}. \quad (3.109)$$

The two approximants are compared in Fig. 3.6, with strikingly better performance of $\mathcal{RA}_1^*(f, v')$ (see (3.108)) originating from the confluent singularities! Thus, we encounter not just a critical state of the composite, but specific, "confluent-singular" state of the composite with perfectly rigid incompressible inclusions embedded into the matrix!

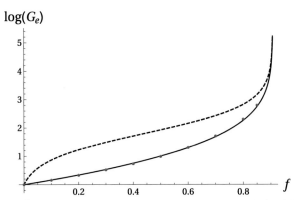

Figure 3.6 Case of $v' = 0.3$. $\log(G_e)$ of the effective shear modulus $G_e(f, v')$ of the composite with perfectly rigid inclusions embedded into the matrix. The results from (3.109) are shown with dashed line. Correspondingly, approximant $\mathcal{RA}_1^*(f, v')$ from (3.108), is shown with solid line. Numerical results from Eischen and Torquato (1993) are presented with dots.

3.3.4 Effective viscosity of 2D suspension

Consider another "material" case from (Eischen and Torquato, 1993), corresponding to fibrous composite material which has a special physical meaning,

$$\frac{G_1}{G} = \infty, \; v' = \frac{1}{2} \text{ and } v_1' = \frac{1}{2}.$$

This "material" case with the incompressible matrix and perfectly rigid, incompressible inclusions represent an elastic analog of the viscous suspension. The effective viscosity μ_e of a suspension of perfectly rigid disks in an incompressible fluid of the viscosity μ under creeping flow conditions, is equivalent to the effective modulus G_e for the material conditions just stated, see Appendix A.1. Therefore all results from this subsection apply to the effective viscosity of the hexagonal array of perfectly rigid particles in an incompressible fluid.

The lowest order $DLog$ additive approximants for the effective shear modulus can be obtained similarly to the case just discussed above,

$$\mathcal{RA}_1^*(f) = \frac{13.1805 e^{-2.75918\sqrt{0.9069-f}}}{\sqrt{0.9069-f}},$$
$$\mathcal{RA}_3^*(f) = \frac{276574. e^{-4.22019(0.9069-f)^{3/2} - 9.38171\sqrt{0.9069-f} - 9.50549 f}}{\sqrt{0.9069-f}}.$$
(3.110)

The approximants (3.110) for the effective shear modulus work well till $f \approx 0.4$. A significant improvement is achieved for the approximant Cor^*, obtained as the $\mathcal{RA}_3^*(f)$ corrected with the non-diagonal Padé approximant. The correcting term is determined from the asymptotic equivalence with the original series, so that corresponding approximants can be found explicitly,

$Cor^*(f) =$

$$\frac{(f(f((-69684.4f - 63494.7)f - 49309.5) - 38476.3) - 16924.4) e^{-4.22019(0.9069-f)^{3/2} - 9.38171\sqrt{0.9069-f} - 9.50549 f}}{(f(f(f(f(f - 0.287107) - 0.229576) - 0.178287) - 0.139117) - 0.0611929)\sqrt{0.9069-f}}.$$
(3.111)

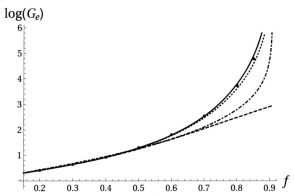

Figure 3.7 Dependence of G_e on concentration for the approximants $\mathcal{R}\mathcal{A}_1^*$ given by (3.110) (dot–dashed), $\mathbf{Cor}^*(f)$ given by (3.112) (dotted), and $\mathbf{Cor}^*(f)$ given by (3.111) (solid). The original truncated power series is shown with dashed line. Numerical results for the "material" case 5 (Eischen and Torquato, 1993) are shown with dots.

The condition on critical amplitude can not be imposed because of a divergence in the dependence on Poisson ratio (Eischen and Torquato, 1993).

Certain improvement at very high concentrations is achieved also for the approximant $\mathcal{R}\mathcal{A}_1^*(f)$ given in (3.110), corrected with Padé approximant determined from the asymptotic equivalence with the original truncated series,

$$\mathbf{Cor}^*(f) = \frac{(f(f((-1.94163f-1.60417)f-1.73157)-1.89862)-0.834311)e^{-2.75918\sqrt{0.9069-f}}}{(f(f(f(f(f-0.244534)-0.208899)-0.175892)-0.144048)-0.063299)\sqrt{0.9069-f}}. \tag{3.112}$$

Various approximants for the effective shear modulus are presented in Fig. 3.7, and compared with the numerical FEM results from Eischen and Torquato (1993), available up to $f = 0.85$. For sake of comparison we also considered the approximation given by the square-root singularity corrected with Padé approximant satisfying exactly same conditions as the approximant (3.111),

$$\mathbf{Cor}^*(f) = \frac{f(f(f(f(29.1174f+3.84744)+3.51806)+3.24466)+3.02139)+1.46196}{(f(f(f(f+0.917655)+0.888492)+0.948722)+1.53517)\sqrt{0.9069-f}}. \tag{3.113}$$

The two approximants (3.113) and (3.111) are compared in Fig. 3.8, with strikingly better performance of the latter approximant. Such performance is due to its ability to take into the confluent singularities. Thus, we encounter not just a critical state of the 2D viscous suspension, but specific "confluent-singular" state.

3.4 Effective shear modulus for soft fibers

For an array of soft fibers used to model circular holes on the cites of the hexagonal lattice, the effective shear modulus is expected to decay in the vicinity of f_c. And

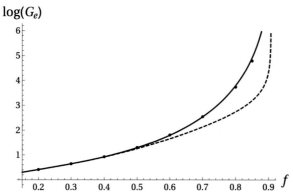

Figure 3.8 $\log(G_e)$ of the effective shear modulus for the composite with perfectly rigid incompressible inclusions embedded into the incompressible matrix. The results from (3.113) are shown with dashed line. Correspondingly, corrected approximant (3.111) is shown with solid line. Numerical results from from Eischen and Torquato (1993) are presented as well with dots.

again, a power-law

$$G_e \simeq A\,(f_c - f)^{\mathcal{T}}$$

holds, with positive critical index $\mathcal{T}$. The phase-interchange symmetry (Torquato, 2002), does not hold in the case of shear modulus for considered array, i.e., it should be two different values for the critical index for holes and inclusions[2].

The holes are punched regularly in the matrix with given elastic properties. The plane strain elastic problem is considered for such composite and the effective elastic modulus is obtained in the form of power series in the inclusions concentration and elastic constants for holes and matrix. For very soft fibers (holes) $G_1 \to 0$, $v_1' = 0$, and without loss of generality we consider G as normalized to unity. In this case, let us expand the expression (3.65) into the series

$$G_e = 1 + \sum_{k=1}^{\infty} c_k(v) f^k. \tag{3.114}$$

The coefficients of the series (3.65) are presented below,

$$c_1(v) = -\frac{4}{v+1}, \quad c_2(v) = -\frac{4(v-3)}{(v+1)^2}, \tag{3.115}$$

$$c_3(v) = -\frac{4(v-3)^2}{(v+1)^3}, \quad c_4(v) = -\frac{4(v-3)^3}{(v+1)^4},$$

$$c_5(v) = \frac{-39.4786v^4 - 93.9144v^3 - 428.8715v^2 + 290.0856v - 359.4786}{(v+1)^5},$$

---

[2] But the phase-interchange symmetry does hold for the bulk modulus (Torquato, 2002).

$$c_6(v) = \frac{1}{(v+1)^6} \Big(1.06837v^5 + 369.1706v^4 + 825.9986v^3$$
$$+2833.656v^2 - 459.3433v + 1260.8971 \Big),$$

$$c_7(v) = \frac{1}{(v+1)^7} \Big(0.585668v^6 + 342.7958v^5 - 957.7785v^4 - 2127.3584v^3$$
$$-12576.2314v^2 + 264.0335v - 4371.1049 \Big),$$

$$c_8(v) = \frac{1}{(v+1)^8} \Big(0.102966v^7 + 319.3172v^6 - 2592.9221v^5 + 291.7547v^4$$
$$-2361.0204v^3 + 46465.5942v^2 + 1373.0612v + 14878.5556 \Big),$$

$$c_9(v) = \frac{1}{(v+1)^9} \Big(-315.0623v^8 - 2218.7259v^7 - 12903.5778v^6$$
$$-8185.6578v^5 - 20068.2625v^4 + 32899.8894v^3$$
$$-165717.1671v^2 - 9601.6475v - 50246.072 \Big),$$

$$c_{10}(v) = \frac{1}{(v+1)^{10}} \Big(403.7295v^9 + 7698.5669v^8 + 39305.3967v^7$$
$$+164423.8549v^6 + 237192.7626v^5 + 341341.564v^4 - 72822.6867v^3$$
$$+622736.0775v^2 + 53406.9803v + 169862.1191 \Big),$$

$$c_{11}(v) = \frac{1}{(v+1)^{11}} \Big(-70.1792v^{10} - 1500.9936v^9 - 49754.1485v^8$$
$$-243231.8994v^7 - 1.05819 \times 10^6 v^6 - 1.76514 \times 10^6 v^5$$
$$-2.61539 \times 10^6 v^4 - 265147.972v^3 - 2.44374 \times 10^6 v^2$$
$$-273879.5068v - 576062.7868 \Big),$$

$$c_{12}(v) = \frac{1}{(v+1)^{12}} \Big(-7.40462v^{11} - 275.2794v^{10}$$
$$-34475.7764v^9 + 56662.9906v^8 + 511267.5111v^7$$
$$+4.12894 \times 10^6 v^6 + 7.96057 \times 10^6 v^5 + 1.37853 \times 10^7 v^4$$
$$+3.00337 \times 10^6 v^3 + 9.56261 \times 10^6 v^2 + 1.27173 \times 10^6 v$$
$$+1.95645 \times 10^6 \Big).$$

In principle, the series (3.114) are required up to the 6-th order inclusively, for sufficiently high numerical accuracy.

Remark 3.1. Formulas (3.115) and (3.90) are different, since they are obtained under the different assumptions for $G_1 \to \infty$ and for $G_1 \to 0$, respectively.

Consider first the case when the value of the index is known, and $\mathcal{T} = \frac{3}{2}$ (Torquato, 2002). We proceed to derivation of an accurate formula for arbitrary f and v'. Let us extract the critical part first, and then apply Padé approximants as the correction. Simple approximant obtained along these lines,

$$
\begin{aligned}
&\mathbf{Cor}^*(f, v') \\
&= \frac{(0.9069 - f)^{3/2}(f(0.156313 - 0.189308v') - 0.289468v' + 0.169774)}{-0.25v' + (v'(v' - 1.3365) + 0.478986)f + 0.146626},
\end{aligned}
$$
(3.116)

brings good accuracy in the two "material" cases relevant to soft fibers (holes) considered numerically in (Eischen and Torquato, 1993).

The critical amplitude dependence on Poisson coefficient can be readily obtained from (3.116),

$$
A(v') = \frac{0.9069(0.156313 - 0.189308v') - 0.289468v' + 0.169774}{0.9069\left(v'^2 - 1.3365v' + 0.478986\right) - 0.25v' + 0.146626}.
$$
(3.117)

For ceramics, metals and polymer matrices, the 3D Poisson ratio is typically in the range $1/4 < v' < 1/3$. The two relevant "material" cases studied by Eischen and Torquato (1993) belong to this interval.

The following formula for the effective modulus was suggested in (Drygaś et al., 2017),

$$
G(f, v') = \sqrt{\frac{1.21585(f - 0.9069)^2(v' - 0.754965) - 0.28372(0.9069 - f)^{3/2}}{(v' - 1)\left(\frac{f(2.4184v' - 1.4184) - 0.9069}{f - 0.9069}\right)^{3/2}}}.
$$
(3.118)

Formula (3.118) is applicable only qualitatively for all v', because of its inconsistency with Hashin–Shtrikman bound at small f. But formula (3.116) is applicable everywhere, including negative Poisson ratios. The two formulas are compared in Fig. 3.9.

3.4.1 Shear modulus in critical regime for arbitrary v'

In the case of holes the standard $DLog$ Padé methodology fails to find reasonable approximation to the critical region and calculate the critical index $\mathcal{T}$. In order to improve convergence of the Padé-method we can try employ the technique of corrected approximants (Gluzman et al., 2017), expressing the correction function in terms of

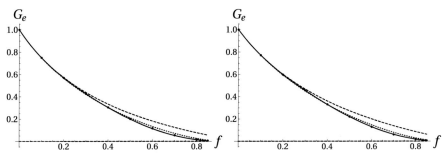

Figure 3.9 Left: Hashin–Shtrikman upper and lower bounds (dashed) compared to formula (3.118) (dotted), and corrected approximant (3.116) (solid) for $\nu' = 1/4$. Numerical results from Eischen and Torquato (1993) are shown as well. Right: Upper Hashin–Shtrikman bound (dashed) is compared to lower Hashin–Shtrikman bound (dashed), formula (3.118) (dotted), and with (3.116) (solid) for $\nu' = 1/3$. Numerical results from Eischen and Torquato (1993) are shown with dots for each case.

the $DLog$ Padé approximants. The standard transformation, $z = \frac{f}{f_c - f} \Leftrightarrow f = \frac{z f_c}{z+1}$ will be applied.

First, let us look for the solution in the form of a factored approximation,

$$G_0(f, \nu') = \left(1 - \frac{f}{f_c}\right)^{\mathcal{T}_0} R\left(f, \nu'\right), \tag{3.119}$$

so that our zero approximation for the critical index is equal to some $\mathcal{T}_0$ and the rest of the approximation is factored into the known regular part $R(f, \nu')$. Let us divide then the original series (3.114) by $G_0(f, \nu')$, express the newly found series in term of variable z defined above, then apply $DLog$ transformation and call the transformed series $L(z, \nu')$. Applying now the Padé approximants $P_{n,n+1}(z, \nu')$ to the transformed series, one can obtain the following sequence of corrected approximations for the critical index,

$$\mathcal{T}_n = \mathcal{T}_0 - \lim_{z \to \infty} (z\, P_{n,n+1}(z, \nu')). \tag{3.120}$$

Such correction technique was discussed in Gluzman et al. (2017). In particular, let us choose the simplest variant, $\mathcal{T}_0 = 1$ and $\mathcal{R}(f, \nu') = 1$.

The effective shear modulus $G_n(f, \nu')$ can be reconstructed using the complete expression for the effective critical index $\mathcal{T}_n$, employing explicitly written approximant $P_{n,n+1}$,

$$G_n(f, \nu') = G_0(f, \nu') \exp\left(\int_0^{\frac{f}{f_c - f}} P_{n,n+1}(z, \nu')\, dz\right).$$

The integral can be found analytically for various particular values of the Poisson ratio. Only the best result is shown for each ν' and the sub-script index is skipped,

$G(f, -1/3)$
$$= \frac{0.300011(1-1.10266f)^{1.60234}(1.f^2-0.124099f+0.106302)^{0.106675}e^{0.110764\tan^{-1}\left(\frac{2.63949f+0.156302}{0.9069-f}\right)}}{(0.807605f+0.174483)^{0.815693}};$$

$G(f, -1/9)$
$$= \frac{0.408109(1-1.10266f)^{1.51069}(f(f-0.0546954)+0.0873759)^{0.090863}e^{0.00147382\tan^{-1}\left(2.98836-\frac{2.92274}{0.9069-f}\right)}}{(0.780623f+0.198953)^{0.692421}};$$

$G_e(f, 1/16)$
$$= \frac{0.503584(1-1.10266f)^{1.44521}(f(f-0.0196543)+0.0735265)^{0.061163}e^{0.051156\tan^{-1}\left(3.31048-\frac{3.24072}{0.9069-f}\right)}}{(0.757195f+0.2202)^{0.567531}};$$

$G(f, 1/4) =$
$$\frac{0.605833(1-1.10266f)^{1.36805}(f(f+0.00387037)+0.0590761)^{0.0270128}e^{0.0578444\tan^{-1}\left(3.73932-\frac{3.64147}{0.9069-f}\right)}}{(0.729525f+0.245293)^{0.422077}};$$

$G(f, 1/3) =$
$$\frac{0.652281(1-1.10266f)^{1.32724}(f^2+0.0107358f+0.0525065)^{0.0153063}e^{-0.0484623\tan^{-1}\left(\frac{3.98231f+0.250457}{0.9069-f}\right)}}{(0.716015f+0.257546)^{0.357849}}.$$

$$(3.121)$$

The approximants (3.121) lead to the following estimates for the critical index,

$$\mathcal{T}(-1/3) = 1.60234, \quad \mathcal{T}(-1/9) = 1.51069, \quad \mathcal{T}(1/16) = 1.44521,$$

$$\mathcal{T}(1/4) = 1.36805, \quad \mathcal{T}(1/3) = 1.32724.$$

The quality of approximations seems to get worse for positive v'.

Consider now the different way to correct the critical index, employed also on pages 37, 193, and in Section 8.1. Let us start again with formulating the initial approximation,

$$G_0(f, v') = \left(1 - \frac{f}{f_c}\right)^{\mathcal{T}_0} R(f, v'), \quad (3.122)$$

where $\mathcal{T}_0$ is the initial guess for the critical index, to be found or fixed in the course of calculations, and $R(f, v')$ stands for the regular part. In what follows we attempt to correct $G_0(f, v')$, assuming instead of a constant $\mathcal{T}_0$ some functional dependence $\mathcal{T}(f, v')$. As $f \to f_c$,

$$\mathcal{T}(f, v') \to \mathcal{T}_c(v'),$$

and supposed to give the corrected value $\mathcal{T}_c(v')$ for the critical index. The function $\mathcal{T}(f, v')$ will be designed in such a way, that it smoothly interpolates between the initial value $\mathcal{T}_0$ and the sought value $\mathcal{T}_c$. The corrected functional form for the shear modulus is now

$$G(f, v') = \left(1 - \frac{f}{f_c}\right)^{\mathcal{T}(f,v')} R(f, v'). \quad (3.123)$$

From the (3.123) one can only formally express $\mathcal{T}(f, \nu')$, since the exact form of $G(f, \nu')$ is not known. But we can use its asymptotic, power-series form with the coefficients (3.115), express $\mathcal{T}(f, \nu')$ as the series for small f and apply some resummation procedure (e.g. the Padé technique). Finally, one can calculate the limit of the diagonal Padé approximants as $f \to f_c$ to obtain the critical index.

In what follows the ratio

$$C\left(f, \nu'\right) = \frac{G\left(f, \nu'\right)}{R\left(f, \nu'\right)}$$

stands for an asymptotic form of the critical part of the solution,

$$\mathcal{T}(f, \nu') \simeq \frac{\log\left(C\left(f, \nu'\right)\right)}{\log\left(1 - \frac{f}{f_c}\right)}, \tag{3.124}$$

which can be easily expanded in powers f and as $f \to 0$, around the value of $\mathcal{T}_0$. Let us also choose simply, $R(f, \nu') = 1$.

From the series for $T(f, \nu')$ one can now construct the diagonal Padé approximants, and find their corresponding limits as $f = f_c$. Thus found values will be our estimates for the critical index. The corresponding Padé approximants need to be holomorphic functions. In such case they do represent not only the critical index, but the whole "index" function $\mathcal{T}(f, \nu')$. The effective modulus is reconstructed from the "index" function in the whole region of concentrations for various values of the Poisson ratio,

$$G(f, -1/3) = (1 - 1.10266\,f)^{\frac{(0.333888\,f+2.23661)\,f}{f(f+0.663112)+0.0904856} + \frac{0.437661}{f(f+0.663112)+0.0904856}};$$

$$G(f, -1/9) = (1 - 1.10266\,f)^{\frac{(0.342008\,f+2.1842)\,f}{f(f+0.68557)+0.08101} + \frac{0.326524}{f(f+0.68557)+0.08101}};$$

$$G(f, 1/16) = (1 - 1.10266\,f)^{\frac{(0.275411\,f+2.2365)\,f}{f(f+0.75285)+0.066762} + \frac{0.227049}{f(f+0.75285)+0.066762}};$$

$$G(f, 1/4)$$
$$= (1 - 1.10266\,f)^{\frac{f((0.571843\,f+2.0282)\,f+0.206795)}{f(f(f+0.823813)+0.0998667)+0.0226939} + \frac{0.0617432}{f(f(f+0.823813)+0.0998667)+0.0226939}};$$

$$G(f, 1/3)$$
$$= (1 - 1.10266\,f)^{\frac{f((0.76389\,f+1.83917)\,f+0.203696)}{f(f(f+0.834883)+0.114455)+0.0341686} + \frac{0.0826332}{f(f(f+0.834883)+0.114455)+0.0341686}}.$$

$$\tag{3.125}$$

Only the best result is shown for each ν' and the sub-script index is skipped again.

We obtain the following estimates for the critical index by simply calculating the value of index functions at $f = f_c$,

$$\mathcal{T}(f_c, -1/3) = 1.80981, \quad \mathcal{T}(f_c, -1/9) = 1.69724, \quad \mathcal{T}(f_c, 1/16) = 1.57879,$$

$$\mathcal{T}(f_c, 1/4) = 1.5253, \quad \mathcal{T}(f_c, 1/3) = 1.49619.$$

Remarkably, there is a crossover in the performance of the two corrective methods around $\nu' = 1/16$. At $\nu' > 1/16$ the second method of correction predicts better values for the critical index. Correspondingly at $\nu' < 1/16$ the first method of correction predicts better values for the index. The corrected index gets fairly close to the exact $\frac{3}{2}$ (Eischen and Torquato, 1993, Torquato, 2002). Corresponding approximants represent the effective modulus in the whole region of concentrations. Thus, we find that qualitatively different types of the functions are needed to describe negative and positive Poisson ratios.

Accuracy and consistency of our calculations for the corresponding critical index is good for the perfectly rigid inclusions and for the soft fibers (holes). The former case probably requires more terms in the series. In the latter case the estimates obtained by various resummation techniques appear to be close to each other and more reliable. For soft fibers (holes) and perfectly rigid inclusions we obtain qualitatively correct general analytical expressions valid for arbitrary matrix material constants. It is also observed that in the case of inclusions it is more difficult to satisfy the Hashin–Shtrikman bounds.

The case of perfectly rigid inclusions is treated by analogy with conductivity problem, but the case of soft fibers is completely independent. We look for the solution in the form extending critical behavior to the whole region of concentrations. But such an extension leads to the necessity to consider instead of critical amplitude and index some functional dependencies. Such dependencies are constructed to satisfy the low-concentration limit. The most intriguing possibility is considered above in the form of concentration-dependent index function.

The results for the composite with perfectly rigid inclusions embedded into incompressible matrix apply to the effective viscosity of the hexagonal array of perfectly rigid particles in an incompressible fluid. In the case of random suspension considered in the next chapter, one needs to combine the analytical approach of the present work with Monte-Carlo simulations. We hope that in the same manner as in (Gluzman and Mityushev, 2015) it would be possible to obtain terms of higher order than linear Einstein's correction in the effective viscosity of $2D$ suspensions.

3.4.2   *DLog additive approximants for soft fibers*

For holes we may also consider an approach based on the high-concentration expansions presented below in general form (3.126). By analogy to the inclusions problem one can consider the following additive ansatz for the effective modulus valid in the vicinity of f_c,

$$
\begin{aligned}
G_n(f, v') &\simeq \sum_{k=0}^{n} A_k(v')(f_c - f)^{\frac{3+k}{2}} \\
&= A_0(v')(f_c - f)^{3/2} \left(1 + \frac{A_1(v')}{A_0(v')}(f_c - f)^{1/2} + \frac{A_2(v')}{A_0(v')}(f_c - f) + \dots \right),
\end{aligned} \tag{3.126}
$$

where $A_0 \equiv A$. In addition to the leading term there are correction terms, or the confluent singularities which tends to zero at the critical point, but can not be ignored away from the critical point. The critical index $\mathcal{T}$ can be estimated from the standard representation for the derivative

$$
\mathcal{B}_a(f, v') = \frac{d}{df} \log(G_e(f, v')) \simeq -\frac{\mathcal{T}}{f_c - f},
$$

as $f \to f_c$, thus defining critical index $\mathcal{T}$ as the residue in the corresponding single pole. Corrected ansatz in the form of (3.126) allows to take into account confluent additive singularities,

$$\mathcal{B}_a(f, v') = \frac{d}{df}\log(G_e(f, v')) \simeq \frac{-\mathcal{T}}{f_c - f} + \sum_{k=1}^{n} \alpha_k \, (f_c - f)^{\frac{k-2}{2}}.$$

In what follows, the critical index is going to be fixed to the correct value $\mathcal{T} = \frac{3}{2}$ and $n \geq 1$.

The effective shear modulus is calculated from the general recursion formula presented in Chapter 1, subsection 1.3.5, and the $DLog$ additive approximants are given as the particular case of the formula (1.128) on page 53. For example,

$$\mathcal{RA}_1^*(f, v') = \frac{(f_c - f)^{3/2} e^{-\frac{(\sqrt{f_c - f} - \sqrt{f_c})(2f_c c_1(v') + 3)}{\sqrt{f_c}}}}{f_c^{3/2}}$$
$$= 0.0164306(0.9069 - f)^{3/2} e^{(7.2552 - 7.6185\sqrt{0.9069 - f})v' + 4.46828\sqrt{0.9069 - f}}.$$

$$(3.127)$$

In the two "material" cases from Eischen and Torquato (1993), we have

$$\mathcal{RA}_1^*(f, 1/4) = 0.10078 e^{2.56365\sqrt{0.9069 - f}}(0.9069 - f)^{3/2},$$
$$\mathcal{RA}_1^*(f, 1/3) = 0.18448 e^{1.92878\sqrt{0.9069 - f}}(0.9069 - f)^{3/2}.$$

$$(3.128)$$

The approximants (3.128) are decreasing monotonously and give correct qualitative description for the shear modulus. Significant improvement at high concentrations is achieved for the approximant $\mathcal{RA}_1^*$ (see (3.128)), multiplicatively corrected with the diagonal Padé approximant determined from the asymptotic equivalence with the original truncated series,

$$Cor^*(f, 1/4) = \frac{0.17571(f(f(f+0.153562)+0.227224)+0.0968943)e^{2.56365\sqrt{0.9069 - f}}(0.9069 - f)^{3/2}}{f(f(f-0.2024)+0.396164)+0.168935},$$
$$Cor^*(f, 1/3)$$
$$= \frac{0.270201(f(f(f+0.0917634)+0.195668)+0.0852695)e^{1.92878\sqrt{0.9069 - f}}(0.9069 - f)^{3/2}}{f(f(f-0.125363)+0.286588)+0.124891}.$$

$$(3.129)$$

In order to discern the contribution from the confluent singularities, we also constructed a simpler approximation. It corresponds to the $(f_c - f)^{3/2}$ function, corrected with the diagonal Padé approximant, satisfying exactly the same conditions as the approximants (3.129),

$$Cor(f, 1/4) = \frac{0.636141(f(f(f+0.822478)+0.165131)+0.0749028)(0.9069 - f)^{3/2}}{f(f(f+0.512013)+0.146115)+0.041152},$$
$$Cor(f, 1/3) = \frac{0.773836(f(f(f+0.718031)+0.154988)+0.070288)(0.9069 - f)^{3/2}}{f(f(f+0.524268)+0.151153)+0.0469753}.$$

$$(3.130)$$

Various approximants for the effective shear modulus are presented in Fig. 3.10, and compared with the numerical FEM results from (Eischen and Torquato, 1993), ex-

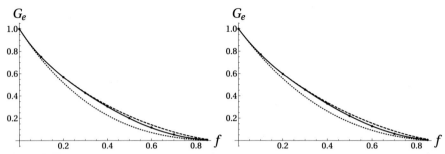

Figure 3.10 Left: Effective shear modulus for $\nu' = 1/4$. Various approximations are compared, such as (3.129) (solid), (3.128) (dotted) and (3.130) (dashed). Numerical results from Eischen and Torquato (1993) are shown as well. Right: For $\nu' = 1/3$. Various approximations are compared, such as (3.129) (solid), (3.128) (dotted) and (3.130) (dashed). Numerical results from Eischen and Torquato (1993) are shown with dots.

tending to $f = 0.85$. Almost perfect agreement with the numerical data of Eischen and Torquato (1993) is achieved by the approximation (3.129).

3.5 Method of contrast expansion for elastic incompressible composites

3.5.1 Method of Schwarz

It follows from Section A.1 that the 2D Stokes equations valid for viscous fluid, is a limit case of the 2D elasticity equations. In particular, this implies that the shear elastic modulus G coincides with the viscosity μ in this case.

In the present section, we apply Schwarz's method based on the contrast parameter expansion for a circular multiply connected domains. Schwarz's method is used in the form of the functional equations method (Mityushev and Rogosin, 2000, Mityushev, 2015, Drygaś, 2016a). Elastic isotropic materials are described through any two independent moduli. We greatly simplify the problem by considering an incompressible materials, with Poisson's ratio ν equals $1/2$. Then, we have only one contrast parameter

$$\varrho_1 = \varrho_2 = \varrho_3 = \frac{\frac{G_1}{G} - 1}{\frac{G_1}{G} + 1}, \tag{3.131}$$

where G_1 and G denote the shear modulus of inclusions and matrix, respectively. Here, the designation $G_1 = \mu_1$ and $G = \mu$ is used for incompressible media in order to stress that the results for elasticity can be applied to viscous fluids following Appendix A.1.

In the present section, we deduce a computationally efficient algorithm. Then, implement it in a symbolic form to compute the local fields in 2D elastic composites and the effective shear modulus G_e for macroscopically isotropic composites. The obtained new analytical formula is valid up to $O(\varrho_1^3)$, and it explicitly demonstrates the dependence on the location of inclusions. Such an approach is advantageous over the pure numerical methods, when dependencies of the effective constants on the mechanical properties of constituents and on geometrical structure are studied. The numerical examples from Section 3.5.6 give sufficiently accurate values of G_e for all admissible ϱ_1, i.e. for $|\varrho_1| \leq 1$, and for f not exceeding 0.4.

Consider a finite number n of inclusions as mutually disjointed, simply connected domains D_k $(k = 1, 2, \ldots, n)$ in the complex plane $\mathbb{C} \equiv \mathbb{R}^2$. It is worth noting that the number n is given in a symbolic form, with an implicit purpose to pass to the limit $n \to \infty$ later. Components of the stress tensor can be determined by the Kolosov–Muskhelishvili formulas (3.7)–(3.10).

The uniform shear stress (3.11) will be considered in this section. Then,

$$\psi_0(z) = iz + \psi(z), \tag{3.132}$$

where $\varphi_0(z)$ and $\psi(z)$ are analytical in D and bounded at infinity. Functions $\varphi_k(z)$ and $\psi_k(z)$ are analytical in D_k and twice differentiable in closures of the considered domains. It follows from (3.131) that

$$\frac{G_1}{G} = \frac{1 + \varrho_1}{1 - \varrho_1}. \tag{3.133}$$

The condition of perfect bonding at the matrix-inclusion interface can be expressed by two equations (Muskhelishvili, 1966)

$$\varphi_k(t) + t\overline{\varphi_k'(t)} + \overline{\psi_k(t)} = \varphi_0(t) + t\overline{\varphi_0'(t)} + \overline{\psi_0(t)}, \tag{3.134}$$

$$\varphi_k(t) - t\overline{\varphi_k'(t)} - \overline{\psi_k(t)} = \frac{1 + \varrho_1}{1 - \varrho_1}\left(\varphi_0(t) - t\overline{\varphi_0'(t)} - \overline{\psi_0(t)}\right), \tag{3.135}$$

$$t \in \partial D_k, \quad k = 1, 2, \ldots, n.$$

The problem (3.134)–(3.135) is the classic boundary value problem of the plane elasticity possessing the unique solution up to additive constants corresponding to rigid shifts of medium. It follows from (Muskhelishvili, 1966) that solutions of (3.134)–(3.135) analytically depend on ϱ_1 for sufficiently small $|\varrho_1|$. One can see also Chapter 2, Section 3 in Grigolyuk and Filishtinskii (1994), where the problem (3.134)–(3.135) is reduced to the Fredholm integral equation shortly written as

$$\Phi = \left(\frac{G_1}{G} - 1\right)\mathcal{K}\Phi + g,$$

where $\mathcal{K}$ is a compact integral operator in the space of the Hölder continuous functions.

We are looking for the complex potentials in form of contrast expansion. For instance, $\varphi_k(z)$ for sufficiently small $|\varrho_1|$ has the form

$$\varphi_k(z) = \sum_{l=0}^{\infty} \varphi_k^{(l)}(z)\varrho_1^l. \tag{3.136}$$

Then, the problem (3.134)–(3.135) is reduced to the following cascade of boundary value problems. Eq. (3.134) becomes

$$\varphi_k^{(l)}(t) + t\overline{\varphi_k^{(l)\prime}(t)} + \overline{\psi_k^{(l)}(t)} = \varphi_0^{(l)}(t) + t\overline{\varphi_0^{(l)\prime}(t)} + \overline{\psi_0^{(l)}(t)}, \tag{3.137}$$
$$t \in \partial D_k, \quad k = 1, 2, \dots, n, \quad l = 0, 1, \dots.$$

Eq. (3.135) yields the cascade

$$\varphi_k^{(0)}(t) - t\overline{\varphi_k^{(0)\prime}(t)} - \overline{\psi_k^{(0)}(t)} = \varphi_0^{(0)}(t) - t\overline{\varphi_0^{(0)\prime}(t)} - \overline{\psi_0^{(0)}(t)}, \tag{3.138}$$
$$\varphi_k^{(l)}(t) - t\overline{\varphi_k^{(l)\prime}(t)} - \overline{\psi_k^{(l)}(t)} = F_{l-1}(t), \; t \in \partial D_k, \; k = 1, 2, \dots, n,$$
$$l = 1, 2, \dots,$$

where

$$F_{l-1}(t) = 2\left[\varphi_0^{(l-1)}(t) - t\overline{\varphi_0^{(l-1)\prime}(t)} - \overline{\psi_0^{(l-1)}(t)}\right]. \tag{3.139}$$

Addition and subtraction of the boundary condition (3.137) for $l = 0$ and of the first condition (3.138) yields

$$\varphi_k^{(0)}(t) = \varphi_0^{(0)}(t), \quad \psi_k^{(0)}(t) = \psi_0^{(0)}(t), \quad t \in \partial D_k, \quad k = 1, 2, \dots, n. \tag{3.140}$$

Application of the principle of analytic continuation to (3.140) implies analytic continuation of $\varphi_k^{(0)}(z)$ and $\varphi_0^{(0)}(z)$ into all the domains D_k. Then, (3.132) gives the exact formulas for the zero approximation

$$\varphi_0^{(0)}(z) = 0, \quad \psi_0^{(0)}(z) = iz, \tag{3.141}$$

and $F_0(t) = 2i\,\bar{t}$.

Ultimately, we arrive at the following cascade. First, we solve n boundary value problems separately for every domain D_k

$$\varphi_k^{(1)}(t) - t\overline{\varphi_k^{(1)\prime}(t)} - \overline{\psi_k^{(1)}(t)} = 2i\,\bar{t}, \quad t \in \partial D_k. \tag{3.142}$$

Next, we solve the problem following from (3.138)

$$\varphi_0^{(1)}(t) + t\overline{\varphi_0^{(1)\prime}(t)} + \overline{\psi_0^{(1)}(t)} =$$
$$\varphi_k^{(1)}(t) + t\overline{\varphi_k^{(1)\prime}(t)} + \overline{\psi_k^{(1)}(t)}, \; t \in \partial D_k, \quad k = 1, 2, \dots, n. \tag{3.143}$$

This is a boundary value problem for the multiply connected domain D, to find $\varphi_0^{(1)}(z)$, $\psi_0^{(1)}(z)$ analytic in D and twice continuously differentiable in the closure of D including infinity. This is the first step of the iterative scheme, when we pass from $\varphi_0^{(0)}(z)$ and $\psi_0^{(0)}(z)$ to $\varphi_0^{(1)}(z)$ and $\psi_0^{(1)}(z)$. The $(l+1)$ step consists in the solution to the problems for every domain D_k

$$\varphi_k^{(l+1)}(t) - t\overline{\varphi_k^{(l+1)'}(t)} - \overline{\psi_k^{(l+1)}(t)} = F_l(t), \; t \in \partial D_k, \; k = 1, 2, \ldots, n, \quad (3.144)$$

and, further, solving the problem for the domain D

$$\varphi_0^{(l+1)}(t) + t\overline{\varphi_0^{(l+1)'}(t)} + \overline{\psi_0^{(l+1)}(t)}$$

$$= \varphi_k^{(l+1)}(t) + t\overline{\varphi_k^{(l+1)'}(t)} + \overline{\psi_k^{(l+1)}(t)}, \; t \in \partial D_k, \quad k = 1, 2, \ldots, n. \quad (3.145)$$

Therefore, the conjugation problem (3.134)–(3.135) is reduced to the sequence of problems to be considered separately for the domains D_k $(k = 1, 2, \ldots, n)$ and D. The described iterative scheme is computationally effective if the inclusions D_k have simple shape. The next subsection is devoted to its explicit realization for circular inclusions.

3.5.2   Method of functional equations

The system of functional equations (3.21)–(3.22) for the incompressible media is presented as follows:

$$\varphi_k(z) = \varrho_1 \sum_{m \neq k} \left[\overline{\Phi_m(z_{(m)}^*)} - (z - a_m) \overline{\varphi_m'(a_m)} \right] - \varrho_1 \overline{\varphi_k'(a_k)} \, (z - a_k) + p_0,$$

$$(3.146)$$

$$\Phi_k(z) = \varrho_1 \sum_{m \neq k} \overline{\varphi_m(z_{(m)}^*)} + \varrho_1 \sum_{m \neq k} \left(\overline{z_{(k)}^*} - \overline{z_{(m)}^*} \right) \left[\left(\overline{\Phi_m(z_{(m)}^*)} \right)' - \overline{\varphi_m'(a_m)} \right]$$

$$+ (1 + \varrho_1) i z - \varrho_1 \sum_{k=1}^{n} \frac{r^2 \overline{\varphi_k'(a_k)}}{z - a_k} + q_0, \; |z - a_k| \leq r, \; k = 1, 2, \ldots, n,$$

$$(3.147)$$

and

$$\psi_k(z) = \varrho_1 \sum_{m \neq k} \overline{\varphi_m(z_{(m)}^*)}$$

$$- \varrho_1 \sum_{m \neq k} \overline{z_{(m)}^*} \left[\overline{\varphi_m'(z_{(m)}^*)} + z \left(\overline{\varphi_m'(z_{(m)}^*)} \right)' + \left(\overline{\psi_m'(z_{(m)}^*)} \right)' - \overline{\varphi_m'(a_m)} \right]$$

$$- \varrho_1 \sum_{m \neq k} \frac{r_m^2}{z - a_m} \overline{\psi_m'(a_m)} + (1 + \varrho_1) i z + \varrho_1 \overline{a_k} \overline{\varphi_k'(a_k)} + q_0,$$

$$|z - a_k| \leq r_k, \ k = 1, 2, ..., N, \tag{3.148}$$

where p_0 and q_0 are constants. One can see that the contrast parameter ϱ_1 plays the role of the spectral parameter. Hence, ϕ can be written in the form of power series in ϱ_1. This implies that the method of successive approximations applied to Eqs. (3.146)–(3.147) converges in $\mathcal{H}^{(2,2)} \times \mathcal{H}^{(2,2)}$ for sufficiently small $|\varrho_1|$, see also page 107.

Functions $\varphi(z)$ and $\psi(z)$ are determined through $\varphi_k(z)$ and $\Phi_k(z)$ up to the additve constants. For instance,

$$\varphi(z) = \frac{\varrho_1}{1 + \varrho_1} \sum_{m=1}^{n} \left[\overline{\Phi_m(z^*_{(m)})} - (z - a_m) \overline{\varphi'_m(a_m)} \right], \quad z \in \mathbb{D}. \tag{3.149}$$

3.5.3 Method of successive approximations in contrast parameter

Application of the method of successive approximations to functional equations is equivalent to Schwarz's method described in the end of Subsection 3.5.1. It follows, for instance, from the uniqueness of the analytic expansion in ϱ_1 near zero. Moreover, each iteration for the functional equations corresponds to an iteration step in Schwarz's method, since the coefficients in the series in ϱ_1 are also uniquely determined. It can be also established directly from formulas written below.

Using the series (3.34) for $\varphi_k(z)$ and analogous series for other functions, and applying the successive approximations to the functional equations we arrive to the following iteration scheme. The zero-th approximation is

$$\varphi_k^{(0)}(z) = 0, \quad \Phi_k^{(0)}(z) = iz. \tag{3.150}$$

The next approximations for $s = 1, 2, \ldots$ are given as

$$\varphi_k^{(s)}(z) = \sum_{m \neq k} \left(\overline{\Phi_m^{(s-1)}\left(z_m^*\right)} - (z - a_m)\overline{(\varphi_m^{(s-1)})'(a_m)} \right) - (z - a_k)\overline{(\varphi_k^{(s-1)})'(a_k)}, \tag{3.151}$$

$$(\varphi_k^{(s)})'(z) = \sum_{m \neq k} \left(\left(\overline{\Phi_m^{(s-1)}\left(z_m^*\right)} \right)' - \overline{(\varphi_m^{(s-1)})'(a_m)} \right) - \overline{(\varphi_k^{(s-1)})'(a_k)},$$

and

$$\Phi_k^{(s)}(z) = \sum_{m \neq k} \overline{\varphi_m^{(s-1)}(z_m^*)} - \sum_{m=1}^{n} \frac{r^2}{z - a_m} \overline{(\varphi_m^{(s-1)})'(a_m)}$$

$$+ \sum_{m \neq k} \left(\overline{z_k^*} - \overline{z_m^*} \right) \left(\left(\overline{\Phi_m^{(s-1)}(z_m^*)} \right)' - \overline{(\varphi_m^{(s-1)})'(a_m)} \right) + (\delta_{0s} + \delta_{1s})iz, \tag{3.152}$$

where $\delta_{j,k}$ denotes the Kronecker symbol.

Introduce the functions

$$F^{(s)}(z) = \sum_{m=1}^{n} \left[\overline{\Phi_m^{(s)}(z_{(m)}^*)} - (z - a_m) \overline{\varphi_m^{(s)\prime}(a_m)} \right]. \tag{3.153}$$

Then,

$$\sum_{m=1}^{n} \left[\overline{\Phi_m(z_{(m)}^*)} - (z - a_m) \overline{\varphi_m'(a_m)} \right] = \sum_{s=0}^{\infty} F^{(s)}(z) \varrho_1^s. \tag{3.154}$$

It follows from (3.149) that $\varphi(z)$ can be written in the form

$$\varphi(z) = \sum_{s=0}^{\infty} \left(\sum_{j=0}^{s-1} (-1)^{s-j-1} F^{(j)}(z) \right) \varrho_1^s, \tag{3.155}$$

where the following expansion is used

$$\frac{\varrho_1}{1 + \varrho_1} = \sum_{l=0}^{\infty} (-1)^l \varrho_1^{l+1}.$$

The sth approximation for $\varphi(z)$ becomes

$$\varphi^{(s)}(z) = \sum_{j=0}^{s-1} (-1)^{s-j-1} F^{(j)}(z), \quad s = 1, 2, \ldots \tag{3.156}$$

and $\varphi^{(0)}(z) = 0$.

The limit $G_e = \lim_{n \to \infty} G_e^{(n)}$, for the effective shear modulus G_e of macroscopically isotropic composites can be calculated from formulas (3.65) and (3.60). In order to find the limit A from (3.60) in the form of contrast expansion, we calculate the integral $\int_{\partial \mathbb{D}_k} \varphi(z) dz$. Eq. (3.155) implies that

$$\int_{\partial \mathbb{D}_k} \varphi(t) dt = \sum_{s=0}^{\infty} \left(\sum_{j=0}^{s-1} (-1)^{s-j-1} \int_{\partial \mathbb{D}_k} F^{(j)}(t) dt \right) \varrho_1^s. \tag{3.157}$$

Application of (3.153) and Cauchy's integral theorem yields

$$\int_{\partial \mathbb{D}_k} F^{(s)}(z) dz = \sum_{m=1}^{n} \int_{\partial \mathbb{D}_k} \overline{\Phi_m^{(s)}(z_{(m)}^*)} dz. \tag{3.158}$$

The described iterative scheme can be easily realized numerically. But we are interested in analytical formulas which can be obtained by symbolic computations. In the next sections, we perform symbolic computations to determine $\varphi(z)$ and the effective shear modulus in the third order approximation.

3.5.4 Second order approximation

We are looking for $\varphi(z)$ in the form

$$\varphi(z) = \varphi^{(1)}(z)\varrho_1 + \varphi^{(2)}(z)\varrho_1^2 + \varphi^{(3)}(z)\varrho_1^3 + O(|\varrho_1|^4). \tag{3.159}$$

It follows from (3.155) that for the third order approximation we need to calculate integrals (3.158) for $s = 0, 1, 2$. Using the second equation (3.150) we have

$$\int_{\partial \mathbb{D}_k} F^{(0)}(z)dz = \sum_{m=1}^{n} \int_{\partial \mathbb{D}_k} \overline{\Phi_m^{(0)}(z_{(m)}^*)}dz = -i\sum_{m=1}^{n} \int_{\partial \mathbb{D}_k} \overline{z_{(m)}^*}dz. \tag{3.160}$$

Functions $\overline{z_{(m)}^*} = \dfrac{r^2}{z - a_m} + \overline{a_m}$ are analytic in $\mathbb{D}_k$ for $m \neq k$. Therefore, integrals with $m \neq k$ vanish in (3.160), as shown by invoking Cauchy's integral theorem; and (3.160) with $m = k$ can be calculated by the residue theorem

$$\int_{\partial \mathbb{D}_k} F^{(0)}(z)dz = -i\int_{\partial \mathbb{D}_k} \overline{z_{(k)}^*}dz = 2\pi r^2.$$

In order to use (3.155) for $s = 1$ we find from (3.152)

$$\Phi_k^{(1)}(z) = \sum_{m \neq k} \overline{\varphi_m^{(0)}(z_m^*)} - \sum_{m=1}^{n} \frac{r^2}{z - a_m} \overline{(\varphi_m^{(0)})'(a_m)}$$

$$+ \sum_{m \neq k} \left(\overline{z_k^*} - \overline{z_m^*} \right) \left(\left(\overline{\Phi_m^{(0)}(z_m^*)} \right)' - \overline{(\varphi_m^{(0)})'(a_m)} \right) + iz$$

$$= -i\sum_{m \neq k} \left(\overline{z_k^*} - \overline{z_m^*} \right) \left(\overline{z_m^*} \right)' + iz = i \left(z + \sum_{m \neq k} \left(\overline{z_k^*} - \overline{z_m^*} \right) \frac{r^2}{(z - a_m)^2} \right)$$

$$+ i \left(z + \sum_{m \neq k} \left((\overline{z_k^*} - \overline{a_k}) - (\overline{z_m^*} - \overline{a_m}) + (\overline{a_k} - \overline{a_m}) \right) \frac{r^2}{(z - a_m)^2} \right). \tag{3.161}$$

Using the expansion for $m \neq k$

$$(z - a_m)^{-l} = \sum_{j=0}^{\infty} (-1)^j \binom{l + j - 1}{j} (a_k - a_m)^{-(l+j)} (z - a_k)^j,$$

we have

$$(\overline{z_k^*} - \overline{a_k}) \frac{r^2}{(z - a_m)^2} = \frac{r^2}{z - a_k} \frac{r^2}{(z - a_m)^2}$$

$$= r^4((a_k - a_m)^{-2}(z - a_k)^{-1} - \sum_{j=0}^{\infty} (-1)^j \binom{j+2}{j+1}(a_k - a_m)^{-(j+3)}(z - a_k)^j)$$

$$= r^4(a_k - a_m)^{-2}(z - a_k)^{-1} - r^4 \sum_{j=0}^{\infty} (-1)^j (j+2)(a_k - a_m)^{-(j+3)}(z - a_k)^j,$$

$$(3.162)$$

and

$$(\overline{z_m^*} - \overline{a_m}) \frac{r^2}{(z - a_m)^2} = \frac{r^4}{2} \sum_{j=0}^{\infty} (-1)^j (j+1)(j+2)(a_k - a_m)^{-(j+3)}(z - a_k)^j.$$

$$(3.163)$$

Along the similar lines we find

$$(\overline{a_k} - \overline{a_m}) \frac{r^2}{(z - a_m)^2}$$

$$= r^2 \sum_{j=0}^{\infty} (-1)^j \binom{j+1}{j}(\overline{a_k} - \overline{a_m})(a_k - a_m)^{-(j+2)}(z - a_k)^j. \qquad (3.164)$$

Subtracting (3.162) from (3.163) we obtain

$$(\overline{z_k^*} - \overline{a_k}) \frac{r^2}{(z - a_m)^2} - (\overline{z_m^*} - \overline{a_m}) \frac{r^2}{(z - a_m)^2}$$

$$= r^4(a_k - a_m)^{-2}(z - a_k)^{-1}$$

$$- \frac{r^4}{2} \sum_{j=0}^{\infty} (-1)^j (j+2)(j+3)(a_k - a_m)^{-(j+3)}(z - a_k)^j. \qquad (3.165)$$

Using the above formulas we obtain from (3.161) the exact formula

$$\Phi_m^{(1)}(z) = i \left(z + \sum_{m_1 \neq m} \left(r^4(a_m - a_{m_1})^{-2}(z - a_m)^{-1} \right. \right.$$

$$- \frac{r^4}{2} \sum_{j=0}^{\infty} (-1)^j (j+2)(j+3)(a_m - a_{m_1})^{-(j+3)}(z - a_m)^j$$

$$\left. \left. + r^2 \sum_{j=0}^{\infty} (-1)^j (j+1)(\overline{a_m} - \overline{a_{m_1}})(a_m - a_{m_1})^{-(j+2)}(z - a_m)^j \right) \right)$$

$$(3.166)$$

We are now ready to calculate the integral

$$
\int_{\partial \mathbb{D}_k} F^{(1)}(z)dz = \int_{\partial \mathbb{D}_k} \sum_{m=1}^{n} \overline{\Phi_m^{(1)}(z_{(m)}^*)}dz
$$

$$
= -i \int_{\partial \mathbb{D}_k} \sum_{m=1}^{n} \left(\overline{z_{(m)}^*} + \sum_{m_1 \neq m} \left(r^4 \overline{(a_m - a_{m_1})}^{-2}(\overline{z_{(m)}^*} - a_m)^{-1} \right. \right.
$$

$$
- \frac{r^4}{2} \sum_{j=0}^{\infty} (-1)^j (j+2)(j+3)\overline{(a_m - a_{m_1})}^{-(j+3)}(\overline{z_{(m)}^*} - a_m)^j
$$

$$
\left. \left. + r^2 \sum_{j=0}^{\infty} (-1)^j (j+1)(a_m - a_{m_1})\overline{(a_m - a_{m_1})}^{-(j+2)}(\overline{z_{(m)}^*} - a_m)^j \right) \right) dz.
$$

$$(3.167)$$

Using the Cauchy integral theorem and the residue theorem, we obtain

$$
\int_{\partial \mathbb{D}_k} \sum_{m=1}^{n} \left(\overline{z_{(m)}^*} - a_m \right)^{-1} dz = \int_{\partial \mathbb{D}_k} \sum_{m=1}^{n} \frac{z - a_m}{r^2}dz = 0,
$$

and

$$
\int_{\partial \mathbb{D}_k} \sum_{m=1}^{n} \left(\overline{z_{(m)}^*} - a_m \right)^{j} dz = \int_{\partial \mathbb{D}_k} \left(\frac{r^2}{z - a_k} \right)^{j} dz = \begin{cases} 2\pi i r^2 \text{ for } j = 1 \\ 0 \text{ otherwise} \end{cases}.
$$

This yields

$$
\int_{\partial \mathbb{D}_k} F^{(1)}(z)dz = 2\pi r^2 + 12\pi r^6 \sum_{m \neq k} \overline{(a_k - a_m)}^{-4}
$$

$$
- 4\pi r^4 \sum_{m \neq k} (a_k - a_m)\overline{(a_k - a_m)}^{-3}.
$$

$$(3.168)$$

3.5.5   *Third order approximation*

In order to calculate the third order approximation $\varphi^{(3)}$ we need the function

$$
\Phi_m^{(2)}(z) = -i \left[\sum_{m_1 \neq m} \sum_{m_2 \neq m_1} \overline{(z_{m_1}^*)_{m_2}^*} - \sum_{m_1 = 1}^{n} \frac{r^2}{z - a_{m_1}} \sum_{m_2 \neq m_1} \frac{r^2}{(a_{m_1} - a_{m_2})^2} \right.
$$

$$
\left. + \sum_{m_1 \neq m} \left((\overline{z_m^*} - \overline{a_m}) - (\overline{z_{m_1}^*} - \overline{a_{m_1}}) + (\overline{a_m} - \overline{a_{m_1}}) \right) \left(\frac{r^2}{(z - a_{m_1})^2} \right. \right.
$$

$$- \sum_{m_2 \neq m_1} \left(\frac{r^2}{\left(\frac{r^2}{z - a_{m_1}} + \overline{a}_{m_1} - \overline{a}_{m_2} \right)^2} + \frac{2r^4}{\left(\frac{r^2}{z - a_{m_1}} + \overline{a}_{m_1} - \overline{a}_{m_2} \right)^3 (z - a_{m_1})} \right.$$

$$- \frac{3r^6}{\left(\frac{r^2}{z - a_{m_1}} + \overline{a}_{m_1} - \overline{a}_{m_2} \right)^4 (z - a_{m_1})^2} + \frac{2r^4 (\overline{a}_{m_1} - \overline{a}_{m_2})}{\left(\frac{r^2}{z - a_{m_1}} + \overline{a}_{m_1} - \overline{a}_{m_2} \right)^3 (z - a_{m_1})^2}$$

$$\left. - \sum_{m_2 \neq m_1} \frac{r^2}{(\overline{a}_{m_1} - \overline{a}_{m_2})^2} \right) \right].$$

It is obtained from (3.152) for $s = 2$ by substituting $\Phi_k^{(1)}$, given by (3.161) and $\varphi_k^{(1)}$ calculated from (3.151) with $s = 1$.

Below, we describe general recurrent formulas for an arbitrary sth approximation, and explicitly write down the third order terms when $s = 3$. The following double series in ϱ_1 and $(z - a_k)$ are used

$$\varphi_k(z) = \sum_{s=0}^{\infty} \left(\sum_{j=0}^{\infty} \varphi_{k,j}^{(s)} (z - a_k)^j \right) \varrho_1^s, \qquad (3.169)$$

$$\Phi_k(z) = \frac{r^2}{z - a_k} \varphi_{m,1} + \sum_{j=0}^{\infty} \gamma_{m,j} (z - a_k)^j. \qquad (3.170)$$

Coefficients in the power series (3.170) are presented as series in ϱ_1

$$\varphi_{m,l} = \sum_{s=0}^{\infty} \varphi_{m,l}^{(s)} \varrho_1^s, \quad \gamma_{m,l} = \sum_{s=0}^{\infty} \gamma_{m,l}^{(s)} \varrho_1^s. \qquad (3.171)$$

Substituting (3.169)–(3.171) into (3.146), and selecting coefficients in the same powers of ϱ_1 and $(z - a_k)$, we obtain

$$\varphi_{k,j}^{(s)} = -\delta_{j,1} \overline{\varphi_{k,1}^{(s-1)}} + (-1)^j \sum_{m \neq k} \sum_{l=1}^{\infty} \overline{\gamma_{m,l}^{(s-1)}} r^{2l} \binom{l+j-1}{j} (a_k - a_m)^{-(l+j)}, \qquad (3.172)$$

with the zeroth approximation $\varphi_{k,j}^{(0)} = 0$. Along the similar lines equation (3.147) yields

$$\gamma_{k,j}^{(s)} = i \delta_{j,1} (\delta_{0,s} + \delta_{1,s}) + \sum_{m \neq k} \left(-(-1)^j r^2 \overline{\varphi_{m,1}^{(s-1)}} (a_k - a_m)^{-(j+1)} \right.$$

$$+ (-1)^j \sum_{l=1}^{\infty} \overline{\varphi_{m,l}^{(s-1)}} r^{2l} \binom{l+j-1}{j} (a_k - a_m)^{-(l+j)}$$

$$+ (-1)^j \sum_{l=1}^{\infty} l \overline{\gamma_{m,l}^{(s-1)}} r^{2(l+1)} \binom{l+j+2}{j+1} (a_k - a_m)^{-(l+j+2)}$$

$$- (-1)^j \overline{(a_k - a_m)} \sum_{l=1}^{\infty} l \overline{\gamma_{m,l}^{(s-1)}} r^{2l} \binom{l+j}{j} (a_k - a_m)^{-(l+j+1)} \Bigg), \quad (3.173)$$

with zeroth approximation $\gamma_{k,j}^{(0)} = i\delta_{1,j}$.

The expanded form of $\varphi(z)$ follows from (3.149).

$$\varphi(z) = \frac{\varrho_1}{1+\varrho_1} \sum_{k=1}^{n} \sum_{s=0}^{\infty} \sum_{j=0}^{\infty} \overline{\gamma_{k,j}^{(s)}} r^{2j} (z - a_k)^{-j} \varrho_1^s. \quad (3.174)$$

Eq. (3.174) is in accordance with (3.155), where

$$F^{(j)}(z) = \sum_{m=1}^{n} \sum_{l=0}^{\infty} \overline{\gamma_{m,l}^{(j)}} r^2 (z - a_m)^{-l}. \quad (3.175)$$

Using (3.175), we calculate the integral

$$\int_{\partial \mathbb{D}_k} F^{(j)}(t) dt = \int_{\partial \mathbb{D}_k} \sum_{m=1}^{n} \sum_{l=0}^{\infty} \overline{\gamma_{m,l}^{(j)}} r^2 (t - a_m)^{-l} dt$$

$$= \int_{\partial \mathbb{D}_k} \sum_{l=0}^{\infty} \overline{\gamma_{k,l}^{(j)}} r^2 (t - a_k)^{-l} dt = 2\pi i \overline{\gamma_{k,1}^{(j)}} r^2. \quad (3.176)$$

Coefficients $\gamma_{m,l}^{(j)}$ for $j = 0, 1, 2$ can be explicitly written by iterations (3.173),

$$\gamma_{k,1}^{(0)} = i,$$

$$\gamma_{k,1}^{(1)} = i - ir^2 \left(2 \sum_{m \neq k} \frac{\overline{a_k - a_m}}{(a_k - a_m)^3} - 6r^2 \sum_{m \neq k} \frac{1}{(a_k - a_m)^4} \right), \quad (3.177)$$

and

$$\gamma_{k,1}^{(2)} = i \left(-r^4 \sum_{k_1 \neq k} \sum_{k_2 \neq k_1} \frac{1}{(a_k - a_{k_1})^2} \frac{1}{(\overline{a_{k_1}} - \overline{a_{k_2}})^2} \right.$$

$$+ 6r^4 \sum_{k_1 \neq k} \frac{1}{(a_k - a_{k_1})^4} - 2r^2 \sum_{k_1 \neq k} \frac{\overline{a_k - a_{k_1}}}{(a_k - a_{k_1})^3}$$

$$+ \sum_{l=1}^{\infty} (-1)^l \left(-r^{2+2l} l \sum_{k_1 \neq k} \sum_{k_2 \neq k_1} \frac{1}{(a_k - a_{k_1})^{l+1}} \frac{1}{(\overline{a_{k_1}} - \overline{a_{k_2}})^{l+1}} \right.$$

$$-\frac{1}{4}r^{2(l+3)}l((l+2)(l+3))^2 \sum_{k_1 \neq k} \frac{1}{(\overline{a_k} - \overline{a_{k_1}})^{l+3}} \sum_{k_1 \neq k} \frac{1}{(a_k - a_{k_1})^{l+3}}$$

$$+\frac{1}{2}r^{2(l+2)}l(l+1)(l+2)(l+3) \sum_{k_1 \neq k} \frac{a_k - a_{k_1}}{(\overline{a_k} - \overline{a_{k_1}})^{l+2}} \sum_{k_1 \neq k} \frac{1}{(a_k - a_{k_1})^{l+3}}$$

$$+\frac{1}{2}r^{2(l+2)}l(l+1)(l+2)(l+3) \sum_{k_1 \neq k} \frac{1}{(\overline{a_k} - \overline{a_{k_1}})^{l+3}} \sum_{k_1 \neq k} \frac{\overline{a_k} - \overline{a_{k_1}}}{(a_k - a_{k_1})^{l+2}}$$

$$-r^{2(l+1)}l(l+1)^2 \sum_{k_1 \neq k} \frac{a_k - a_{k_1}}{(\overline{a_k} - \overline{a_{k_1}})^{l+2}} \sum_{k_1 \neq k} \frac{\overline{a_k} - \overline{a_{k_1}}}{(a_k - a_{k_1})^{l+2}} \Bigg)\Bigg).$$

The limit (3.60) ought to be found. Using (3.157) and (3.176) we express A through $\gamma_{k,1}^{(j)}$,

$$A = \sum_{s=0}^{\infty} \left(\sum_{j=0}^{s-1} (-1)^{s-j-1} \pi i r^2 \lim_{n \to \infty} \frac{1}{|Q_n|} \sum_{k=1}^{n} \overline{\gamma_{k,1}^{(j)}} \right) \varrho_1^s.$$

Taking into account terms up to $O(\varrho_1^4)$, we receive $\gamma_{k,1}^{(j)}$ for $j = 0, 1, 2$. Hence,

$$A = \pi i r^2 \left(\lim_{n \to \infty} \frac{1}{|Q_n|} \sum_{k=1}^{n} \overline{\gamma_{k,1}^{(0)}} \right) \varrho_1 + \pi i r^2 \lim_{n \to \infty} \frac{1}{|Q_n|} \sum_{k=1}^{n} \left(\overline{\gamma_{k,1}^{(1)}} - \overline{\gamma_{k,1}^{(0)}} \right) \varrho_1^2$$

$$+ \pi i r^2 \lim_{n \to \infty} \frac{1}{|Q_n|} \sum_{k=1}^{n} \left(\overline{\gamma_{k,1}^{(2)}} - \overline{\gamma_{k,1}^{(1)}} + \overline{\gamma_{k,1}^{(0)}} \right) \varrho_1^3 + O(\varrho_1^4). \quad (3.178)$$

The above limits are calculated by use of the formalism developed in (Drygaś and Mityushev, 2016, Drygaś, 2016a). It was supposed that N non-overlapping disks belong to the fundamental cell (parallelogram). Next this parallelogram is periodically extended to the complex plane by two linearly independent translation vectors. As an example, we present the transformation of the term of $\gamma_{k,1}^{(1)}$ given by (3.177), multiplied by $\pi i r^2$ and complexly conjugated in accordance with (3.178)

$$T_n = 12\pi r^6 \sum_{m \neq k} \frac{1}{(a_k - a_m)^4}.$$

We are interested in the limit associated with the Eisenstein summation, see (Drygaś and Mityushev, 2016, Drygaś, 2016a),

$$T = \lim_{n \to \infty} \frac{T_n}{2|Q_n|}. \quad (3.179)$$

The normalized structural sums are introduced following (Drygaś and Mityushev, 2016, Drygaś, 2016a), by means of the Eisenstein summation

$$e_p \equiv e_p^{(0)(0)} = \lim_{n \to \infty} \frac{1}{N^{1+\frac{p}{2}}|Q_n|^{\frac{p}{2}}} \sum_{m \neq k} \frac{1}{(a_k' - a_m')^p}, \tag{3.180}$$

$$e_{p,p} \equiv e_{p,p}^{(0,0)(0,1)} = \lim_{n \to \infty} \frac{1}{N^{1+p}|Q_n|^p} \sum_{m \neq k} \sum_{k \neq k_1} \frac{1}{(a_k' - a_m')^p (a_k' - a_{k_1}')^p}, \tag{3.181}$$

$$e_p^{(1)(0)} = \lim_{n \to \infty} \frac{1}{N^{1+\frac{p-1}{2}}|Q_n|^{\frac{p-1}{2}}} \sum_{m \neq k} \frac{\overline{a_k' - a_m'}}{(a_k' - a_m')^p}, \tag{3.182}$$

where $a_k' = |Q_n|^{-\frac{1}{2}} a_k$. This means that the rectangle Q_n is normalized by the $\mathbb{C}$-linear transformation $z \to z' = |Q_n|^{-\frac{1}{2}} z$ to Q_n' having the unit area. Therefore, $T = 6 \frac{f^3}{\pi^2} e_4^{(0)(1)}$. Other terms are transformed by the same method. As a result we arrive at the formula

$$
\begin{aligned}
A = f\varrho_1 &+ 2f \left(3\frac{f^2}{\pi^2} e_4^{(0)(1)} - \frac{f}{\pi} e_3^{(1)(1)} \right) \varrho_1^2 \\
&+ f \left(-\frac{f^2}{\pi^2} e_{2,2}^{(0,0)(0,1)} + \sum_{l=1}^{\infty} (-1)^l \left(-\frac{f}{\pi} \right)^{(l+1)} l e_{l+1,l+1}^{(0,0)(0,1)} \right. \\
&\quad - \frac{1}{4} \left(\frac{f}{\pi} \right)^{(l+3)} l((l+2)(l+3))^2 \left| e_{l+3}^{(0)(0)} \right|^2 \\
&\quad + \frac{1}{2} \left(\frac{f}{\pi} \right)^{(l+2)} l(l+1)(l+2)(l+3) e_{l+2}^{(1)(1)} e_{l+3}^{(0)(0)} \\
&\quad + \frac{1}{2} \left(\frac{f}{\pi} \right)^{(l+2)} l(l+1)(l+2)(l+3) e_{l+3}^{(0)(1)} e_{l+2}^{(1)(0)} \\
&\quad \left. - \left(\frac{f}{\pi} \right)^{(l+1)} l(l+1)^2 \left| e_{l+2}^{(1)(0)} \right|^2 \right) \varrho_1^3 + O(\varrho_1^4), \tag{3.183}
\end{aligned}
$$

where

$$f = \lim_{n \to \infty} \frac{N\pi r^2}{|Q_n|},$$

denotes the concentration. Here, we use designations (1.18) for the structural e-sums. The effective shear modulus is calculated from (3.65),

$$\frac{G_e}{G} = \frac{1 + f\varrho_1 + R_2\varrho_1^2 + R_3\varrho_1^3}{1 - f\varrho_1 - R_2\varrho_1^2 - R_3\varrho_1^3}, \tag{3.184}$$

where

$$R_2 = 2\left(3\frac{e_4^{(0)(1)}}{\pi^2}f^3 - \frac{e_3^{(1)(1)}}{\pi}f^2\right),$$ (3.185)

and

$$R_3 = f\left[\frac{f^2}{\pi^2}e_{2,2}^{(0,0)(0,1)} + \sum_{l=1}^{\infty}(-1)^l\left(-\left(\frac{f}{\pi}\right)^{(l+1)}le_{l+1,l+1}^{(0,0)(0,1)}\right.\right.$$

$$-\frac{1}{4}\left(\frac{f}{\pi}\right)^{(l+3)}l((l+2)(l+3))^2\left|e_{l+3}^{(0)(0)}\right|^2$$

$$+\frac{1}{2}\left(\frac{f}{\pi}\right)^{(l+2)}l(l+1)(l+2)(l+3)e_{l+2}^{(1)(1)}e_{l+3}^{(0)(0)}$$

$$+\frac{1}{2}\left(\frac{f}{\pi}\right)^{(l+2)}l(l+1)(l+2)(l+3)e_{l+3}^{(0)(1)}e_{l+2}^{(1)(0)}$$

$$\left.\left.-\left(\frac{f}{\pi}\right)^{(l+1)}l(l+1)^2\left|e_{l+2}^{(1)(0)}\right|^2\right)\right].$$ (3.186)

Absolute convergence of series in R_3 follows from the standard root test (Cauchy's criterion).

Expansion of (3.184) in ϱ_1 yields

$$\frac{G_e}{G} = 1 + 2f\varrho_1 + 2\left(f^2 + R_2\right)\varrho_1^2 + 2\left(f^3 + 2fR_2 + R_3\right)\varrho_1^3 + O(\varrho_1^4).$$ (3.187)

3.5.6 Numerical simulations

The asymptotic formula (3.187) is novel. Computationally effective formulas and algorithms for the absolutely convergent sums $e_p^{(0)(0)}$ and $e_{p,p}^{(0,0)(0,1)}$ ($p > 2$) were developed in (Czapla et al., 2012b, Mityushev and Nawalaniec, 2015). However, the numerical implementation of the conditionally convergent series (3.180)–(3.181) for $p = 2$ and (3.182) for $p = 3$ requires further investigation.

Formula (3.65) is similar to the famous Clausius–Mossotti approximation (Maxwell's formula) (Milton, 2002, Section 10.4) applied for calculation of the effective conductivity of dilute composites. It is surprising that the Eisenstein summation and Maxwell's self-consistent formalism are based on the different summation definitions of the conditionally convergent series (McPhedran and Movchan, 1994, Mityushev, 1997a). Using an analogy with the conductivity problem we now justify the limit value of $e_3^{(1)(0)}$, i.e., the formula (3.182) for $p = 3$. The Hashin–Shtrikman bounds (Milton, 2002, Ch. 23) for 2D incompressible elastic media become as follows:

$$h^+ = \frac{1 + f\varrho_1}{1 - f\varrho_1}, \quad h^- = \frac{(\varrho_1 + 1)((f - 1)\varrho_1 + 1)}{(\varrho_1 - 1)((f - 1)\varrho_1 - 1)} \quad \text{for } \varrho_1 < 0.$$ (3.188)

Table 3.1 The nonzero values of the lattice sums for the hexagonal array.

k	S_{k+1}	$S_k^{(1)}$
5	3.80815	4.2426
11	2.53532	2.93754
17	1.64402	1.89818
23	1.06787	1.23308
29	0.693603	0.800903
35	0.450508	0.520202
41	0.292614	0.337881
47	0.190058	0.21946
53	0.123446	0.142544
59	0.0801808	0.0925848

In the case of $\varrho_1 > 0$, the bounds h^+ and h^- should be intercahnged. One can check that the upper and lower bounds (3.188) coincide up to $O(\varrho^3)$,

$$h^{\pm} = 1 + 2f\varrho_1 + 2f^2\varrho_1^2 + O(\varrho_1^3). \tag{3.189}$$

This implies the equality of coefficients in the term $f^2\varrho_1^2$ of expressions (3.187) and (3.189). It follows from (3.185) that $e_3^{(1)(0)}$ must vanish in the framework of the considered Maxwell's formalism. The definition of $e_{2,2}^{(0,0)(0,1)}$ is not essential in the final formulas, since the terms with $e_{2,2}^{(0,0)(0,1)}$ are getting canceled in (3.186). The above demonstration is based on the formal, pure mathematical arguments. Its physical interpretation is not clear yet.

Consider the numerical example with the regular hexagonal lattice, when the e-sums become the Eisenstein–Rayleigh lattice sums $e_k^{(1)(0)} = S_k^{(1)}$ and $e_k^{(0)(0)} = S_k$. Only nonzero values of S_k and $S_k^{(1)}$ are presented in Table 3.1.

The effective shear modulus for the regular hexagonal lattice takes the following form,

$$\frac{G_e}{G} = \frac{1 + f\varrho_1 + H_3\varrho_1^3}{1 - f\varrho_1 - H_3\varrho_1^3} + O(\varrho_1^4), \tag{3.190}$$

where

$$H_3 = f\left[f^2 + \sum_{l=1}^{\infty} (-1)^l \left(-\left(\frac{f}{\pi}\right)^{(l+1)} lS_{l+1}^2 - \frac{1}{4}\left(\frac{f}{\pi}\right)^{(l+3)} l((l+2)(l+3))^2 S_{l+3}^2 \right. \right.$$

$$+ \left(\frac{f}{\pi}\right)^{(l+2)} l(l+1)(l+2)(l+3) S_{l+2}^{(1)} S_{l+3}$$

$$\left. \left. - \left(\frac{f}{\pi}\right)^{(l+1)} l(l+1)^2 \left(S_{l+2}^{(1)}\right)^2 \right) \right].$$

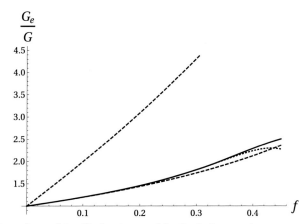

Figure 3.11 Dependence of the effective shear modulus for the hexagonal array on concentration f for $\varrho_1 = 0.9$. The data are computed from (3.190) (solid line) and from its polynomial expansion up to $O(\varrho_1^4)$ (dotted line). The Hashin–Shtrikman bounds (3.188) are shown by dashed lines.

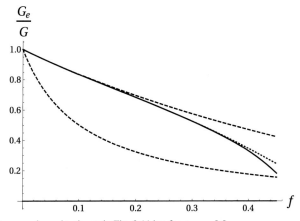

Figure 3.12 The same dependencies as in Fig. 3.11 but for $\varrho_1 = -0.9$.

The typical dependencies of the effective modulus (3.190) on f are displayed in Fig. 3.11 for ϱ_1 and in Fig. 3.12 for negative ϱ_1. The dependence on ϱ_1 is shown in Fig. 3.13. The figures demonstrate sufficiently good precision of (3.190) for moderate $f < 0.4$ and for arbitrary $|\varrho_1|$.

Thus, Schwarz's alternating method is applied to the 2D elastic problem for dispersed composites. It is realized for circular inclusions in a symbolic form. Exact and approximate formulas for the local fields and for the effective shear modulus are established. In general, Schwarz's method can be realized as expansion on the concentration f and on the contrast parameter ϱ_1. The concentration expansion was recently developed by Drygaś (2016a). In the present work, we use the expansion in contrast parameter. These two expansions in f and ϱ_1 yield two different computa-

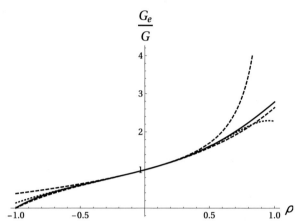

Figure 3.13 Dependence of the effective shear modulus for the hexagonal array on ϱ_1 for $f = 0.3$. The designations of lines are the same as in Fig. 3.11. (3.190) and its polynomial expansion coincide (solid line).

tional schemes (Mityushev, 2015). Formulas for the effective modulus are the same in the second order but start to differ in the third order terms in f and ϱ_1, respectively. Schwarz's method is used in the form of the functional equations method for circular inclusions (Mityushev and Rogosin, 2000, Mityushev, 2015, Drygaś, 2016a).

We develop a computationally efficient algorithm, implemented in a symbolic form, to compute the local fields in 2D elastic incompressible composites and the effective shear modulus for macroscopically isotropic composites. The new analytical formula (3.184) contains the third order term in ϱ_1 and explicitly demonstrates the dependence on locations of inclusions. The theory is supplemented by the numerical example for the hexagonal array of inclusions. Figs. 3.11–3.13 illustrate the dependence of the effective shear modulus on f and $\varrho_1 < 0.5$. These numerical examples give sufficiently accurate values of G_e for $|\varrho_1| \leq 1$, and for $f < 0.4$.

One can expect that the precision of G_e will increase, if the next approximation terms ϱ_1^m ($m > 3$) are obtained. Schwarz's method can be based on two different expansions. The expansion in ϱ_1 is used here, and the expansion in f was used in (Drygaś and Mityushev, 2016). Both expansions contain the locations of inclusions in a symbolic form. Expansions in f were obtained for small f, but for an arbitrary ϱ_1 (Drygaś and Mityushev, 2016). At the present time, we are inclined to use the contrast parameter expansions. However, we expect that the choice between different variants of Schwarz's method will depend on the further implementation of high order symbolic-numerical codes.

3.6   Practice of asymptotic methods

The effective constants are presented in the form of polynomials. The polynomial (3.87) gives the shear modulus and (3.88) expresses the bulk modulus. In our study of the elastic properties we are going to dwell on the intuition developed for the high-contrast conductivity problem (Gluzman et al., 2017). The following strategy was advanced.

Consider first the case of a perfect, ideally-conducting inclusions. We ought to necessarily consider the percolation regime of $f \to f_c$. Our strategy dictates to take into account the effect of singularity explicitly. It could be accomplished through formulation of the corresponding critical form ansatz. As soon as the non-perfection of the real conductors is taken into account, the contrast parameter decreases and the singularity starts moving to the positive nonphysical region. For high-contrast situations the critical form ansatz could be preserved while true threshold being replaced with the effective threshold, dependent on contrast parameter. For low-contrast composites one should be concerned only with faithful description of the physical region of concentrations, which is now very weakly affected by the singularity. Correspondingly, the ansatz should be corrected with the Padé approximations, till to the point of a complete neglect of the singularity.

In application to the elasticity problem we are going to follow similar strategy. Mind that analytical and asymptotic formulas show more details and trends than numerical results computed for special cases.

3.6.1   Glass-resin composite

Consider fibrous composite material composed of resin matrix and glass fibers with the 3D elastic moduli (where prime denotes 3D moduli) (Czaplinski et al., 2018),

$$E' = 3\text{GPa}, \ v' = 0.38 \quad \text{and} \quad E_1' = 73\text{GPa}, \ v_1' = 0.2. \tag{3.191}$$

For 3D moduli the following relations hold (Jun and Jasiuk, 1993, Eischen and Torquato, 1993)

$$k' = \frac{E'}{3(1 - 2v')}, \quad G' = \frac{E'}{2(1 + v')}.$$

It is noted on page 103 that

$$k = k' + G'/3, \quad E = \frac{E'}{1 - v'^2}, \quad v = \frac{v'}{1 - v'},$$

and the parameter $\kappa = 3 - 4v'$.

Thus, corresponding bulk modulus for the matrix $k = 4.53$ GPa, and for the inclusions $k_1 = 50.69$ GPa. The other $2D$ elastic moduli can be calculated as well,

$$v = 0.61, \quad v_1 = 0.25, \quad E = 3.51 \text{ GPa}, \quad E_1 = 76.0 \text{ GPa}.$$

Since $G = G'$, we obtain for the matrix $G = 1.09$ GPa, and for the inclusions $G_1 = 30.42$ GPa.

For the transverse 2D Young modulus we use the formula $E = \frac{4kG}{G+k}$. The latter formula could be applied for the effective 2D properties for the plane strain problem, such as transverse effective 2D Young modulus E_e^{2D}, or just E_e:

$$E_e = \frac{4k_e G_e}{k_e + G_e}. \tag{3.192}$$

The last formula is exploited as a source of asymptotic expansion as $f \to 0$,

$$E_e \simeq 3.50631 + 5.21146 f + 4.74723 f^2 + 4.43883 f^3 + 4.11625 f^4$$
$$+ 19.8413 f^5 - 3.53222 f^6 + 1.24937 f^7 + \dots, \tag{3.193}$$

deduced from rigorously obtained expansions for the two other moduli

$$G_e \simeq 1.08696 + 1.71489 f + 1.61462 f^2 + 1.52022 f^3 + 1.43133 f^4$$
$$+ 7.5035 f^5 - 0.330421 f^6 - 0.100326 f^7 + 0.103992 f^8, \tag{3.194}$$

$$k_e \simeq 4.52899 + 5.00687 f + 4.46385 f^2 + 3.97972 f^3 + 3.5481 f^4$$
$$+ 3.16329 f^5 + 2.82022 f^6 + 2.61716 f^7 + \dots, \tag{3.195}$$

near $f = 0$.

Let G_1 tend to ∞ and $v_1' = v' = 1/2$, as discussed above in this chapter. Such "material" case with the incompressible matrix and inclusions is expected to serve as an elastic analog of the viscous suspension. The effective viscosity of a suspension of perfectly rigid particles in an incompressible fluid under creeping flow conditions, is equivalent to the effective modulus G_e for the material conditions just stated (Torquato, 2002). There is a square-root singularity in the effective shear modulus as $f \to f_c = \frac{\pi}{\sqrt{12}} \approx 0.9069$.

Since we are interested in high-contrast cases, close to the ideal case, one can first solve the corresponding critical problem with perfectly rigid inclusions, and then simply modify parameters of the solution to move away to non-critical situations. Conductivity problem is rigorously analogous to the one-component (anti-plane) elasticity problem (Torquato, 2002), so that all results concerning effective conductivity of high-contrast composites can be applied qualitatively to the effective shear modulus G_e. In this case the contrast parameter is given as follows:

$$\varrho = \frac{G_1 - G}{G_1 + G}. \tag{3.196}$$

In our particular case $\varrho = 0.931$.

Consider now the simplest possible dependence of the effective threshold leading to the correct value of f_c as $\varrho \to 1$ and

$$f_c^*(\varrho) = \frac{f_c}{\varrho},$$

where $f_c^* = 0.974$. Such dependence is motivated by the celebrated Clausius–Mossotti approximation (CMA) (Mityushev and Rylko, 2013, Torquato, 2002), which also includes a singular behavior at $f_c^*(\varrho)$. Moving singularity to the non-physical values of f allows to preserve the form typical for critical regime for all values of ϱ but to suppress the original singularity at $f = f_c$ (Gluzman et al., 2017).

Using the analogy with the problem of effective conductivity, the effective shear (bulk) modulus is expected to diverge as a power-law in the vicinity of the singularity

$$G_e \sim (f_c^* - f)^{-\mathcal{S}}, \quad \text{as } f \to f_c^*.$$

Here the superelasticity index $\mathcal{S}$ is positive, and $\mathcal{S} = \frac{1}{2}$ (Torquato, 2002). The next order correction to the modulus is usually assumed to be a constant. By analogy to the case of perfectly rigid inclusions one can consider the following, more general additive ansatz for the shear effective modulus in the vicinity of f_c^*,

$$G_e(f) \simeq \sum_{k=0}^{n} A_k \left(f_c^* - f \right)^{\frac{k-1}{2}}, \quad n = 0, 1 \ldots N, \tag{3.197}$$

where the unknown amplitudes A_k have to be calculated from the series at small f^3. The dependence on material parameters is also hidden within the amplitudes A_k. Similar expressions can be written for $k_e(f)$, $E_e(f)$.

From the asymptotic equivalence with the series for corresponding moduli, we obtain the sequence of additive approximants $\mathcal{A}_i^*(f)$, rapidly converging to the following expressions (with $n = 3$ in (3.197))

$$G_e(f) \equiv \mathcal{A}_3^*(f) = -6.45376 - 0.655188(1 - 1.02657f)$$

$$+ 3.08259\sqrt{1 - 1.02657f} + \frac{5.11336}{\sqrt{1 - 1.02657f}},$$

$$k_e(f) = -0.425381(1 - 1.02637f)$$

$$\tag{3.198}$$

$$+ 3.59137\sqrt{1 - 1.02637f} + \frac{12.4975}{\sqrt{1 - 1.02637f}}.$$

The effective 2D Young modulus $E_e(f)$ is calculated by adapting the formula (3.192). As $f = 0$ this expression produces the 2D Young modulus of the matrix, and as $f \to f_c^*$ the effective modulus $E_e(f)$ diverges as square-root.

It is feasible to employ again the technique of $DLog$ additive approximants, to incorporate known value of the index, as well as the modified expression for the threshold (Czaplinski et al., 2018). Such approach allows to guarantee an always positive effective moduli for all concentrations. The critical index $\mathcal{S}$ can be estimated also from a standard representation for the derivative

$$\mathcal{B}_a(f) = \partial_f \log(G_e(f)) \simeq \frac{\mathcal{S}}{f_c^* - f}, \quad \text{as } f \to f_c^*,$$

---

[3] The index n in the right-hand side of (3.197) may remain "silent", its value then should be specified.

thus defining critical index $\mathcal{S}$ as the residue in the corresponding single pole. Standard techniques for the residue estimation allow also to determine G_e for arbitrary f, but in such approach the correct form of the corrections to scaling is missed.

Let us consider an additively corrected $\mathcal{B}_a(f)$, when "single-pole" approximation is complemented with correction to scaling terms,

$$\mathcal{B}_a(f) = \partial_f \log(G_e(f)) \simeq \frac{\mathcal{S}}{f_c^* - f} + \sum_{k=1}^{n} B_k \left(f_c^* - f\right)^{\frac{k-2}{2}}, \tag{3.199}$$

where the critical index is going to be fixed to the correct value $\mathcal{S} = 1/2$ (Torquato, 2002). With

$$G_0^*(f) = \frac{1}{\sqrt{1 - \frac{f}{f_c^*}}},$$

for all $k \geq 1$, one can find the following recursion for $G_k^*(f) \equiv \mathcal{R}\mathcal{A}_k^*(f)$,

$$G_k^*(f) = G_{k-1}^*(f) \exp\left(\frac{2B_k\left((f_c^*)^{k/2} - (f_c^* - f)^{k/2}\right)}{k}\right). \tag{3.200}$$

For example, when normalized to unity at $f = 0$,

$$G_1^*(f) = \frac{e^{2B_1\left(\sqrt{f_c^*} - \sqrt{f_c^* - f}\right)}}{\sqrt{1 - \frac{f}{f_c^*}}},$$

$$G_2^*(f) = \frac{e^{2B_1\left(\sqrt{f_c^*} - \sqrt{f_c^* - f}\right) + B_2 f}}{\sqrt{1 - \frac{f}{f_c^*}}}, \tag{3.201}$$

$$G_3^*(f) = \frac{\exp\left(2B_1\left(\sqrt{f_c^*} - \sqrt{f_c^* - f}\right) + \frac{2}{3}B_3\left((f_c^*)^{3/2} - (f_c^* - f)^{3/2}\right) + B_2 f\right)}{\sqrt{1 - \frac{f}{f_c^*}}}.$$

To complete the procedure one has to find B_k explicitly, from the asymptotic equivalence with the weak-coupling expansion. In the case when the effective threshold $f_c^*(\varrho)$ is known, the problem has a unique solution, as it is reduced to the linear system.

For the parameters of interest, the lowest order approximants for the effective shear modulus corresponding to $n = 1, 2, 3, 4$, are written below,

$$G_1^*(f) = \frac{8.53385e^{-2.10111\sqrt{0.974114 - f}}}{\sqrt{0.974114 - f}},$$

$$G_2^*(f) = \frac{0.903861e^{0.17365\sqrt{0.974114 - f} + 1.15239 f}}{\sqrt{0.974114 - f}},$$

$$G_3^*(f) = \frac{291.858e^{-1.50229(0.974114 - f)^{3/2} - 4.21655\sqrt{0.974114 - f} - 3.29576 f}}{\sqrt{0.974114 - f}},$$

$$G_4^*(f) = \frac{0.00013981 \exp\left(f\left(-6.06539\sqrt{0.974114 - f} - 1.91689 f + 11.6424\right) + 9.06361\sqrt{0.974114 - f}\right)}{\sqrt{0.974114 - f}}.$$

$$\tag{3.202}$$

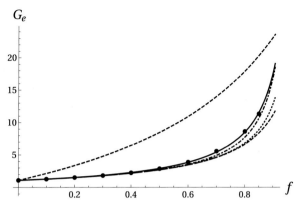

Figure 3.14 Glass-resin composite. The effective shear modulus $G_e(f)/G$ for the composite with rigid inclusions embedded into the matrix. The lower and upper Hashin–Shtrikman bounds are shown with dashed lines. The results for G_1^* are shown with solid line. They are compared with G_3^* (dotted) and additive approximant G_3 shown with dot–dashed line. The numerical FEM results are displayed as dots.

Note, that in (3.202) we returned to the original units. One can find directly that these expressions are asymptotically equivalent to the original power series (3.87). The higher order approximants become progressively closer to the lower bound and will not be discussed further.

In Fig. 3.14, the effective shear modulus of the composite with inclusions embedded into the matrix is compared with the FEM numerical results for different theoretical formulas. The approximant G_1^* is in a good agreement with FEM results. The approximants G_2^* and G_4^* do not satisfy the Hashin–Shtrikman bounds.

The approximants for the effective bulk modulus, $k_i^*(f)$, can be written explicitly along similar lines with the same critical index and effective threshold. The expressions for k_e will be different,

$$
\begin{aligned}
k_1^*(f) &= \frac{14.179 e^{-1.16936\sqrt{0.974306-f}}}{\sqrt{0.974306-f}}, \\[4pt]
k_2^*(f) &= \frac{10.4005 e^{0.15904 f - 0.855393\sqrt{0.974306-f}}}{\sqrt{0.974306-f}}, \\[4pt]
k_3^*(f) &= \frac{11.1093 e^{-0.0171372(0.974306-f)^{3/2} - 0.905484\sqrt{0.974306-f}+0.108294 f}}{\sqrt{0.974306-f}}, \\[4pt]
k_4^*(f) &= \frac{1.15803 \exp\left(f\left(-1.15841\sqrt{0.974306-f}-0.297736 f+2.42898\right)+1.3685\sqrt{0.974306-f}\right)}{\sqrt{0.974306-f}}, \\[4pt]
k_5^*(f) &= \\
&\frac{0.0237431 \exp\left(f\left(f\left(0.259284\sqrt{0.974306-f}-1.57739\right)-4.18987\sqrt{0.974306-f}+7.41609\right)+5.3066\sqrt{0.974306-f}\right)}{\sqrt{0.974306-f}}.
\end{aligned}
$$

$$(3.203)$$

The effective bulk modulus of the composite with inclusions embedded into the matrix obtained from different theoretical formulas is compared in Fig. 3.15 with the Hashin–Shtrikman upper and lower bounds. There is clear convergence within the sequence of approximants (3.203) to the approximant k_5^*.

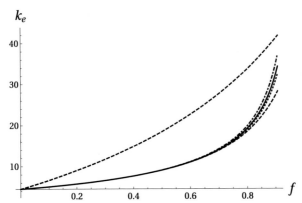

Figure 3.15 Glass-resin composite. The effective 2D bulk modulus $k_e(f)$ of the composite with rigid inclusions embedded into the matrix. The lower and upper Hashin–Shtrikman bounds are shown with dashed lines. The results for k_5^* are shown with dotted line. They are compared with k_4^* (solid) and with additive approximant k_3 shown with dot–dashed line.

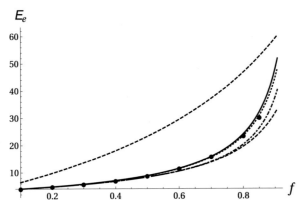

Figure 3.16 Glass-resin composite. Dependencies of E_e on concentration are constructed and illustrated based on various approximants for G_e and k_e. Formulas for upper and lower bounds are both shown with dashed lines. The results for E_1^* are shown with solid line. They are compared with E_5^* (dotted) and with additive approximant E_3 deduced from (3.198), shown with dot–dashed line.

The effective Young modulus for these approximants is calculated from the formula analogous to (3.192), which reads in this case as follows:

$$E_i^*(f) = \frac{4G_1^*(f)\,k_i^*(f)}{k_i^*(f)+G_1^*(f)}, \qquad i = 1, 3, 5\dots. \tag{3.204}$$

In Fig. 3.16, the effective Young modulus of the composite with inclusions embedded into the matrix is reconstructed from various approximations for G_e and k_e. Different theoretical formulas are compared. Formulas for upper and lower bounds are derived from the corresponding bounds for the shear and bulk moduli. Again we observe a good agreement of E_1^* and E_5^* with FEM data. Alternatively, one can obtain low-order approximants for the effective Young modulus directly from the expansion (3.193) and conditioning on square-root singularity. Such calculated approximants are in a reasonably good agreement with FEM data.

3.6.2 Comparison with numerical FEM results

Typically for our study, the obtained analytical formulas are compared with numerical FEM results, in the spirit of Eischen and Torquato (1993). Czaplinski et al. (2018) performed FEM computations for the fibrous material specifically designed for applications composed of glass fibers and resin matrix. For computations Czaplinski et al. (2018) used the ANSYS package. Good agreement is achieved for all available elastic parameters and concentrations.

In the present subsection, some pertaining information on the finite element method (FEM) is presented. Consider circular in cross-section fibers arranged in the hexagonal array (Fig. 3.2). The length in the x_3-direction is finite and is taken as $\frac{1}{2}\omega_1 = \frac{1}{\sqrt[4]{12}}$. Therefore, instead of the 2D periodicity cell Q of unit area we consider a 3D representative element $\mathcal{Q}$. Moreover, the double RVE enclosed by solid rectangle is considered for computations (see Fig. 3.2). The effective material properties combine averaged stresses and averaged strains over $\mathcal{Q}$

$$\langle \sigma_{ij} \rangle = C_{ijkl} \langle \epsilon_{kl} \rangle, \quad \langle \sigma_{ij} \rangle = \int_{\mathcal{Q}} \sigma_{ij} dx_1 dx_2 dx_3, \quad \langle \epsilon_{ij} \rangle = \int_{\mathcal{Q}} \epsilon_{ij} dx_1 dx_2 x_3,$$

where C_{ijkl} denote the effective elastic constants. Introduce the designations $\sigma_1 = \sigma_{11}$, $\sigma_2 = \sigma_{22}$, $\sigma_3 = \sigma_{33}$, $\sigma_4 = \sigma_{23}$, $\sigma_5 = \sigma_{13}$, $\sigma_6 = \sigma_{12}$. Strains ϵ_{ij} are ordered in the same way. Then, Hooke's law for transversely isotropic materials can be written in matrix form (Lekhnitskii, 1981)

$$
\begin{bmatrix}
\langle \sigma_1 \rangle \\
\langle \sigma_2 \rangle \\
\langle \sigma_3 \rangle \\
\langle \sigma_4 \rangle \\
\langle \sigma_5 \rangle \\
\langle \sigma_6 \rangle
\end{bmatrix}
=
\begin{bmatrix}
C_{11} & C_{12} & C_{13} & 0 & 0 & 0 \\
 & C_{11} & C_{13} & 0 & 0 & 0 \\
 & & C_{33} & 0 & 0 & 0 \\
 & sym & & C_{44} & 0 & 0 \\
 & & & & C_{44} & 0 \\
 & & & & & \frac{C_{11}-C_{12}}{2}
\end{bmatrix}
\begin{bmatrix}
\langle \epsilon_1 \rangle \\
\langle \epsilon_2 \rangle \\
\langle \epsilon_3 \rangle \\
\langle \epsilon_4 \rangle \\
\langle \epsilon_5 \rangle \\
\langle \epsilon_6 \rangle
\end{bmatrix}.
$$

In order to determine the first column of the stiffness matrix C_{ijkl}, we solve the boundary conditions which forces the ϵ_{11} strain equal to 1. The remaining strain tensor components are equal to zero, i.e.

$$\langle \epsilon_1 \rangle = 1, \langle \epsilon_2 \rangle = \langle \epsilon_3 \rangle = \langle \epsilon_4 \rangle = \langle \epsilon_5 \rangle = \langle \epsilon_6 \rangle = 0. \quad (3.205)$$

Solving this problem, we can find the stresses and the first column of the stiffness matrix. Its components denoted by C_{i1} are equal to the average stresses $\langle \sigma_{i1} \rangle$ in the periodicity cell at the given unit strains. Further application of the same procedure yields the components of the stiffness matrix. The ith column is determined from conditions $\langle \epsilon_i \rangle = 1$ and $\langle \epsilon_j \rangle = 0$ for $j \neq i$ $(i, j = 1, 2, \ldots, 6)$.

The effective elastic constants are determined by computed components of the stiffness matrix

$$E^L = C_{11} - \frac{2C_{12}^2}{C_{11} + C_{13}}, \quad E_e = \frac{(C_{11}(C_{11} + C_{13}) - 2C_{12}^2)(C_{11} - C_{13})}{C_{11}^2 - C_{12}^2},$$

$$(3.206)$$

$$G_e = C_{44}, \quad G^L = \frac{C_{11} - C_{12}}{2}, \quad v_e = \frac{C_{12}}{C_{22} + C_{23}}, \quad v^L = \frac{C_{11}C_{13} - C_{12}^2}{C_{11}^2 - 2C_{12}^2}.$$

$$(3.207)$$

This procedure is used to calculate effective properties of composite yarn by FEM, using the finite element code ANSYS v17.0. The representative unit cell is modeled with relative dimensions. It means that the diameter of fiber is fixed and overall dimensions of unit cell are calculated in order to achieve assumed degrree of reinforcement. The FE model consists of 296310 solid 186 elements and 1229966 nodes. The next step is to compute average components of stresses and strain over the volume of double RVE. This procedure is implemented in the ANSYS package with the help of APDL (Ansys Parametric Design Language) and performed automatically. The homogenization procedure is used to determine the effective properties for the numerical data (3.191). Results of computations are shown by dots in Figs. 3.14 and 3.16.

The 2D elastic problem considered above in the chapter is classical in the theory of composites. It is difficult to measure effective elastic properties of the fibrous composites and the challenge still remains to perform more reliable body of experimental work. Thus, theoretical investigations by analytical and numerical techniques are paramount. In particular, they are important for the regime with high-concentration of fibers. Numerical approaches to computations of the effective properties of elastic media based on the integral equations and the series method, are briefly described in Introduction.

Various analytical formulas have been obtained by means of the self-consistent methods. It was rigorously demonstrated by Mityushev and Rylko (2013) for isotropic composites, that any self-consistent method, without using of the additional geometrical assumptions, can give formulas for the effective conductivity valid only to $O(f^2)$.

It was shown by Drygaś and Mityushev (2016) that the terms on f^2 in the 2D effective elastic constants include conditionally convergent sums (3.86). The Eisenstein summation (3.79) on page 119 was applied to these sums and the local elastic fields were described in analytical form. Surprisingly, it turns out that the effective elastic constants must be calculated by the symmetric summation (3.80) on page 119. Therefore, the higher order terms for the hexagonal array from Supplement to (Drygaś and Mityushev, 2016) are wrong. In the book we correct them and express the effective constants in the form of rational and polynomial functions (3.86), (3.87) and (3.88). The dependence on the elastic parameters of components G_1, G, κ_1, κ and on the concentration f is explicitly presented in latter formulas.

This is in sharp contrast with (Guinovart-Diaz et al., 2001), where the term "closed form expressions" means really a numerical solution to an infinite system of linear algebraic equations. To finalize, one has to implement a purely numerical procedure. Our results can be compared with (Guinovart-Diaz et al., 2001) in the following way. Let the constants G_1, G, κ_1 and κ be fixed. For such parameters let us numerically solve an infinite system described in Guinovart-Diaz et al. (2001)[4] by the truncation method. The obtained constants have to coincide with the numerical values of the coefficients in f from (3.87) after substitution of the numerical data.

Therefore, despite of numerous preceding claims of "closed form solutions"[5], we assert that completely new approximate analytical formulas (3.86), (3.87) and (3.88) are deduced for the effective elastic constants of the hexagonal array. The latter formulas are written up to $O(f^7)$. It is interesting that the effective shear modulus G_e does not depend on κ_1, i.e., on the Poisson coefficient of inclusions v_1, up to $O(f^7)$. This approximation can be extended up to $O(f^{15})$, as it was done for holes in the same manner (Drygaś et al., 2017). One can see that starting from f^7, the coefficients do depend on κ_1.

In present, we choose not to compute higher order terms, since they do not impact the effective constants, at least up to moderate concentrations (f about 0.5). Moreover, higher order terms do not resolve the problem of high concentrations as shown in Fig. 3.3. The problem of the percolation regime $f \to f_c$ is studied by conditioning the formulas on the square root singularity for rigid fibers as $f \to f_c$. Simultaneously, the polynomial approximation near $f = 0$ is considered. Using asymptotically equivalent transformations we find compact analytical expressions for the effective moduli for all concentrations in Section 3.6.

Our theoretical considerations are designed to include two regimes, of low and high concentrations of inclusions. The former regime is controlled by approximations (3.86), (3.87) and (3.88), while the latter is controlled by singularity. In realistic situations the singularity gets blunted through application of the specially extended analytical technique. Our theoretical considerations are designed to match the two regimes, and derive formulas valid for all concentrations. Comparison with numerical FEM is good overall.

3.6.3 Boron-aluminum composite

Consider yet another "material" case from (Eischen and Torquato, 1993). It corresponds to the fibrous composite material, composed of boron inclusions embedded into the aluminum matrix, with material parameters

$$\frac{G_1}{G} = 6.75,\ v' = 0.35,\quad v_1' = 0.2. \tag{3.208}$$

---

[4] Their system, in fact, coincides with the system from Natanson (1935) and Grigolyuk and Filishtinskii (1970).

[5] See critical comments by Andrianov and Mityushev (2018)

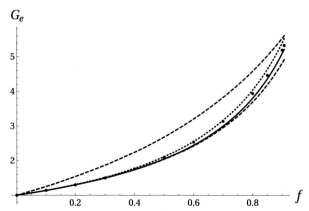

Figure 3.17 Boron-aluminum composite. Dependence of G_e on concentration for the approximants G_1^* (dotted) and G_3^* (solid). Hashin–Shtrikman bounds are shown with dashed line. Numerical results from Helsing (1995) are shown with dots.

Let us follow the approach of Subsection 3.6.1. The lowest order $DLog$ additive approximants for the effective shear modulus, can be obtained readily,

$$G_1^*(f) = \frac{9.01769e^{-1.89834\sqrt{1.22234-f}}}{\sqrt{1.22234-f}},$$
$$G_3^*(f) = \frac{424.712e^{-1.0623(1.22234-f)^{3/2}-4.08415\sqrt{1.22234-f}-2.75023f}}{\sqrt{1.22234-f}}. \tag{3.209}$$

As above we use the effective threshold instead of the true threshold. The approximants for the effective shear modulus are presented in Fig. 3.17, and compared with the Hashin–Shtrikman bounds. Numerical results from Helsing (1995) extending to very high concentrations of inclusions are shown with dots. Numerical approach of Helsing (1995) is based on the integral equations and allows to extend the results to arbitrary high concentrations.

Further improvement is achieved for the approximant **Cor**$^*(f)$, obtained as the G_1^* is corrected with the Padé approximant. The Padé approximant is determined from the asymptotic equivalence with the original series, so that

$$\mathbf{Cor}^*(f) = \frac{(f(f(518.962f+62.4595)+49.999)+30.251)e^{-1.89834\sqrt{1.22234-f}}}{(f(f(f(1f+58.3136)+7.45683)+5.54455)+3.35463)\sqrt{1.22234-f}}, \tag{3.210}$$

as shown in Fig. 3.18.

The approximants for bulk modulus are obtained along similar lines,

$$k_2^*(f) = \frac{4.83457e^{0.320055f-0.245509\sqrt{1.22234-f}}}{\sqrt{1.22234-f}},$$
$$k_3^*(f) = \frac{3.00307e^{0.0880848(1.22234-f)^{3/2}+0.0775003\sqrt{1.22234-f}+0.612214f}}{\sqrt{1.22234-f}},$$
$$k_4^*(f) = \frac{0.0846566\exp\left(f\left(-1.40847\sqrt{1.22234-f}-0.298569f+3.53184\right)+3.4131\sqrt{1.22234-f}\right)}{\sqrt{1.22234-f}}.$$

$$\tag{3.211}$$

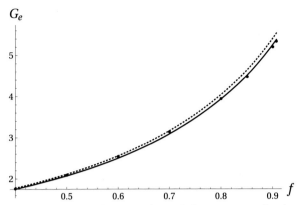

Figure 3.18 Boron-aluminum composite. Dependence of G_e on concentration for the approximants $Cor^*(f)$ (solid), G_1^* (dotted). Numerical results from Helsing (1995) are shown with dots.

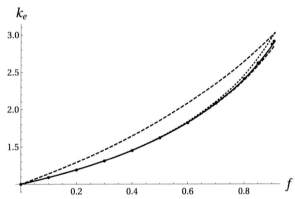

Figure 3.19 Boron-aluminum composite. Dependence of k_e on concentration for the approximants k_2^* (dotted) and k_4^* (solid). Hashin–Shtrikman bounds are shown with dashed line. Numerical results from Helsing (1995) are shown with dots. The approximants, bounds and data are normalized to unity as $f = 0$.

The approximants for the 2D effective bulk modulus normalized to unity as $f = 0$, are presented in Fig. 3.19. The are compared with the Hashin–Shtrikman bounds. Numerical results from Helsing (1995) extend to the very high concentrations of inclusions. They are shown with dots.

For both moduli we observe good agreement with the numerical data.

3.6.4 Boron-epoxy composite

Consider another "material" case from (Eischen and Torquato, 1993, Helsing, 1995), corresponding to fibrous composite material made with boron inclusions embedded into the epoxy matrix. It is characterized by the material parameters

$$\frac{G_1}{G} = 135, \quad v' = 0.35, \quad v'_1 = 0.2. \tag{3.212}$$

The lowest order $DLog$ additive approximants for the effective shear modulus, can be obtained similarly to the other cases

$$G_1^*(f) = \frac{6.79336e^{-2.04025\sqrt{0.920435-f}}}{\sqrt{0.920435-f}},$$
$$G_3^*(f) = \frac{99.8744e^{-1.31184(0.920435-f)^{3/2}-3.63452\sqrt{0.920435-f}-2.71873f}}{\sqrt{0.920435-f}}. \tag{3.213}$$

We again use the effective threshold instead of the true threshold. The approximants (3.213) for the effective shear modulus work well till $f \approx 0.9$. Further improvement in the region of very high f is achieved for the approximant Cor^*, obtained as the G_3^* from (3.213) multiplicatively corrected with the Padé approximant. The Padé approximant is determined from the asymptotic equivalence with the original series, so that

$$Cor^*(f) = \frac{e^{-1.31184(0.920435-f)^{3/2}-3.63452\sqrt{0.920435-f}-2.71873f}}{\sqrt{0.920435-f}} \times$$
$$\frac{(f(f(f(f(196.216f+71.8411)+54.4748)+43.6043)+34.9402)+18.7073)}{(f(f(f(f(f+0.742627)+0.545433)+0.436591)+0.349841)+0.187309)}. \tag{3.214}$$

Let us impose the condition on the known critical amplitude, but at the effective threshold f_c^* instead of the true threshold f_c, where

$$A(v) = \frac{\sqrt{3}\pi(3-v)\sqrt{f_c^*}}{4(1-v)},$$

and $f_c^* = \frac{f_c}{\varrho}$. Corresponding approximants can be found explicitly,

$$G_1^*(f) = \frac{6.9795e^{-2.06842\sqrt{0.920435-f}}}{\sqrt{0.920435-f}},$$
$$G_2^*(f) = \frac{7.17075e^{-2.0966\sqrt{0.920435-f}-0.0293688f}}{\sqrt{0.920435-f}},$$
$$G_3^*(f) = \frac{52.9763e^{-1.13233(0.920435-f)^{3/2}-3.13884\sqrt{0.920435-f}-2.20207f}}{\sqrt{0.920435-f}}. \tag{3.215}$$

Approximants $G_1^*(f), G_2^*(f)$ given by (3.215), work slightly better than approximants (3.213) obtained only from knowledge of the low-concentration series. But still, both of them give much smaller values for the last two points computed in Helsing (1995). Certain improvement at very high concentrations is achieved for the approximant G_3^* (from (3.215)), corrected with the Padé approximant determined from the asymptotic equivalence with the original truncated series,

$$Cor^*(f) = \frac{e^{-1.13233(0.920435-f)^{3/2}-3.13884\sqrt{0.920435-f}-2.20207f}}{\sqrt{0.920435-f}}$$
$$\times \frac{(f(f(f(f(112.177f+39.1167)+30.2812)+25.0322)+20.6394)+11.2434)}{(f(f(f(f(f+0.750502)+0.566207)+0.472516)+0.389597)+0.212235)}. \tag{3.216}$$

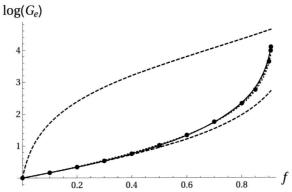

Figure 3.20 Boron-epoxy composite. Dependence of G_e on concentration for the approximants $G_1^*(f)$ given by (3.215) (dotted), $Cor^*(f)$ as given in (3.214) (solid), and $Cor^*(f)$ given by (3.216) (dot–dashed). Hashin–Shtrikman bounds are shown with dashed line. Numerical results from Helsing (1995) are shown with dots.

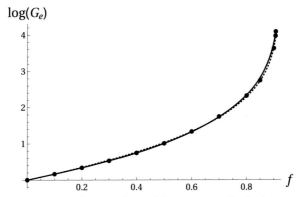

Figure 3.21 Boron-epoxy composite. Dependence of G_e on concentration for the approximants $Cor^*(f)$ (3.214) (solid) and fit G_f (3.217) (dotted). Numerical results from Helsing (1995) are shown with dots.

Various approximants for the effective shear modulus are presented in Fig. 3.20, and compared with the Hashin–Shtrikman bounds. Numerical results from Helsing (1995), extending to the very high concentrations of inclusions are shown with dots.

In order to evaluate better the quality of corrected approximant (3.214), we compare it with a simple fit to available data. Consider the effective threshold f_c^* as a free parameter to be found by fitting to the data, while the rest of parameters are still defined from the asymptotic conditions. Good agreement with data is achieved by a simple approximant

$$G(f) = \frac{7.29336 e^{-2.12204\sqrt{0.915924-f}-0.0480238f}}{\sqrt{0.915924-f}}, \tag{3.217}$$

and the approximant (3.214) compares with it quite well, as shown in Fig. 3.21.

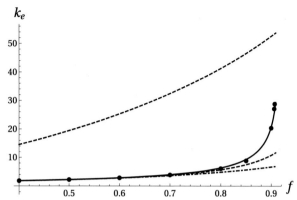

Figure 3.22 Boron-epoxy composite. Dependence of k_e on concentration for the approximant k_3^* (solid) and the truncated, 8th order series k_e (dot–dashed). Hashin–Shtrikman bounds are shown with dashed line. Numerical results from Helsing (1995) are shown with dots. The approximants, bounds and data are normalized to unity as $f = 0$.

The approximations for bulk modulus, k_i^*, are obtained along similar lines based only on the low-concentration truncated series,

$$k_1^*(f) = \frac{12.302 e^{-1.40427\sqrt{0.920435-f}}}{\sqrt{0.920435-f}},$$
$$k_3^*(f) = \frac{18.5946 e^{-0.224691(0.920435-f)^{3/2}-1.62804\sqrt{0.920435-f}-0.439973f}}{\sqrt{0.920435-f}}. \tag{3.218}$$

They are in a good agreement with numerical data of Eischen and Torquato (1993), Helsing (1995), practically for all f. Approximants for the 2D effective bulk modulus are presented in Fig. 3.22. The approximant k_3^* is compared with the Hashin–Shtrikman bounds and with the original truncated series for the bulk modulus. Numerical results from Helsing (1995) are shown with dots.

Random elastic composites with circular inclusions

4

4.1 Introduction

Based on the MMM principle by Hashin and the theory of homogenization, we apply the method of random constructive homogenization developed in (Gluzman et al., 2017). In the present chapter we derive general analytical formulas for the local fields and for the effective elastic constants in 2D random composites. The complete methodology, beginning from the construction of representative volume element (RVE) by a given image, is described in Gluzman et al. (2017), Kurtyka and Rylko (2017). In the present chapter we consider several examples of simulated random media and application of the derived symbolic-numerical algorithms.

We simulate two types of macroscopically isotropic composites. The first composite is displayed in Fig. 4.2, where a fundamental triangle with 6 inclusions is continued by reflections to the infinite plane. Next, the minimal hexagonal array with 288 inclusions is constructed. Such a symmetric construction yields an ideal macroscopically isotropic structure.

The second composite is displayed in Fig. 4.4, where $N = 81$ inclusions are randomly positioned inside the hexagonal cell. The obtained structure is not ideally isotropic. But it will become ideally isotropic as N tends to infinity. More precisely, we have to introduce the deviation (an error ϵ) from the ideal isotropy and investigate the dependence $\epsilon = \epsilon(N)$ in the considered class of random composites. In the present section we describe the way to determine $\epsilon(N)$ at the point $N = 81$ for the class of random composites generated by random walks, as explained in Chapter 9 of Gluzman et al. (2017).

Composites with polydisperse structure are shown in Fig. 4.6 and Fig. 4.8. They can be seen as modifications of the first and second types when inclusions of different radii are taken.

4.2 Method of functional equations for local fields

We begin our study with a finite number n of inclusions on the infinite plane. This number n is given in a symbolic form, allowing to pass to the limit $n \to \infty$. Mutually disjointed disks $D_k := \{z \in \mathbb{C} : |z - a_k| < r_k\}$ $(k = 1, 2, ..., n > 1)$ are considered in the complex plane $\mathbb{C}$ of the variable $z = x + iy$. Let $D := \mathbb{C} \cup \{\infty\} \setminus \left(\cup_{k=1}^{n} D_k \cup \partial D_k \right)$, where $\partial D_k := \{t \in \mathbb{C} : |t - a_k| = r_k\}$. We assume that ∂D_k are orientated clockwise. In the present section, we use letter t for a complex variable on a curve, and z for the variable in a domain.

Applied Analysis of Composite Media. https://doi.org/10.1016/B978-0-08-102670-0.00013-5

The stress and strain tensors, displacement, stresses at infinity and the perfect contact conditions are given by the general equations (3.7)–(3.17) in Chapter 3, based on the Kolosov–Muskhelishvili formulas. We will use the same designations for the elastic parameters and complex potentials as in Chapter 3. We follow the programme advanced in Section 3.1 of Chapter 3, devoted to a finite number of inclusions on the plane.

The principle difference between Chapter 3 and the present chapter lies in the passage to the limit $n \to \infty$. In Chapter 3 it is assumed that the target limit-structure is a regular hexagonal array. Here, the limit-structure is a random set of inclusions on the plane. The stress tensor in the domains D, D_k and at infinity is expressed by means of the complex potentials through the same formulas (3.7)–(3.12). Formulas for the strain tensor (3.13) and for the displacement (3.17) are slightly modified by replacing constants G_1 and κ_1 by G_k and κ_k in D_k.

The condition of perfect bonding at the matrix-inclusion interface is also slightly modified, so that

$$\varphi_k(t) + t\overline{\varphi'_k(t)} + \overline{\psi_k(t)} = \varphi_0(t) + t\overline{\varphi'_0(t)} + \overline{\psi_0(t)}, \tag{4.1}$$

$$\kappa_k \varphi_k(t) - t\overline{\varphi'_k(t)} - \overline{\psi_k(t)} = \frac{G_k}{G}\left(\kappa\varphi_0(t) - t\overline{\varphi'_0(t)} - \overline{\psi_0(t)}\right). \tag{4.2}$$

Further study will result in much more complicated formulas and mathematical expressions, than in Chapter 3. Introduce the following nondimensional contrast parameters

$$\begin{aligned}
\varrho_{1k,m} &= \frac{G_m - G}{G_k + \kappa_k G} \cdot \frac{G_k}{G_m}, \\
\varrho_{2k,m} &= \frac{\kappa G_m - \kappa_m G}{\kappa G_k + G} \cdot \frac{G_k}{G_m}, \\
\varrho_{3k,m} &= \frac{G_m - G}{\kappa G_k + G} \cdot \frac{G_k}{G_m}.
\end{aligned} \tag{4.3}$$

In the case of $k = m$ for brevity we use notations $\varrho_{ik,k} = \varrho_{ik}$ ($i = 1, 2, 3$), and it coincides with ϱ_i ($i = 1, 2, 3$) introduced in previous sections. Introduce also the new unknown functions

$$\Phi_k(z) = \left(\frac{r_k^2}{z - a_k} + \overline{a_k}\right)\varphi'_k(z) + \psi_k(z), \quad |z - a_k| \le r_k,$$

analytic in D_k, except the point a_k, where its principal part has the form $r_k^2 (z - a_k)^{-1} \varphi'_k(a_k)$.

Let $z^*_{(k)} = r_k^2 \left(\overline{z - a_k}\right)^{-1} + a_k$ denote the inversion with respect to the circle ∂D_k. The problem (4.1), (4.2) was reduced by Mityushev and Rogosin (2000) (see Eqs. (5.6.11) and (5.6.16) in Chapter 5), and by Mityushev (2000) to the system of functional equations

$$\varphi_k(z) = \sum_{m \neq k} \varrho_{1k,m} \left[\overline{\Phi_m(z^*_{(m)})} - (z - a_m) \overline{\varphi'_m(a_m)} \right]$$

$$- \varrho_{1k} \overline{\varphi'_k(a_k)} (z - a_k) + \varrho_{1k}(1 + \varrho_{3k}^{-1}) B_0 z + p_0, \ |z - a_k| \leq r_k,$$

$$k = 1, 2, ..., n, \tag{4.4}$$

$$\Phi_k(z) = \sum_{m \neq k} \varrho_{2k,m} \overline{\varphi_m(z^*_{(m)})}$$

$$+ \sum_{m \neq k} \varrho_{3k,m} \left(\overline{z^*_{(k)}} - \overline{z^*_{(m)}} \right) \left[\left(\overline{\Phi_m(z^*_{(m)})} \right)' - \overline{\varphi'_m(a_m)} \right]$$

$$+ (1 + \varrho_{3k}) B_0 \overline{z^*_{(k)}} + (1 + \varrho_{3k}) \Gamma_0 z + \omega(z), \ |z - a_k| \leq r_k,$$

$$k = 1, 2, ..., n, \tag{4.5}$$

where

$$\omega(z) = \sum_{k=1}^{n} \frac{r_k^2 q_k}{z - a_k} + q_0, \tag{4.6}$$

$q_{0,k}$ is a constant and

$$q_k = (\varrho_{2k} - \varrho_{3k}) \varphi'_k(a_k) - \varrho_{3k} \overline{\varphi'_k(a_k)}. \tag{4.7}$$

The unknown functions, $\varphi_k(z)$ and $\Phi_k(z)$ ($k = 1, 2, ..., n$), are related by $2n$ Eqs. (4.4)–(4.5). One can see that the functional equations do not contain integral operators, but do contain compositions of $\varphi_k(z)$ and $\Phi_k(z)$ combined with inversions. Such compositions define the compact operators in Banach space (Mityushev and Rogosin, 2000).

Functions $\varphi(z)$ and $\psi(z)$ are expressed through $\varphi_k(z)$ and $\psi_k(z)$ by formulas

$$\varphi(z) = \sum_{m=1}^{n} \frac{\varrho_{3m}}{1 + \varrho_{3m}} \left[\overline{\Phi_m(z^*_{(m)})} - (z - a_m) \overline{\varphi'_k(a_k)} \right] + p'_0, \ z \in D, \tag{4.8}$$

$$\psi(z) = \omega(z) - \sum_{m=1}^{n} \frac{\varrho_{3m}}{1 + \varrho_{3m}} \left(\frac{r_m^2}{z - a_m} + \overline{a_m} \right) \left[\left(\overline{\Phi_m(z^*_{(m)})} \right)' - \overline{\varphi'_m(a_m)} \right]$$

$$+ \sum_{m=1}^{n} \frac{\varrho_{2m}}{1 + \varrho_{3m}} \overline{\varphi_m(z^*_{(m)})}, \ z \in D. \tag{4.9}$$

Introduce the averaged square radius proportional to concentration f

$$r^2 = \frac{1}{n}\sum_{k=1}^{n} r_k^2.$$

Introduce also the dimensionless constants R_k in such a way that

$$r_k^2 = R_k r^2.$$

We are looking for the complex potentials φ_k and ψ_k, up to terms of $O(r^6)$, in the form

$$\varphi_k(z) = \varphi_k^{(0)}(z) + r^2 \varphi_k^{(1)}(z) + r^4 \varphi_k^{(2)}(z) + \ldots + r^{2p} \varphi_k^{(p)} + O(r^{2(p+1)}), \quad (4.10)$$

and

$$\psi_k(z) = \psi_k^{(0)}(z) + r^2 \psi_k^{(1)}(z) + r^4 \psi_k^{(2)}(z) + \ldots + r^{2p} \psi_k^{(p)} + O(r^{2(p+1)}). \quad (4.11)$$

Functions $\varphi_k^{(s)}$ and $\psi_k^{(s)}$ ($s = 0, 1, 2, \ldots, p$) inside of each inclusion are presented by their Taylor series. It is sufficient to take only first p terms

$$\varphi_k^{(s)}(z) = \alpha_{k,0}^{(s)} + \alpha_{k,1}^{(s)}(z - a_k) + \alpha_{k,2}^{(s)}(z - a_k)^2 + \ldots + \alpha_{k,2}^{(s)}(z - a_k)^p + O((z - a_k)^{p+1}), \tag{4.12}$$

$$\psi_k^{(s)}(z) = \beta_{k,0}^{(s)} + \beta_{k,1}^{(s)}(z - a_k) + \beta_{k,2}^{(s)}(z - a_k)^2 + \ldots + \beta_{k,2}^{(s)}(z - a_k)^p + O((z - a_k)^{p+1}), \tag{4.13}$$

as $z \to a_k$. The precision $O((z - a_k)^{p+1})$ is taken here for the reason explained below, after Eqs. (4.14)–(4.16).

For brevity let us introduce the auxiliary constants $\gamma_{m,l}^{(s)}$, so that

$$\gamma_{m,l}^{(s)} = (l + 2) R_m r^2 \alpha_{m,l+2}^{(s)} + \overline{a_m}(l + 1)\alpha_{m,l+1}^{(s)} + \beta_{m,l}^{(s)} \quad s, l = 1, 2, \ldots, p. \quad (4.14)$$

Substitution of (4.10) and (4.11) to (4.4) and (4.5) yields

$$\sum_{s=0}^{p} r^{2s} \varphi_k^{(s)}(z) = \sum_{m \neq k} \sum_{l+s \leq p} \varrho_{1k,m} r^{2(s+l)} \overline{\gamma}_{m,l}^{(s)}(z - a_m)^{-l}$$

$$+ \left(\varrho_{1k}(1 + \varrho_{3k}^{-1}) B_0 - \varrho_{1k} \sum_{s=0}^{\infty} r^{2s} \overline{\alpha}_{k,1}^{(s)} \right)(z - a_k)$$

$$+ p_1 + O(r^{2(p+1)}), \tag{4.15}$$

and

$$\sum_{s=0}^{p} r^{2s} \left(\overline{a_k} \left(\varphi_k^{(s)}(z) \right)' + \psi_k^{(s)}(z) \right)$$

$$= -\sum_{s=0}^{p-1} R_k r^{2(s+1)} \left(\varphi_k^{(s)}(z) \right)' (z - a_k)^{-1}$$

$$+ \sum_{m \neq k} \sum_{l+s \leq p} \varrho_{2k,m} R_m^l r^{2(l+s)} \overline{\alpha}_{m,l}^{(s)}(z - a_m)^{-l}$$

$$- \sum_{m \neq k} \varrho_{3k,m} R_k R_m r^4 \overline{\gamma}_{m,1}^{(0)}(z - a_k)^{-1}(z - a_m)^{-2}$$

$$+ \sum_{m \neq k} \sum_{l+s \leq p-1} \varrho_{3k,m} l R_m^{l+1} r^{2(l+s+1)} \overline{\gamma}_{m,l}^{(s)}(z - a_m)^{-l-2}$$

$$- \sum_{m \neq k} \sum_{l+s \leq p} \varrho_{3k,m} R_m^l l r^{2(l+s)} \overline{\gamma}_{m,l}^{(s)} \overline{(a_k - a_m)}(z - a_m)^{-l-1}$$

$$+ (1 + \varrho_{3k}) B_0 R_k r^2 (z - a_k)^{-1} + (1 + \varrho_{3k}) \Gamma_0 (z - a_k)$$

$$+ \sum_{m \neq k} R_m r^2 q_m (z - a_m)^{-1} + q_0 + O(r^{2(p+1)}), \qquad (4.16)$$

where the sum $\sum_{l+s \leq p}$ contains terms with $l = 1, 2, ..., p$ and $s = 0, 1, ..., p - 1$, satisfying the inequality $l + s \leq p$. The sum $\sum_{m \neq k}$ means that m runs over $\{1, 2, ..., n\}$, except k. Other sums are defined analogously.

One can check that the higher-order terms $(z - a_k)^l$ (for $l \geq 3$) in (4.12)–(4.13) produce terms of order $O(r^6)$ in (4.15)–(4.16). Such a rule holds in general case, when the terms $(z - a_k)^l$ yield $O(r^{2l})$. The constants $p_1, q_0, \alpha_{m,0}^{(s)}$ and $\beta_{m,0}^{(s)}$ ($m = 1, .., n$, $s = 0, 1, 2$) in (4.15)–(4.16) are not relevant, since they determine parallel translations not important for the stress and deformation fields.

Selecting the terms with the same powers r^{2p}, we arrive at the following iteration scheme for Eqs. (4.15)–(4.16). The zero terms ($p = 0$) are calculated as follows:

$$\varphi_k^{(0)}(z) = B_0 \frac{\varrho_{1k}(1 + \varrho_{3k})}{\varrho_{3k}(1 + \varrho_{1k})} (z - a_k), \qquad (4.17)$$

$$\psi_k^{(0)}(z) = \Gamma_0 (1 + \varrho_{3k})(z - a_k). \qquad (4.18)$$

It follows from (4.4)–(4.8) that

$$\varphi(z) = r^2 \varphi^{(1)}(z) + r^4 \varphi^{(2)}(z) + ... + r^{2p} \varphi^{(p)} + O(r^{2(p+1)}),$$

where, for instance,

$$\varphi^{(1)}(z) = \overline{\Gamma_0} \sum_{k=1}^{n} R_k \varrho_{3k}(z - a_k)^{-1}. \qquad (4.19)$$

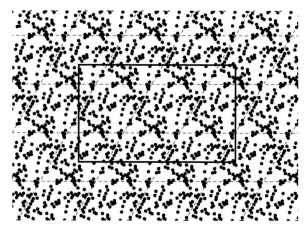

Figure 4.1 The rectangle Q_n with n inclusions enclosed (compare to Fig. 3.1).

Analogously, the function ψ has the form

$$\psi(z) = r^2 \psi^{(1)}(z) + r^4 \psi^{(2)}(z) + \ldots + r^{2p} \psi^{(p)} + O(r^{2(p+1)}),$$

where, for instance,

$$\psi^{(1)}(z) = \sum_{k=1}^{n} \left(2B_0 R_k \frac{\varrho_{1k} - \varrho_{3k}}{\varrho_{3k}(1 + \varrho_{1k})} (z - a_k)^{-1} + R_k \varrho_{3k} \overline{\Gamma_0 a_k} (z - a_k)^{-2} \right). \quad (4.20)$$

This computational scheme can be easily extended to higher orders by means of symbolic computations with *Mathematica*.

4.3 Averaged fields in composites and effective shear modulus

The averaged shear modulus $G_e^{(n)}$ of the finite composite with n inclusions and the effective shear modulus G_e for macroscopically isotropic composites are computed by the method described in Chapter 3. The rectangle Q_n is displayed in Fig. 4.1. First, $G_e^{(n)}$ is determined. Next, the limit $G_e = \lim_{n \to \infty} G_e^{(n)}$ is explicitly calculated.

The macroscopic shear modulus can be computed from the same formula as in Chapter 3, namely

$$\frac{G_e}{G} = \frac{1 + \operatorname{Re} A}{1 - \kappa \operatorname{Re} A}, \quad (4.21)$$

where

$$A = \lim_{n \to \infty} \frac{1}{2|Q_n|} \sum_{k=1}^{n} \int_{\partial D_k} \varphi(t) dt. \tag{4.22}$$

Though formulas (4.21)–(4.22) are formally the same, the explicit expression for the constant A is much more complicated. We have

$$\Phi_k(z) = \frac{r^2}{z - a_k} \varphi_{m,1} + \sum_{j=0}^{\infty} \gamma_{m,j}(z - a_k)^j, \tag{4.23}$$

and

$$\varphi(z) = \sum_{m=1}^{n} \sum_{l=0}^{\infty} \frac{\varrho_{3m}}{1 + \varrho_{3m}} \overline{\gamma}_{m,l} R_m^l r^{2l} (z - a_m)^{-l}.$$

Using the residue theorem we obtain

$$\int_{\partial D_k} \varphi(z) dz = \int_{\partial D_k} \sum_{m=1}^{n} \sum_{l=0}^{\infty} \frac{\varrho_{3m}}{1 + \varrho_{3m}} \overline{\gamma}_{m,l} R_m^l r^{2l} (z - a_m)^{-l} dz$$

$$= 2\pi i \frac{\varrho_{3k}}{1 + \varrho_{3k}} \overline{\gamma}_{k,1} R_k r^2,$$

and

$$A = i \frac{f}{n} \sum_{k=1}^{n} R_k \frac{\varrho_{3k}}{1 + \varrho_{3k}} \overline{\gamma}_{k,1},$$

where

$$\gamma_{m,1} = 3 R_m r^2 \alpha_{m,3} + 2\overline{a_m} \alpha_{m,2} + \beta_{m,1}, \tag{4.24}$$

by (4.14). The constant A can be expanded into the series

$$A = \sum_{s=1}^{\infty} A^{(s)} f^s, \tag{4.25}$$

where

$$A^{(s)} = \frac{i}{n^s \pi^{s-1}} \sum_{k=1}^{n} R_k \frac{\varrho_{3k}}{1 + \varrho_{3k}} \left(3 R_k \overline{\alpha}_{k,3}^{(s-2)} + 2 a_k \overline{\alpha}_{k,2}^{(s-1)} + \overline{\beta}_{k,1}^{(s-1)} \right). \tag{4.26}$$

We have

$$A^{(1)} = \frac{1}{n} \sum_{k=1}^{n} R_k \varrho_{3k},$$

and

$$A^{(2)} = -\frac{2}{n^2\pi}\sum_{k=1}^{n}\sum_{m=1}^{n} R_k R_m \varrho_{3k}\varrho_{3k,m}\frac{1+\varrho_{3m}}{1+\varrho_{3k}}\overline{E_3^{(1)}(a_k-a_m)}.$$

Using the relation

$$\frac{1+\varrho_{3m}}{1+\varrho_{3k}}\varrho_{3k,m}=\varrho_{3m}, \tag{4.27}$$

we arrive at the formula

$$A^{(2)} = -\frac{2}{n^2\pi}\sum_{k=1}^{n}\sum_{m=1}^{n} R_k R_m \varrho_{3k}\varrho_{3m}\overline{E_3^{(1)}(a_k-a_m)}.$$

Using (4.27) and the following relation

$$\frac{(\varrho_{2m_1,m_2}-\varrho_{3m_1,m_2})\varrho_{1m_2,m_3}(1+\varrho_{3m_3})}{1+\varrho_{3m_1}}=\frac{(\varrho_{1m_2}-\varrho_{3m_2})\varrho_{3m_3}}{\varrho_{3m_2}}, \tag{4.28}$$

we have

$$A^{(3)} = \frac{1}{n^3\pi^2}\left(\sum_{m_1=1}^{n}\sum_{m_2=1}^{n}\sum_{m_3=1}^{n}\left[4R_{m_1}R_{m_2}R_{m_3}\varrho_{3m_1}\varrho_{3m_2}\varrho_{3m_3}\right.\right.$$

$$\times \overline{E_3^{(1)}(a_{m_1}-a_{m_2})}E_3^{(1)}(a_{m_2}-a_{m_3})$$

$$+R_{m_1}R_{m_2}R_{m_3}\frac{\varrho_{3m_1}(\varrho_{1m_2}-\varrho_{3m_2})\varrho_{3m_3}}{(1+\varrho_{1m_2})\varrho_{3m_2}}\overline{E_2(a_{m_1}-a_{m_2})}E_2(a_{m_2}-a_{m_3})$$

$$\left.-R_{m_1}R_{m_2}R_{m_3}\frac{\varrho_{3m_1}(\varrho_{1m_2}-\varrho_{3m_2})\varrho_{3m_3}}{(1+\varrho_{1m_2})\varrho_{3m_2}}\overline{E_2(a_{m_1}-a_{m_2})}E_2(a_{m_2}-a_{m_3})\right]$$

$$\left.+3\sum_{m_1=1}^{n}\sum_{m_2=1}^{n} R_{m_1}R_{m_2}(R_{m_1}+R_{m_2})\varrho_{3m_1}\varrho_{3m_2}\overline{E_4(a_{m_1}-a_{m_2})}\right). \tag{4.29}$$

Analogously, we obtain

$$A^{(4)} = \frac{2}{n^4\pi^3}\left(\sum_{m_1=1}^{n}\sum_{m_2=1}^{n}\sum_{m_3=1}^{n}\sum_{m_4=1}^{n}\right.$$

$$\left[R_{m_1}R_{m_2}R_{m_3}R_{m_4}\frac{\varrho_{3m_1}\varrho_{3m_2}(\varrho_{1m_3}-\varrho_{3m_3})\varrho_{3m_4}}{(1+\varrho_{1m_3})\varrho_{3m_3}}\right.$$

$$\times \overline{E_3^{(1)}(a_{m_1}-a_{m_2})}E_2(a_{m_2}-a_{m_3})E_2(a_{m_3}-a_{m_4})$$

$$-R_{m_1}R_{m_2}R_{m_3}R_{m_4}\frac{\varrho_{3m_1}\varrho_{3m_2}(\varrho_{1m_3}-\varrho_{3m_3})\varrho_{3m_4}}{(1+\varrho_{1m_3})\varrho_{3m_3}}$$

$$\times \overline{E_3^{(1)}(a_{m_1} - a_{m_2})} E_2(a_{m_2} - a_{m_3}) \overline{E_2(a_{m_3} - a_{m_4})}$$

$$- R_{m_1} R_{m_2} R_{m_3} R_{m_4} \frac{\varrho_{3m_1}(\varrho_{1m_2} - \varrho_{3m_2})\varrho_{3m_3}\varrho_{3m_4}}{(1 + \varrho_{1m_2})\varrho_{3m_2}}$$

$$\times \overline{E_2(a_{m_1} - a_{m_2})} E_2(a_{m_2} - a_{m_3}) E_3^{(1)}(a_{m_3} - a_{m_4})$$

$$- 4 R_{m_1} R_{m_2} R_{m_3} R_{m_4} \varrho_{3m_1}\varrho_{3m_2}\varrho_{3m_3}\varrho_{3m_4}$$

$$\times \overline{E_3^{(1)}(a_{m_1} - a_{m_2})} E_3^{(1)}(a_{m_2} - a_{m_3}) \overline{E_3^{(1)}(a_{m_3} - a_{m_4})}$$

$$R_{m_1} R_{m_2} R_{m_3} R_{m_4} \frac{\varrho_{3m_1}(\varrho_{1m_2} - \varrho_{3m_2})\varrho_{3m_3}\varrho_{3m_4}}{(1 + \varrho_{1m_2})\varrho_{3m_2}}$$

$$\left. \times \overline{E_2(a_{m_1} - a_{m_2})} E_2(a_{m_2} - a_{m_3}) E_3^{(1)}(a_{m_3} - a_{m_4}) \right]$$

$$+ \sum_{m_1=1}^{n} \sum_{m_2=1}^{n} \sum_{m_3=1}^{n} \left[- R_{m_1} R_{m_2}^2 R_{m_3} \frac{\varrho_{3m_1}\varrho_{1m_2}\varrho_{2m_2}\varrho_{3m_3}}{\varrho_{3m_2}} \right.$$

$$\times \overline{E_3(a_{m_1} - a_{m_2})} E_3(a_{m_2} - a_{m_3})$$

$$- 3 R_{m_1} R_{m_2} \left(R_{m_1} + R_{m_2} \right) R_{m_3} \varrho_{3m_1}\varrho_{2m_2}\varrho_{3m_3}$$

$$\times \overline{E_4(a_{m_1} - a_{m_2})} E_3^{(1)}(a_{m_2} - a_{m_3})$$

$$- 3 R_{m_1} R_{m_2} \left(R_{m_1} + R_{m_2} \right) R_{m_3} \varrho_{3m_1}\varrho_{2m_2}\varrho_{3m_3}$$

$$\times \overline{E_3^{(1)}(a_{m_1} - a_{m_2})} E_4(a_{m_2} - a_{m_3})$$

$$\left. - 9 R_{m_1} R_{m_2}^2 R_{m_3} \varrho_{3m_1}\varrho_{2m_2}\varrho_{3m_3} \overline{E_4^{(1)}(a_{m_1} - a_{m_2})} E_4^{(1)}(a_{m_2} - a_{m_3}) \right].$$

$$(4.30)$$

Therefore, the general formula for the effective shear modulus of random composites takes the form

$$\frac{G_e}{G} = \frac{1 + \text{Re} \sum_{s=1}^{\infty} A^{(s)} f^s}{1 - \kappa \, \text{Re} \sum_{s=1}^{\infty} A^{(s)} f^s}, \tag{4.31}$$

with exact coefficients $A^{(s)}$ ($s = 1, 2, 3, 4$). Finding coefficients involves taking the limit of the conditionally convergent sums, which have to be precisely described as n tends to infinity

$$e_2^{(0)(0)}(n) = \sum_{l=1}^{n} \sum_{m \neq l} \frac{1}{(a_m - a_l)^2}, \quad e_3^{(1)(1)}(n) = \sum_{l=1}^{n} \sum_{m \neq l} \frac{a_m - a_l}{\left(a_m - a_l\right)^3}, \tag{4.32}$$

$$e_2^{(0)(0)} = \lim_{n \to \infty} e_2(n), \quad e_3^{(1)(1)} = \lim_{n \to \infty} e_3^{(1)(1)}(n). \tag{4.33}$$

Here, we use designations (1.18).

It follows from Section 3.1.3 and Appendix A.4 that the Eisenstein–Rayleigh approach has to be used in formulas for the local fields derived in Section 4.2. In this case, we find similarly to (3.81) that

$$e_2^{(0)(0)} = \pi, \quad e_3^{(1)(1)} = \frac{\pi}{2}. \tag{4.34}$$

The effective elastic constants have to be calculated from Maxwell's approach, similarly to (3.82) we find

$$e_2^{(0)(0)} = 0, \quad e_3^{(1)(1)} = 0. \tag{4.35}$$

Coefficients $A^{(s)}$ are calculated from the Maxwell approach, when formulas (4.35) hold.

4.4 Identical circular inclusions

Let all inclusions in the representative cell $Q_{0,0}$ be identical, i.e. $R_k = 1$ and $\varrho_{i,k} = \varrho_i$, for $i = 1, 2, 3$, $k = 1, ..., n$. Using (4.35) we write (4.25)–(4.26) in the form

$$A^{(1)} = \varrho_3, \quad A^{(2)} = -\frac{2}{\pi} \varrho_3^2 e_3^{(1)(1)} = 0, \tag{4.36}$$

$$A^{(3)} = \frac{1}{\pi^2} \varrho_3 \left[4\varrho_3^2 e_{3,3}^{(1,1)(1,0)} + 6\varrho_3 e_4^{(0)(1)} \right.$$
$$\left. + \frac{\varrho_1 - \varrho_3}{1 + \varrho_1} \left(e_{2,2}^{(0,0)(1,0)} - e_{2,2}^{(0,0)(1,1)} \right) \right]. \tag{4.37}$$

Using Lemma 2.3 we get

$$A^{(4)} = \frac{1}{\pi^3} \left[-2\varrho_1\varrho_2\varrho_3 e_{3,3}^{(0,0)(1,0)} - 12\varrho_3^3 e_{4,3}^{(0,1)(1,0)} - 12\varrho_3^3 e_{3,4}^{(1,0)(1,0)} \right.$$
$$- 18\varrho_3^3 e_{4,4}^{(1,1)(1,0)} - 8\varrho_3^4 e_{3,3,3}^{(1,1,1)(1,0,1)}$$
$$\left. - 2\varrho_3^2 \frac{\varrho_3 - \varrho_1}{1 + \varrho_1} \left(e_{2,2,3}^{(0,0,1)(0,0,1)} + e_{2,2,3}^{(0,0,1)(1,1,0)} + 2e_{2,2,3}^{(0,0,1)(1,0,1)} \right) \right], \tag{4.38}$$

$$A^{(5)} = \frac{1}{\pi^4} \left[3\varrho_3(\varrho_1\varrho_2 + 12\varrho_3^2) e_{4,4}^{(0,0)(1,0)} \right.$$
$$+ 24\varrho_3^3 \left(5\mathrm{Re}\left(e_{5,4}^{(0,1)(1,0)} \right) + 2e_{5,5}^{(1,1)(1,0)} \right)$$
$$+ 8\varrho_1\varrho_2\varrho_3^2 e_{3,3,3}^{(0,0,1)(1,0,1)} + 16\varrho_3^5 e_{3,3,3,3}^{(1,1,1,1)(1,0,1,0)}$$
$$\left. + 24\varrho_3^4 \left(2e_{4,3,3}^{(0,1,1)(1,0,1)} + e_{3,4,3}^{(1,0,1)(1,0,1)} + 3e_{4,4,3}^{(1,1,1)(1,0,1)} \right) \right.$$

$$+ 12\varrho_3^2 \frac{\varrho_1 - \varrho_3}{1 + \varrho_1} \left(e_{2,2,4}^{(0,0,0)(1,0,1)} + e_{2,3,4}^{(0,0,1)(1,0,1)} \right.$$

$$\left. - \mathrm{Re} \left(e_{2,2,4}^{(0,0,0)(1,1,0)} + e_{2,3,4}^{(0,0,1)(1,1,0)} \right) \right)$$

$$+ \varrho_3 \left(\frac{\varrho_1 - \varrho_3}{1 + \varrho_1} \right)^2 \left(e_{2,2,2,2}^{(0,0,0,0)(1,0,1,0)} - 2 e_{2,2,2,2}^{(0,0,0,0)(1,0,1,1)} + e_{2,2,2,2}^{(0,0,0,0)(1,1,0,0)} \right)$$

$$+ 4\varrho_3^3 \frac{\varrho_1 - \varrho_3}{1 + \varrho_1} \left(2\mathrm{Re} \left(e_{2,2,3,3}^{(0,0,1,1)(1,0,1,0)} \right) - 2 e_{2,2,3,3}^{(0,0,1,1)(1,1,0,1)} \right.$$

$$\left. \left. - e_{3,2,2,3}^{(1,0,0,1)(1,0,0,1)} + e_{3,2,2,3}^{(1,0,0,1)(1,0,1,0)} \right) \right], \tag{4.39}$$

$$A^{(6)} = \frac{1}{\pi^5} \left[-4\varrho_3(\varrho_1 \varrho_2 + 50\varrho_3^2) e_{5,5}^{(0,0)(1,0)} \right.$$

$$- \varrho_3^3 \left(360\mathrm{Re} \left(e_{6,5}^{(0,1)(1,0)} \right) + e_{6,6}^{(1,1)(1,0)} \right)$$

$$+ \varrho_3^2 \frac{\varrho_1 - \varrho_3}{1 + \varrho_1} \left(-40 e_{2,3,5}^{(0,0,0)(1,0,1)} + 40\mathrm{Re} \left(e_{2,3,5}^{(0,0,0)(1,1,0)} \right) \right.$$

$$\left. - 24 e_{2,4,5}^{(0,0,1)(1,0,1)} + 24\mathrm{Re} \left(e_{2,4,5}^{(0,0,1)(1,1,0)} \right) \right)$$

$$+ \varrho_1 \varrho_2 \varrho_3^2 \left(-24 e_{3,3,4}^{(0,0,0)(1,0,1)} - 36 e_{3,4,4}^{(0,0,1)(1,0,1)} \right)$$

$$+ 12\varrho_3^2 (\varrho_1 \varrho_2 + 12\varrho_3^2) e_{4,4,3}^{(0,0,1)(1,0,1)}$$

$$+ \varrho_3^4 \left(-72 e_{4,3,4}^{(0,1,0)(1,0,1)} - 216 e_{4,4,4}^{(0,1,1)(1,0,1)} - 240 e_{5,4,3}^{(0,1,1)(1,0,1)} \right.$$

$$\left. - 240 e_{3,5,4}^{(1,0,1)(1,0,1)} - 192 e_{3,5,5}^{(1,1,1)(1,0,1)} - 216 e_{4,5,4}^{(1,1,1)(1,0,1)} \right)$$

$$- 4\varrho_1 \varrho_2 \varrho_3 \frac{\varrho_1 - \varrho_3}{1 + \varrho_1} \left(\mathrm{Re} \left(e_{2,2,3,3}^{(0,0,0,0)(1,0,1,0)} \right) - e_{2,2,3,3}^{(0,0,0,0)(1,1,0,1)} \right)$$

$$+ 12\varrho_3^3 \frac{\varrho_1 - \varrho_3}{1 + \varrho_1} \left(\mathrm{Re} \left(-2 e_{2,2,4,3}^{(0,0,0,1)(1,0,1,0)} - 2 e_{2,2,3,4}^{(0,0,1,0)(1,0,1,0)} \right. \right.$$

$$- 3 e_{2,2,4,4}^{(0,0,1,1)(1,0,1,0)}$$

$$\left. - 2 e_{2,3,4,3}^{(0,0,1,1)(1,0,1,0)} - 2 e_{4,2,2,3}^{(0,0,0,1)(1,0,1,0)} - 2 e_{3,2,3,4}^{(1,0,0,1)(1,0,1,0)} \right)$$

$$+ 2 e_{2,2,4,3}^{(0,0,0,1)(1,1,0,1)} + 2 e_{2,2,3,4}^{(0,0,1,0)(1,1,0,1)} + 3 e_{2,2,4,4}^{(0,0,1,1)(1,1,0,1)}$$

$$\left. + 2 e_{2,3,4,3}^{(0,0,1,1)(1,1,0,1)} + 2 e_{4,2,2,3}^{(0,0,0,1)(1,0,0,1)} + 2 e_{3,2,3,4}^{(1,0,0,1)(1,0,0,1)} \right)$$

$$- 2\varrho_3 \left(\frac{\varrho_1 - \varrho_3}{1 + \varrho_1} \right)^2 \left(e_{2,3,3,2}^{(0,0,0,0)(1,0,1,0)} + e_{2,3,3,2}^{(0,0,0,0)(1,1,0,0)} - e_{2,3,3,2}^{(0,0,0,0)(1,1,0,1)} \right)$$

$$- 8\varrho_1 \varrho_2 \varrho_3^3 \left(2\mathrm{Re} \left(e_{3,3,3,3}^{(0,0,1,1)(1,0,1,0)} \right) + e_{3,3,3,3}^{(1,0,0,1)(1,0,1,0)} \right)$$

$$- 24\varrho_3^5 \left(4 e_{4,3,3,3}^{(0,1,1,1)(1,0,1,0)} + 3 e_{3,4,4,3}^{(1,1,1,1)(1,0,1,0)} \right.$$

$$\left. + \mathrm{Re} \left(4 e_{3,4,3,3}^{(1,0,1,1)(1,0,1,0)} + 6 e_{3,3,4,4}^{(1,1,1,1)(1,0,1,0)} \right) \right)$$

$$
\begin{aligned}
&+ 2\varrho_3^2 \left(\frac{\varrho_1 - \varrho_3}{1 + \varrho_1}\right)^2 \left(2\mathrm{Re}\left(e_{2,2,2,2,3}^{(0,0,0,0,1)(1,0,1,1,0)} + e_{2,2,3,2,2}^{(0,0,1,0,0)(1,0,1,0,0)}\right.\right.\\
&+ e_{2,2,2,2,3}^{(0,0,0,0,1)(1,1,0,1,0,0)}\bigg) - 2e_{2,2,2,2,3}^{(0,0,0,0,1)(1,0,1,0,1)} - e_{2,2,3,2,2}^{(0,0,1,0,0)(1,0,1,0,1)}\\
&- e_{2,2,3,2,2}^{(0,0,1,0,0)(1,1,0,1,1)} - e_{3,2,2,2,2}^{(1,0,0,0,0)(1,0,0,1,1)}\bigg)\\
&+ 16\varrho_3^4 \frac{\varrho_1 - \varrho_3}{1 + \varrho_1} \left(\mathrm{Re}\left(e_{2,2,3,3,3}^{(0,0,1,1,1)(1,1,0,1,0)} + e_{3,2,2,3,3}^{(1,0,0,1,1)(1,0,0,1,0)}\right.\right.\\
&- e_{3,2,2,3,3}^{(1,0,0,1,1)(1,0,1,0,1)} - e_{3,3,3,2,2}^{(1,1,1,0,0)(1,0,1,0,1)}\bigg) - 32\varrho_3^6 e_{3,3,3,3,3}^{(1,1,1,1,1)(1,0,1,0,1)}\bigg].
\end{aligned}
$$

$$(4.40)$$

The effective shear modulus of random composites with identical disks is calculated from (4.31), with exact coefficients $A^{(s)}$ ($s = 1, \ldots, 6$).

Following Section 3.3, particularly Subsection 3.3.4, we can apply (4.31) to calculate the effective viscosity of random 2D suspension. Setting $\kappa = 1$ and $\varrho_1 = \varrho_2 = \varrho_3 = 1$, we arrive at the formula

$$
\frac{\mu_e}{\mu} = \frac{1 + \mathrm{Re}\sum_{s=1}^{\infty} A^{(s)} f^s}{1 - \mathrm{Re}\sum_{s=1}^{\infty} A^{(s)} f^s},
$$

$$(4.41)$$

with simplified coefficients (4.36)–(4.40),

$$
A^{(1)} = 1, \quad A^{(2)} = 0, \quad A^{(3)} = \frac{1}{\pi^2}\left(4e_{3,3}^{(1,1)(1,0)} + 6e_4^{(0)(1)}\right),
$$

$$(4.42)$$

$$
\begin{aligned}
A^{(4)} = \frac{1}{\pi^3}\bigg(&-2e_{3,3}^{(0,0)(1,0)} - 12e_{4,3}^{(0,1)(1,0)} - 12e_{3,4}^{(1,0)(1,0)}\\
&-18e_{4,4}^{(1,1)(1,0)} - 8e_{3,3,3}^{(1,1,1)(1,0,1)}\bigg),
\end{aligned}
$$

$$(4.43)$$

$$
\begin{aligned}
A^{(5)} = \frac{1}{\pi^4}\bigg[&39e_{4,4}^{(0,0)(1,0)} + 24\left(5\mathrm{Re}\left(e_{5,4}^{(0,1)(1,0)}\right) + 2e_{5,5}^{(1,1)(1,0)}\right)\\
&+ 8e_{3,3,3}^{(0,0,1)(1,0,1)} + 16e_{3,3,3,3}^{(1,1,1,1)(1,0,1,0)}\\
&+ 24\left(2e_{4,3,3}^{(0,1,1)(1,0,1)} + e_{3,4,3}^{(1,0,1)(1,0,1)} + 3e_{4,4,3}^{(1,1,1)(1,0,1)}\right)\bigg],
\end{aligned}
$$

$$(4.44)$$

$$
\begin{aligned}
A^{(6)} = \frac{1}{\pi^5}\bigg[&-204e_{5,5}^{(0,0)(1,0)} - 360\,\mathrm{Re}\left(e_{6,5}^{(0,1)(1,0)}\right) - e_{6,6}^{(1,1)(1,0)} - 24e_{3,3,4}^{(0,0,0)(1,0,1)}\\
&- 36e_{3,4,4}^{(0,0,1)(1,0,1)} + 156e_{4,4,3}^{(0,0,1)(1,0,1)} - 72e_{4,3,4}^{(0,1,0)(1,0,1)} - 216e_{4,4,4}^{(0,1,1)(1,0,1)}\\
&- 240e_{5,4,3}^{(0,1,1)(1,0,1)} - 240e_{3,5,4}^{(1,0,1)(1,0,1)} - 192e_{3,5,5}^{(1,1,1)(1,0,1)}\\
&- 216e_{4,5,4}^{(1,1,1)(1,0,1)} - 96e_{4,3,3,3}^{(0,1,1,1)(1,0,1,0)} - 72e_{3,4,4,3}^{(1,1,1,1)(1,0,1,0)}
\end{aligned}
$$

$$- 48\text{Re} \left(2e_{3,4,3,3}^{(1,0,1,1)(1,0,1,0)} + 3e_{3,3,4,4}^{(1,1,1,1)(1,0,1,0)} \right)$$

$$- 16\text{Re} \left(e_{3,3,3,3}^{(0,0,1,1)(1,0,1,0)} \right) - 8e_{3,3,3,3}^{(1,0,0,1)(1,0,1,0)} - 32e_{3,3,3,3,3}^{(1,1,1,1,1)(1,0,1,0,1)} \Big].$$

$$(4.45)$$

Eq. (4.41) can be written in the form

$$\frac{\mu_e}{\mu} = \frac{1+f}{1-f} + 2A^{(3)}f^3 + 2\left(2A^{(3)}\right) + A^{(4)}\right) f^4 + 2\left(3A^{(3)} + 2A^{(4)} + A^{(5)}\right) f^5$$

$$+ 2\left(4A^{(3)} + \left(A^{(3)}\right)^2 + 3A^{(4)} + 2A^{(5)} + A^{(6)}\right) f^6 + O(f^7). \quad (4.46)$$

4.5   Numerical examples

Consider below four numerical examples of random composites generated by different protocols.

4.5.1   Symmetric location of inclusions with equal radii

In order to get a macroscopically isotropic composite let us take 6 disks of radius $r = 0.023$, randomly located at the basic triangle (see Fig. 4.2).[1] One of the statistical realization is shown in Fig. 4.2. Other triangles are obtained by invoking symmetries with respect to the sides of the basic and generated triangles. As the result we obtain the hexagonal periodicity cell containing 288 disks. Such constructed structure has three lines of symmetry (dashed lines in Fig. 4.2). The latter observation implies that the composite is macroscopically isotropic. The e-sums of the considered structure are calculated using the procedure developed in Chapter 2. It yields the result up to $O(f^7)$,

$$A^{(1)} = \varrho_3, \quad A^{(2)} = 0, \quad\quad\quad\quad\quad\quad\quad\quad\quad\quad (4.47)$$

$$A^{(3)} = \frac{\varrho_3}{1 + \varrho_1} \left(0.169694\varrho_1\varrho_2 - 0.169694\varrho_1\varrho_3 \right.$$

$$\left. + 0.905683\varrho_3^2 + 0.905683\varrho_1\varrho_3^2 \right),$$

$$A^{(4)} = \frac{\varrho_3}{1 + \varrho_1} \left(0.392387\varrho_1\varrho_2 + 0.392387\varrho_1^2\varrho_2 - 0.104308\varrho_1\varrho_2\varrho_3 \right.$$

$$\left. + 0.932775\varrho_3^2 + 1.03708\varrho_1\varrho_3^2 \right),$$

---

[1] Here,"randomly" means the uniform non-overlapping distribution in the considered triangle.

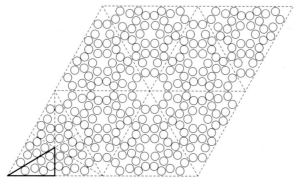

Figure 4.2 The hexagonal cell with 288 inclusions symmetrically generated by 6 inclusions in marked triangle. Coordinates of inclusions are $\{0.09967 + 0.02783i, 0.1552 + 0.04116i, 0.1907 + 0.08114i, 0.2042 + 0.02764i, 0.239 + 0.1074i, 0.2448 + 0.05229i\}$. The concentration is equal to 0.48.

$$A^{(5)} = \frac{\varrho_3}{(1+\varrho_1)^2}\left(0.174541\varrho_1\varrho_2 + 0.349081\varrho_1^2\varrho_2 + 0.174541\varrho_1^3\varrho_2 \right.$$
$$+0.189247\varrho_1^2\varrho_2^2 + 0.241034\varrho_1\varrho_2\varrho_3 - 0.13746\varrho_1^2\varrho_2\varrho_3 - 6.09762\varrho_3^2$$
$$-12.4363\varrho_1\varrho_3^2 - 6.14941\varrho_1^2\varrho_3^2 + 0.884204\varrho_1\varrho_2\varrho_3^2 + 0.884204\varrho_1^2\varrho_2\varrho_3^2$$
$$- 0.884204\varrho_1\varrho_3^3 - 0.884204\varrho_1^2\varrho_3^3$$
$$\left. +2.22295\varrho_3^4 + 4.4459\varrho_1\varrho_3^4 + 2.22295\varrho_1^2\varrho_3^4\right),$$

$$A^{(6)} = \frac{\varrho_3}{(1+\varrho_1)^2}\left(0.186219\varrho_1\varrho_2 + 0.372438\varrho_1^2\varrho_2 + 0.186219\varrho_1^3\varrho_2 \right.$$
$$+0.337272\varrho_1^2\varrho_2^2 + 0.155548\varrho_1^3\varrho_2^2 + 0.380899\varrho_1\varrho_2\varrho_3$$
$$-0.138096\varrho_1^2\varrho_2\varrho_3 - 0.155548\varrho_1^3\varrho_2\varrho_3 + 0.100653\varrho_1^2\varrho_2^2\varrho_3$$
$$+6.30024\varrho_3^2 + 12.2196\varrho_1\varrho_3^2 + 6.10107\varrho_1^2\varrho_3^2 - 0.0736084\varrho_1\varrho_2\varrho_3^2$$
$$+0.385084\varrho_1^2\varrho_2\varrho_3^2 + 0.659998\varrho_1^3\varrho_2\varrho_3^2 + 0.733606\varrho_1\varrho_3^3$$
$$+0.834259\varrho_1^2\varrho_3^3 + 0.138438\varrho_1\varrho_2\varrho_3^3 + 0.138438\varrho_1^2\varrho_2\varrho_3^3$$
$$\left. +1.09012\varrho_3^4 + 2.04181\varrho_1\varrho_3^4 + 0.951685\varrho_1^2\varrho_3^4\right). \tag{4.48}$$

It is worth noting that the constants $A^{(j)}$ for the considered symmetric locations are real. Such a rule should be observed for all symmetric structures.

The obtained result and the Hashin–Shtrikman bounds for boron-aluminum composite considered on page 165, are presented on Fig. 4.3. The effective shear modulus for such composite is given by the polynomial

$$\frac{G_e}{G} = 1 + 1.26695f + 0.987791f^2 + 1.06292f^3 + 1.56428f^4$$
$$+ 0.262378f^5 + 1.70226f^6 + O(f^7). \tag{4.49}$$

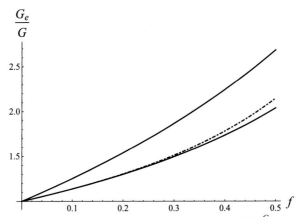

Figure 4.3 Boron-aluminum composite with parameters given on page 165; $\frac{G_e}{G}$ calculated from (4.49) (dashed) and the Hashin–Shtrikman bounds (solid lines).

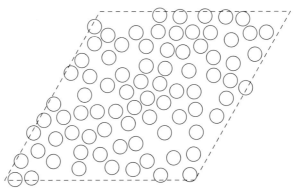

Figure 4.4 The hexagonal cell with 81 inclusions generated by random walks. The volume fraction is equal to 0.5.

4.5.2 Random location of inclusions with equal radii

Consider yet different random composite material with 81 inclusions per cell, obtained from regular configuration by random walks described in Chapter 2. Centers of inclusions are displayed in Fig. 4.4. The structural sums are calculated up to $O(f^7)$, by the procedure described in Chapter 2, and

$$A_1 = \varrho_3, \quad \mathrm{Re}\, A_2 = -0.107677\varrho_3^2,$$

$$\mathrm{Re}\, A_3 = \frac{\varrho_3}{1+\varrho_1}\left(0.10982\varrho_1\varrho_2 + \varrho_1\varrho_3(0.0551957 + 1.21313\varrho_3)\right.$$
$$\left. +\varrho_3(0.165016 + 1.21313\varrho_3)\right),$$

$$\mathrm{Re}\, A_4 = \frac{\varrho_3}{1+\varrho_1}\left(0.162317\varrho_1^2\varrho_2 + \varrho_1\varrho_2(0.162317 + 0.337078\varrho_3)\right.$$
$$\left. +\varrho_1(0.764601 - 0.428199\varrho_3)\varrho_3^2 + (1.10168 - 0.428199\varrho_3)\varrho_3^2\right),$$

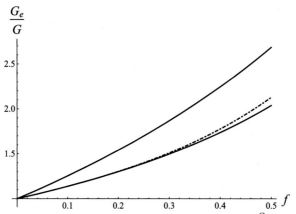

Figure 4.5 Boron-aluminum composite with parameters given on page 165; $\frac{G_e}{G}$ calculated from (4.31) and (4.51) (dashed) and the Hashin–Shtrikman bounds (solid lines).

$$\operatorname{Re} A_5 = \frac{\varrho_3}{(1+\varrho_1)^2}\Big(0.174871\varrho_1\varrho_2 + 0.349741\varrho_1^2\varrho_2 + 0.174871\varrho_1^3\varrho_2$$
$$+0.20535\varrho_1^2\varrho_2^2 + 0.159165\varrho_1\varrho_2\varrho_3 - 0.277665\varrho_1^2\varrho_2\varrho_3$$
$$-0.0261315\varrho_1^3\varrho_2\varrho_3 - 4.2815\varrho_3^2 - 8.74815\varrho_1\varrho_3^2$$
$$-4.26076\varrho_1^2\varrho_3^2 + 1.40073\varrho_1\varrho_2\varrho_3^2 + 1.40073\varrho_1^2\varrho_2\varrho_3^2 + 1.66391\varrho_3^3$$
$$+1.92708\varrho_1\varrho_3^3 + 0.263174\varrho_1^2\varrho_3^3 + 3.90406\varrho_3^4$$
$$+7.80811\varrho_1\varrho_3^4 + 3.90406\varrho_1^2\varrho_3^4\Big),$$

$$\operatorname{Re} A_6 = \frac{\varrho_3}{(1+\varrho_1)^2}\Big(0.0635485\varrho_1\varrho_2 + 0.267969\varrho_1^2\varrho_2 + 0.166858\varrho_1^3\varrho_2$$
$$+0.187028\varrho_1^2\varrho_2^2 + 0.0638303\varrho_1^3\varrho_2^2 - 0.483168\varrho_1\varrho_2\varrho_3$$
$$-0.693732\varrho_1^2\varrho_2\varrho_3 - 0.0869364\varrho_1^3\varrho_2\varrho_3 + 0.907818\varrho_1^2\varrho_2^2\varrho_3$$
$$-4.24245\varrho_3^2 + 0.239362\varrho_1\varrho_3^2 - 3.82879\varrho_1^2\varrho_3^2 + 0.236331\varrho_1\varrho_2\varrho_3^2$$
$$-1.04259\varrho_1^2\varrho_2\varrho_3^2 + 0.548376\varrho_1^3\varrho_2\varrho_3^2 - 3.45364\varrho_3^3 + 2.3754\varrho_1\varrho_3^3$$
$$-2.23941\varrho_1^2\varrho_3^3 + 2.77872\varrho_1\varrho_2\varrho_3^3 + 2.77872\varrho_1^2\varrho_2\varrho_3^3 - 4.33347\varrho_3^4$$
$$+1.11654\varrho_1\varrho_3^4 - 7.11219\varrho_1^2\varrho_3^4 - 2.15453\varrho_3^5 - 2.15453\varrho_1^2\varrho_3^5\Big).$$

$$(4.50)$$

For the boron-aluminum composite the effective shear moduli is given by the polynomial

$$\frac{G_e}{G} = 1 + 1.26695 f + 0.921315 f^2 + 1.14646 f^3$$
$$+ 1.61498 f^4 + 1.16812 f^5 - 1.05377 f^6 + O(f^7), \qquad (4.51)$$

shown in Fig. 4.5.

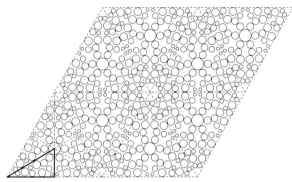

Figure 4.6 The hexagonal cell with 480 inclusions, symmetrically generated by 10 inclusions in marked triangle. The concentration is equal to 0.5. The radii of disks are taken between $r = 0.00869$ and $r = 0.02196$.

Remark 4.1. Formula $A^{(2)} = -\frac{2}{\pi}\varrho_3^2 e_3^{(1)(1)} = 0$, follows from (4.35). It is valid for any macroscopically isotropic composite, i.e., for all uniform non-overlapping distributions of centers $\{a_1, a_2, \cdots, a_N\}$. In particular, this result is valid for symmetric locations, displayed in Fig. 4.2. However, for the composite displayed in Fig. 4.4, we calculate Re $A_2 = -0.107677\varrho_3^2$. This discrepancy arises from the numerical simulation of centers $\{a_1, a_2, \cdots, a_{81}\}$ in Fig. 4.2. It shows that certain degree of anisotropy remains in the simulation. The problem requires special numerical simulations with increasing number of inclusions N per cell. It can be performed along the line of Kurtyka and Rylko (2013, 2017), dedicated to the conductive composites.

4.5.3 Symmetric location of inclusions with different radii

We consider an isotropic material with different radii of inclusions, generated by 10 inclusions per cell, displayed in Fig. 4.6. The e-sums for the considered structure are calculated by the procedure described in Chapter 2. The results are shown up to $O(f^7)$,

$$A_1 = \varrho_3, \quad A_2 = 0,$$

$$A_3 = \frac{\varrho_3}{1 + \varrho_1}\left(0.192529\varrho_1\varrho_2 + 1.3067\varrho_3^2 + \varrho_1\varrho_3(-0.192529 + 1.3067\varrho_3)\right),$$

$$A_4 = \frac{\varrho_3}{1 + \varrho_1}\left(0.17503\varrho_1\varrho_2 + 0.17503\varrho_1^2\varrho_2 + 0.164061\varrho_1\varrho_2\varrho_3 \right.$$
$$\left. -0.343448\varrho_3^2 - 0.50751\varrho_1\varrho_3^2\right),$$

$$A_5 = \frac{\varrho_3}{(1 + \varrho_1)^2}\left(0.162622\varrho_1\varrho_2 + 0.325244\varrho_1^2\varrho_2 + 0.162622\varrho_1^3\varrho_2 \right.$$
$$\left. +0.345226\varrho_1^2\varrho_2^2 + 0.562223\varrho_1\varrho_2\varrho_3 - 0.12823\varrho_1^2\varrho_2\varrho_3 - 0.467195\varrho_3^2\right.$$

$$-1.49661\varrho_1\varrho_3^2 - 0.684191\varrho_1^2\varrho_3^2 + 2.57713\varrho_1\varrho_2\varrho_3^2 + 2.57713\varrho_1^2\varrho_2\varrho_3^2$$
$$-2.57713\varrho_1\varrho_3^3 - 2.57713\varrho_1^2\varrho_3^3 + 5.67553\varrho_3^4 + 11.3511\varrho_1\varrho_3^4$$
$$+5.67553\varrho_1^2\varrho_3^4\Big),$$

$$A_6 = \frac{\varrho_3}{(1+\varrho_1)^2}\Big(0.11274\varrho_1\varrho_2 + 0.22548\varrho_1^2\varrho_2 + 0.11274\varrho_1^3\varrho_2$$
$$+0.157091\varrho_1^2\varrho_2^2 + 0.034094\varrho_1^3\varrho_2^2 + 0.558661\varrho_1\varrho_2\varrho_3 + 0.278573\varrho_1^2\varrho_2\varrho_3$$
$$-0.034094\varrho_1^3\varrho_2\varrho_3 + 1.33079\varrho_1^2\varrho_2^2\varrho_3 - 7.9673\varrho_3^2 - 16.4933\varrho_1\varrho_3^2$$
$$-8.40296\varrho_1^2\varrho_3^2 - 4.47595\varrho_1\varrho_2\varrho_3^2 - 6.83072\varrho_1^2\varrho_2\varrho_3^2$$
$$+0.306809\varrho_1^3\varrho_2\varrho_3^2 + 4.78275\varrho_1\varrho_3^3$$
$$+6.11354\varrho_1^2\varrho_3^3 + 5.24069\varrho_1\varrho_2\varrho_3^3 + 5.24069\varrho_1^2\varrho_2\varrho_3^3 - 17.4647\varrho_3^4$$
$$-40.1702\varrho_1\varrho_3^4 - 22.7054\varrho_1^2\varrho_3^4\Big).$$

The effective shear modulus of random composites with different radii is calculated by (4.31).

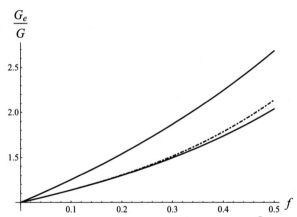

Figure 4.7 Boron-aluminum composite with parameters given on page 165; $\frac{G_e}{G}$ calculated by (4.31) and (4.52) (dashed) and the Hashin–Shtrikman bounds (solid lines).

For the boron-aluminum composite the effective shear modulus is given by the polynomial

$$\frac{G_e}{G} = 1 + 1.26695 f + 0.987791 f^2 + 1.18629 f^3 + 1.25944 f^4 + 1.71405 f^5$$
$$- 1.54071 f^6 + O(f^7), \tag{4.52}$$

see also Fig. 4.7.

4.5.4 Random location of inclusions with different radii

Consider yet different random composite displayed in Fig. 4.8, with different radii of inclusions. Locations of inclusions are simulated by random walks, see Chapter 2. Representative set of centers is generated numerically and displayed in Fig. 4.8.

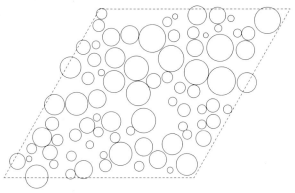

Figure 4.8 Hexagonal cell with 81 inclusions generated by random walks. The concentration equals 0.5. The radii of disks are uniformly distributed in the interval $(0.0141636, 0.0781273)$.

The structural sums for the considered structure are calculated by the algorithm described in Chapter 2. Results are presented up to $O(f^7)$

$$A_1 = \varrho_3, \quad \mathrm{Re}\, A_2 = -0.191758\varrho_3^2,$$

$$
\mathrm{Re}\, A_3 = \frac{\varrho_3}{1+\varrho_1}\Big(0.1367\varrho_1\varrho_2 + 0.277468\varrho_3 + 0.140769\varrho_1\varrho_3 \\
+ 1.60755\varrho_3^2 + 1.60755\varrho_1\varrho_3^2\Big),
$$

$$
\mathrm{Re}\, A_4 = \frac{\varrho_3}{1+\varrho_1}\Big(0.184996\varrho_1\varrho_2 + 0.184996\varrho_1^2\varrho_2 + 0.339696\varrho_1\varrho_2\varrho_3 \\
- 1.27745\varrho_3^2 - 1.61715\varrho_1\varrho_3^2 - 0.747524\varrho_3^3 - 0.747524\varrho_1\varrho_3^3\Big),
$$

$$
\mathrm{Re}\, A_5 = \frac{\varrho_3}{(1+\varrho_1)^2}\Big(0.192193\varrho_1\varrho_2 + 0.384387\varrho_1^2\varrho_2 + 0.192193\varrho_1^3\varrho_2 \\
+ 0.199804\varrho_1^2\varrho_2^2 + 0.0739847\varrho_1\varrho_2\varrho_3 - 0.287961\varrho_1^2\varrho_2\varrho_3 \\
+ 0.0376625\varrho_1^3\varrho_2\varrho_3 - 0.36429\varrho_3^2 - 0.764901\varrho_1\varrho_3^2 - 0.200808\varrho_1^2\varrho_3^2 \\
+ 1.63774\varrho_1\varrho_2\varrho_3^2 + 1.63774\varrho_1^2\varrho_2\varrho_3^2 + 1.72979\varrho_3^3 + 1.82183\varrho_1\varrho_3^3 \\
+ 0.0920414\varrho_1^2\varrho_3^3 + 6.38858\varrho_3^4 + 12.7772\varrho_1\varrho_3^4 + 6.38858\varrho_1^2\varrho_3^4\Big),
$$

$$
\mathrm{Re}\, A_6 = \frac{\varrho_3}{(1+\varrho_1)^2}\Big(0.116459\varrho_1\varrho_2 + 0.232918\varrho_1^2\varrho_2 + 0.116459\varrho_1^3\varrho_2
$$

$$+0.186737\varrho_1^2\varrho_2^2 + 0.0637164\varrho_1^3\varrho_2^2 + 0.0257226\varrho_1\varrho_2\varrho_3$$
$$-0.501007\varrho_1^2\varrho_2\varrho_3 - 0.280688\varrho_1^3\varrho_2\varrho_3 + 0.890573\varrho_1^2\varrho_2^2\varrho_3$$
$$-6.34991\varrho_3^2 - 12.9425\varrho_1\varrho_3^2 - 6.46958\varrho_1^2\varrho_3^2 - 1.40218\varrho_1\varrho_2\varrho_3^2$$
$$-2.38067\varrho_1^2\varrho_2\varrho_3^2 + 0.802658\varrho_1^3\varrho_2\varrho_3^2 - 0.784987\varrho_3^3$$
$$+0.634868\varrho_1\varrho_3^3 + 2.31043\varrho_1^2\varrho_3^3 + 3.21987\varrho_1\varrho_2\varrho_3^3 + 3.21987\varrho_1^2\varrho_2\varrho_3^3$$
$$-17.0029\varrho_3^4 - 37.2257\varrho_1\varrho_3^4 - 20.2228\varrho_1^2\varrho_3^4 - 3.26753\varrho_3^5$$
$$\left.-6.53505\varrho_1\varrho_3^5 - 3.26753\varrho_1^2\varrho_3^5\right).$$

For the boron-aluminum composite the effective shear moduli is given by the polynomial

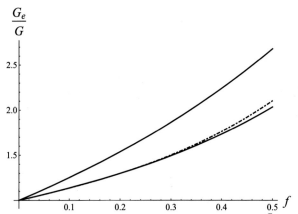

Figure 4.9 Boron-aluminum composite with parameters given on page 165; $\frac{G_e}{G}$ calculated by (4.53) (dashed) and the Hashin–Shtrikman bounds (solid lines).

$$\frac{G_e}{G} = 1 + 1.26695f + 0.869405f^2 + 1.25681f^3$$
$$+ 1.07559f^4 + 1.59829f^5 - 1.13112f^6, \tag{4.53}$$

shown in Fig. 4.9.

4.6 Critical index for the viscosity of 2D random suspension

Consider special composite material with the incompressible matrix and perfectly rigid, incompressible inclusions. It represents an elastic analog of the viscous sus-

Table 4.1 Case of $\frac{G_1}{G} = \infty$, $v' = \frac{1}{2}$ and $v'_1 = \frac{1}{2}$. Critical indices for the $\mathcal{S}_k$ for the viscosity problem obtained from optimization conditions on $\Delta_{kn}(\mathcal{S}_k)$.

$\mathcal{S}_k$	$\min_{\mathcal{S}_k} \mid \Delta_{k\,k+1}(\mathcal{S}_k) \mid$	$\min_{\mathcal{S}_k} \mid \Delta_{k\,6}(\mathcal{S}_k) \mid$
$\mathcal{S}_1$	0.907	1.37
$\mathcal{S}_2$	n.a	n.a
$\mathcal{S}_3$	n.a	1.19
$\mathcal{S}_4$	1.23	1.317
$\mathcal{S}_5$	1.16	1.16

pension as explained in Appendix A.1, with

$$\frac{G_1}{G} = \infty, \ v' = \frac{1}{2} \text{ and } v'_1 = \frac{1}{2}. \tag{4.54}$$

Eqs. (4.54) imply that $\kappa = \kappa_1 = 1$ and $\varrho_1 = \varrho_2 = \varrho_3 = 1$. Mind that the effective viscosity μ_e of a random suspension of perfectly rigid disks in an incompressible fluid of the viscosity μ under the creeping flow conditions, is equivalent to the effective modulus G_e for the material conditions just stated. Therefore all results apply to the effective viscosity of the random array of perfectly rigid particles in an incompressible fluid.

Consider the random medium displayed in Fig. 4.2. Substitution of (4.54) into the corresponding formulas (4.47), (4.25) and (4.21) yields

$$\frac{\mu_e}{\mu} \simeq 1 + 2f + 2f^2 + 3.81137 f^3 + 8.27305 f^4 + 5.33447 f^5 + 20.5096 f^6. \tag{4.55}$$

The critical index $\mathcal{S}$ was introduced in Chapter 3. Here we consider that $\mu_e(f) \simeq A(f_c - f)^{-\mathcal{S}}$, as $f \to f_c$. In order to calculate the critical index $\mathcal{S}$, we analyze the differences defined in (1.108) on page 45, in terms of critical amplitides.

$$\Delta_{kn}(\mathcal{S}_k) = A_k(\mathcal{S}_k) - A_n(\mathcal{S}_k). \tag{4.56}$$

In this way, a set $\mathcal{S}_k$ of control parameters is defined from the minimal difference condition written as

$$\mid \Delta_{kn}(\mathcal{S}_k) \mid = \min_{\mathcal{S}} \mid \Delta_{kn}(\mathcal{S}) \mid. \tag{4.57}$$

Composing the sequences $\min_{\mathcal{S}_k} \mid \Delta_{kn}(\mathcal{S}_k) \mid$, we find the related approximate values $\mathcal{S}_k$ for the critical index. It is possible to investigate different sequences of the conditions on Δ_{kn}. The most logical are the sequences of $\mid \Delta_{k\,k+1} \mid$ and of $\mid \Delta_{k\,6} \mid$, with $k = 1, 2, 3, 4, 5$. The results from the third column, presented in Table 4.1, give rather reasonable numerical estimates of the approximate critical indices $\mathcal{S}_k$ in our case.

Their average value equals $\mathcal{S} = 1.26 \pm 0.09$, with the error bars estimated as the standard deviation. The estimate agrees both with experimental results of 1.24

(Torquato, 2002), and 1.3 (Belzons et al., 1981, Bergman, 2004). The theory of Bergman (2004), also anticipates that in 2D $\mathcal{S} = \mathsf{s}$. Note, that only the value of $\mathcal{S} = 1.317$, chosen from the second column, corresponds to zero of $\Delta_{k6}(\mathcal{S}_k)$, and may be expected to be the best estimate within the range.

The following approximant

$$\mathbf{Cor}^*(f) = \frac{1.25404(f(f-0.777544)+0.303005)^{0.25381}(f(f+0.629195)+0.145546)^{0.148273}}{(0.9069-f)^{1.30417}}$$
$$\times \exp\left(-0.182591\tan^{-1}\left(1.32957 + \frac{1.07858}{f-0.9069}\right)\right. \tag{4.58}$$
$$\left. -0.227575\tan^{-1}\left(5.66002 + \frac{7.12951}{f-0.9069}\right)\right),$$

describes the effective modulus at arbitrary f, and gives $\mathcal{S} \approx 1.304$. To obtain it, we choose as the starting approximation the following simple form $(1 - \frac{f}{f_c})^{-1/2}$, with the critical index $\mathcal{S} = 1/2$ of the regular 2D model, just like in the case of conductivity (Gluzman et al., 2017). The multiplicative correction is sought in the form of $DLog$ Padé approximant, with $\mathcal{B}(z)$ defined as in (1.66) on page 28, and approximated by $P_{3,4}(z)$, where $z = \frac{f}{f_c-f}$.

Remark 4.2. Concerning the infinity used in (4.54), we have to give the following clarification. Of course, the expression $\frac{\mu_1}{\mu} \equiv \frac{G_1}{G}$ tends ∞ means that G_1 is much larger than G. We consider this expressions $\frac{G_e}{G}$ as the first coefficient in the Taylor expansion of the initial function $F(G) = \frac{G_e}{G} = F_1 + F_2 G + F_3 G^3 \ldots$. More precisely, in the present section we consider the asymptotic formula $G_e = F_1 G + O(G^2)$ as G tends to zero.

4.7 Critical behavior of random holes

Holes are punched at random in the matrix with given elastic properties. Plane strain elastic problem is considered for such composite, and the effective elastic modulus is obtained in the form of power series in the inclusions concentration and elastic constants for holes and matrix. Consider the random medium displayed in Fig. 4.2. In the considered case we have $G_1 = 0$ and $v'_1 = 0$ for holes. For matrix we take G normalized to unity and $v' = \frac{1}{4}$. Then, $\kappa = 2$, $\kappa_1 = 3$ and $\varrho_1 = -\frac{1}{3}$ $\varrho_2 = -3$ $\varrho_3 = -1$. Substituting such parameters into the corresponding formulas (4.47), we arrive to the series

$$G_e(f) \simeq 1 - 3f + 6f^2 - 15.2261 f^3 + 32.6161 f^4 - 60.9567 f^5 + 98.6404 f^6. \tag{4.59}$$

As always, we are interested in the formula for all f. The most consistent results for various v' are achieved by the real-valued factor approximant with only real and complex-conjugated pairs of amplitudes and indices. In our case it takes the following form,

$$\mathcal{F}_6^*(f) = (1 - 1.10266 f)^{3.50863}(1 + (0.888456 - 0.659293i) f)^{3.59871-4.19067i}$$
$$\times (1 + (0.888456 + 0.659293i) f)^{3.59871+4.19067i}, \tag{4.60}$$

giving $\mathcal{T} \approx 3.5$. The critical index $\mathcal{T}$ was introduced in Section 3.4. The value of index is close to the experimental estimate of 3.5 ± 0.4 (Benguigui, 1984), and to the accepted numerical estimate of 3.75 (Torquato, 2002).

The method of index function developed in preceding chapters, can be applied to calculate $\mathcal{T}$, as discussed for the regular arrays of holes in Chapter 3. Construct first the simplest approximant, also known as Krieger–Dougherty (KD) formula (Krieger and Dougherty, 1959)

$$G_e(f) = \left(1 - \frac{f}{f_c}\right)^{3f_c},$$

concretized here for $c_1 = -3$. We see that the critical index can be expressed as

$$\mathcal{T} = \lim_{f \to f_c} \frac{\log(G_e(f))}{\log(1 - f/f_c)}. \tag{4.61}$$

Let us look for the solution in more general form,

$$G_e(f) = \left(1 - \frac{f}{f_c}\right)^{\mathcal{T}(f)}.$$

The index function $\mathcal{T}(f)$ will be constructed in such a way that as $f \to 0$, $\mathcal{T}(f) \to -c_1 f_c$, the KD-type estimate for the critical index. And as $f \to f_c$, $\mathcal{T}(f) \to \mathcal{T}$. To perform actual calculations one has to expand the function $\frac{\log(G_e(f))}{\log(1 - f/f_c)}$ for small f, using the known series for the effective shear modulus. When only linear term in the expansion is available, we simply have a constant index function, and return to the KD-type formula. With more terms available, we turn to the *Log* Padé approximants,

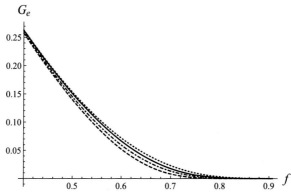

Figure 4.10 Holes, $G_1 \to 0$, $v_1' = 0$, $v' = 1/4$. Dependence of different approximants for the shear modulus as a function of the volume fraction f. The dotted line corresponds to the upper bound from (4.62). The dashed line corresponds to the lower bound from (4.62). The solid line corresponds to the tight upper bound from (4.63), with $\mathcal{T} \approx 3.59$. The dot-dashed line corresponds to the tight lower bound from (4.63), with $\mathcal{T} \approx 4.15$.

as explained on page 37. Very important is to establish bounds for the index function $\mathcal{T}(f)$, employing the upper and lower Padé bounds (Baker and Graves-Moris, 1996), for the sought index function and critical index. In our case they are given as follows:

$$P_{3,2}(f) = \frac{0.251812(f-4.9578)(f+0.636741)}{(f-0.996872)(f^2+1.00334f+0.293096)},$$

$$P_{1,2}(f) = \frac{3.87214(f^2-0.109887f+0.772852)}{f+1.09993}.$$

$$(4.62)$$

The upper and lower bounds for the effective modulus can be found by substituting (4.62) to the general formula, and the upper bound practically coincides with the factor approximant (4.60). Even tighter bounds are given by the following Padé approximants

$$P_{4,0}(f) = 3.81311\left(f^2 - 1.59197f + 1.24169\right)\left(f^2 + 0.132613f + 0.574632\right),$$

$$P_{5,1}(f) = \frac{4.71758(f+0.0779878)(f^2-1.39356f+0.998257)(f^2+0.199039f+0.577729)}{f+0.0779887}.$$

$$(4.63)$$

Correspondingly, the tighter bounds for the critical index follow from (4.63), i.e., $\mathcal{T} = 3.58951$, and $\mathcal{T} = 4.15289$. Various bounds are shown in Fig. 4.10.

Effective conductivity of fibrous composites with cracks on interface

5.1 Deterministic and random cracks on interface

Cracks and voids can be formed on the interface of different media when the condition of perfect contact between the components are violated. Consider a unidirectional fiber-reinforced composite with cracks/voids at the interface of the components. The location and size of de-bonded regions at the boundary of fibers have a strong effect on the transverse effective conductivity and the mechanical properties. In the present chapter we introduce the geometrical measure of cracks/voids and investigate its influence on the effective properties of fibrous composites.

The local 2D fields near cracks located at the interface/surface dividing two different media were described in Muskhelishvili (1966) and developed in Mishuris and Kuhn (2002) and by others. The fundamental book (Grigolyuk and Filishtinskii, 1994) contains systematical theoretical investigation of 2D fields with cracks by the method of integral equations and by series. In particular, the local stresses for the anti-plane problem with one circular inclusion with cracks along the interface are found in closed form.

Previous investigations of the macroscopic properties were based on spoiling of the perfect to an imperfect contact at the boundary by considering a uniform imperfect boundary condition.

The most advanced studies were performed in Bigon et al. (1998), Mishuris et al. (2012), Piccolroaz et al. (2012) and works cited therein. Cracks and voids were considered as obstacles to the heat flux. Analogous problems were discussed for the flow through porous media (Berkowitz, 2002, Adler and Thovert, 2012), where the opposite-to-obstacles case was considered on physical grounds. In the porous media, cracks and voids are considered as high permeability tubes in the surrounding medium of small permeability. Various numerical models are applied to estimate the influence of cracks and voids on the permeability, e.g. (Mourzenko et al., 2011, Adler and Thovert, 2012), as well as analytical, exact and approximate formulas (Mityushev and Adler, 2006, Pesetskaya et al., 2018).

In the present chapter we discuss fiber-reinforced composites with crack at the interface of two different components, and calculate (heat, electric etc) conductivity. Such a crack can be quantified only by the central angle $\psi = \theta_2 - \theta_1$ which spans the crack (see Fig. 5.1). A crack can be considered as an insulation or as a highly conducting medium. In both cases the crack can be modeled by an arc of vanishing width. The perfect contact between the components of media is supposed for the remaining part of interface.

Below, we restrict the discussion to results on closed form solution to the mixed boundary value problems (the Riemann–Hilbert problems), and Maxwell's approach

Applied Analysis of Composite Media. https://doi.org/10.1016/B978-0-08-102670-0.00014-7

to derivation of analytical formulas for the effective properties of composites. Such an approach was first applied in Rylko et al. (2013) to fibrous composite with circular inclusions of infinite conductivity and of concentration f, randomly distributed in a medium of finite conductivity σ_0. Let the central angle $\psi = \theta_2 - \theta_1$ span the crack (see Fig. 5.1). The following analytical formula for the transverse effective conductivity σ_e was established (Rylko et al., 2013)

$$\frac{\sigma_e}{\sigma_0} = \frac{1 + f\langle \cos \frac{\psi}{2} \rangle}{1 - f\langle \cos \frac{\psi}{2} \rangle} + O(f^2), \qquad (5.1)$$

where $\langle \cos \frac{\psi}{2} \rangle$ denotes the mean value over all inclusions of $\cos \frac{\psi}{2}$. This formula is similar to the famous Clausius–Mossotti (Maxwell) approximation for a dilute media with perfect contact between circular inclusions of conductivity σ_1 and matrix of conductivity σ_0

$$\frac{\sigma_e}{\sigma_0} = \frac{1 + f\varrho}{1 - f\varrho} + O(f^3), \qquad (5.2)$$

with the contrast parameter

$$\varrho = \frac{\sigma_1 - \sigma_0}{\sigma_1 + \sigma_0}. \qquad (5.3)$$

The precision up to $O(f^3)$ of (5.2) was established by Mityushev and Rylko (2013) for macroscopically isotropic 2D media. In fact, only the precision up to $O(f^2)$, as for all self-consistent approximations, holds (Mityushev and Rylko, 2013). Eq. (5.2) actually is written in a non-dimensional form. We follow the dimensionless formulation below.

The effective conductivity of composites with arbitrary contrast parameter was found in Vilchevskaya and Sevostianov (2015). The obtained deterministic formula was derived by Maxwell's approach from the local fields obtained in Chao and Shen (1993). Alternatively, the randomization method (Mityushev and Rylko, 2013) can be applied to the local fields obtained in the closed form in Dolgih and Filshtinsky (1980), Chao and Shen (1993). This yields the following analytical formula for the effective conductivity of circular inclusions of conductivity σ_1, randomly distributed in matrix of conductivity σ_0

$$\frac{\sigma_e}{\sigma_0} = \frac{1 + f(\rho\langle \cos^2 \frac{\psi}{4} \rangle - \langle \sin^2 \frac{\psi}{4} \rangle)}{1 - f(\rho\langle \cos^2 \frac{\psi}{4} \rangle - \langle \sin^2 \frac{\psi}{4} \rangle)} + O(f^2). \qquad (5.4)$$

This relation includes formulas (5.1)–(5.2) as particular cases. We write here (5.4) instead of Vilchevskaya and Sevostianov (2015, formula with the same number (5.4)), because the latter formula contains a misleading term of order f^2 (see explanations in Mityushev and Rylko (2013), Mityushev (2018)). One can omit the term f^2 in Vilchevskaya and Sevostianov (2015) and apply the randomization method (Rylko et al., 2013) to get the same result (5.4). It is worth noting that unjustified extension of precision can lead to a seeming contradiction between different formulas (Mityushev et al., 2018b).

The formula (5.4), after application of Keller's identity (Keller, 1964), yields also the effective conductivity σ'_e in the case when ψ stands for the span angle of the high conductivity interface crack/void,

$$\frac{\sigma'_e}{\sigma_0} = \frac{1 + f(\rho\langle\cos^2\frac{\psi}{4}\rangle + \langle\sin^2\frac{\psi}{4}\rangle)}{1 - f(\rho\langle\cos^2\frac{\psi}{4}\rangle + \langle\sin^2\frac{\psi}{4}\rangle)} + O(f^2). \qquad (5.5)$$

In the next sections we derive the formulas (5.4)–(5.5) by the method of complex potentials. First, the corresponding boundary value problem is written as a vector-matrix $\mathbb{C}$-linear problem. After its factorization we arrive at the scalar boundary value problems solved in Muskhelishvili (1966), Gakhov (1966). Such approach yields a potent method to determine the potential (temperature) distributions in fractured media. Application of Maxwell's approach (Maxwell, 1873) is based on the dipole method (Bigon et al., 1998, Piccolroaz et al., 2012, Mityushev et al., 2018a), and yields analytical approximate formulas for the macroscopic conductivity.

5.2 Boundary value problem

In order to describe the local fields let us introduce a complex variable z and consider the disk $|z| < r$, whose boundary is divided onto two arcs L and L' as displayed in Fig. 5.1.

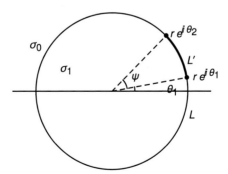

Figure 5.1 Inclusion with crack L' along the interface.

Let $u_1(z)$ and $u(z)$ denote the potential, e.g. the temperature distribution, in $|z| < r$ and in $|z| > r$, respectively. The perfect contact along the arc L is expressed by two equations (Gluzman et al., 2017)

$$u(z) = u_1(z), \qquad \sigma_0 \frac{\partial u}{\partial \mathbf{n}}(z) = \sigma_1 \frac{\partial u_1}{\partial \mathbf{n}}(z), \qquad z \in L, \qquad (5.6)$$

where $\frac{\partial}{\partial \mathbf{n}}$ denotes the normal derivative to the circle $|z| = r$. In addition, we consider the impermeable crack condition

$$\frac{\partial u}{\partial \mathbf{n}}(z) = \frac{\partial u_1}{\partial \mathbf{n}}(z) = 0, \qquad z \in L'. \tag{5.7}$$

Following (Gluzman et al., 2017) let us introduce the complex potentials $\varphi_1(z)$ in $|z| \leq r$ and $\varphi(z)$ in $|z| \geq r$ in such a way that

$$u(z) = \mathrm{Re}[\varphi(z) + z], \ u_1(z) = \frac{2\sigma_0}{\sigma_0 + \sigma_1} \mathrm{Re}\,\varphi_1(z), \tag{5.8}$$

where Re stand for the real part. The term Re z models the external flux applied at infinity parallel to the real axis. The functions $\varphi_1(z)$ and $\varphi(z)$ are analytic in $|z| < r$ and $|z| > r$, respectively, and continuously differentiable in the closures of the considered domains. Excepting points

$$a = r \exp(i\theta_1), \quad b = r \exp(i\theta_2), \tag{5.9}$$

where they are bounded.

The complex potentials satisfy the following equivalent boundary value problem. The condition (5.6) becomes the $\mathbb{R}$-linear condition (Gluzman et al., 2017)

$$\varphi(t) = \varphi_1(t) - \varrho \overline{\varphi_1(t)} - t, \quad t \in L. \tag{5.10}$$

The impermeable condition (5.7) takes the form

$$\varphi(t) = \overline{\varphi(t)} - t + \frac{r^2}{t}, \quad \varphi_1(t) = \overline{\varphi_1(t)}, \quad t \in L'. \tag{5.11}$$

Introduce the functions analytic outside the circle $|z| = r$

$$\Phi_1(z) = \begin{cases} \varphi_1(z), & |z| \leq r, \\ \overline{\varphi_1\left(\frac{r^2}{\overline{z}}\right)} & |z| \geq r, \end{cases} \quad \Phi_2(z) = \begin{cases} \overline{\varphi\left(\frac{r^2}{\overline{z}}\right)}, & |z| \leq r, \\ \varphi(z) & |z| \geq r. \end{cases} \tag{5.12}$$

Let D denote the domain on the complex plane outside of L. It follows from (5.11) that the function $\Phi_1(z)$ is analytic in D and continuous in the closure of D in the two-sided topology of the arc L. It is convenient to consider the functions (5.12) as one vector-function $\Phi(z) = (\Phi_1(z), \Phi_2(z))^\top$.

Introduce the limit values on the different sides on the circle

$$\Phi^+(t) = \lim_{\substack{t \to z \\ |z| < r}} \Phi(z), \quad \Phi^-(t) = \lim_{\substack{t \to z \\ |z| > r}} \Phi(z). \tag{5.13}$$

Then, the problem (5.10)–(5.11) is reduced to the vector-matrix problem

$$\Phi^+(t) = G(t)\Phi^-(t) + g(t), \quad t \in L \cup L', \tag{5.14}$$

where the matrix $G(t)$ is piecewise constant. Namely, $G(t)$ is the identity matrix on L' and

$$G(t) = \begin{pmatrix} \varrho & 1 \\ 1 - \varrho^2 & -\varrho \end{pmatrix}, \quad t \in L. \tag{5.15}$$

The vector-function $g(t)$ is defined as follows:

$$g(t) = \begin{pmatrix} t \\ -(\varrho t + \frac{r^2}{t}) \end{pmatrix}, \quad t \in L, \quad g(t) = \begin{pmatrix} 0 \\ t - \frac{r^2}{t} \end{pmatrix}, \quad t \in L'. \tag{5.16}$$

Introduce the matrix constructed from the eigenvalues of G

$$S = \begin{pmatrix} 1 & 1 \\ -1 - \varrho & 1 - \varrho \end{pmatrix}. \tag{5.17}$$

The inverse matrix is given by the formula

$$S^{-1} = \frac{1}{2} \begin{pmatrix} 1 - \varrho & -1 \\ 1 + \varrho & 1 \end{pmatrix}. \tag{5.18}$$

Introduce the vector-function

$$\Psi(z) = S^{-1}\Phi(z). \tag{5.19}$$

Then, (5.14) becomes

$$\Psi^+(t) = \begin{pmatrix} -1 & 0 \\ 0 & 1 \end{pmatrix} \Psi^-(t) + \frac{1}{2} \begin{pmatrix} t + r^2 t^{-1} \\ t - r^2 t^{-1} \end{pmatrix}, \quad t \in L, \tag{5.20}$$

$$\Psi^+(t) = \begin{pmatrix} 1 & 0 \\ 0 & 1 \end{pmatrix} \Psi^-(t) + \frac{1}{2} \left(t - \frac{r^2}{t} \right) \begin{pmatrix} -1 \\ 1 \end{pmatrix}, \quad t \in L'. \tag{5.21}$$

The vector-matrix problem is decomposed onto two scalar $\mathbb{C}$-linear problems

$$\Psi_1^+(t) = -\Psi_1^-(t) + \frac{1}{2} \left(t + \frac{r^2}{t} \right), \; t \in L, \; \Psi_1^+(t) = \Psi_1^-(t) - \frac{1}{2} \left(t - \frac{r^2}{t} \right), \; t \in L', \tag{5.22}$$

and

$$\Psi_2^+(t) = \Psi_2^-(t) + \frac{1}{2} \left(t - \frac{r^2}{t} \right), \quad |t| = r. \tag{5.23}$$

The second problem is trivial. Its solution has the form up to an additive arbitrary constant (Gakhov, 1966)

$$\Psi_2(z) = \begin{cases} \frac{z}{2} & |z| \leq r, \\ \frac{r^2}{2z} & |z| \geq r. \end{cases} \tag{5.24}$$

In order to solve the problem (5.22) introduce the function

$$R(z) = \sqrt{(z-a)(z-b)}, \quad z \in D, \tag{5.25}$$

analytic in D except at infinity. The square root branch is chosen in such a way that $R^+(t) = -R^-(t)$ on L and $R^+(t) = R^-(t)$ on L'. This function is called the factorization function in the theory of boundary value problems (Gakhov, 1966). The problem (5.22) has the unique solution given by formula (Gakhov, 1966), see page 443,

$$\Psi_1(z) = \frac{1}{2}R(z)\left[f_1(z) + f_2(z)\right], \quad |z| > r, \tag{5.26}$$

where

$$f_1(z) = \frac{1}{2\pi i}\int_{|t|=r}\frac{r^2\,dt}{t\,R^+(t)(t-z)}, \quad f_2(z) = -\frac{1}{2\pi i}\int_{|t|=r}\frac{t\,dt}{R^-(t)(t-z)}. \tag{5.27}$$

The function $\Phi_2(z) = \varphi(z)$ for $|z| > r$ can be calculated by using of (5.24), (5.19) and (5.17)

$$\varphi(z) = -(1+\varrho)\Psi_1(z) + (1-\varrho)\frac{r^2}{2z}, \quad |z| > r. \tag{5.28}$$

The function $\Psi_1(z)$ is analytic at infinity, hence,

$$\Psi_1(z) - \Psi_1(\infty) = \frac{c}{z} + O(|z|^{-2}),$$

for some constant c. Then, (5.28) implies that the asymptotic representation of $\varphi(z)$ at infinity is given by the expression

$$\varphi(z) - \varphi(\infty) = -(1+\varrho)\frac{c}{z} + (1-\varrho)\frac{r^2}{2z} + \ldots, \quad \text{as } z \to \infty. \tag{5.29}$$

The integrals (5.27) can be obtained from the residue theorem. The first integral is calculated by means of the residue at the point $t = 0$

$$f_1(z) = -\frac{r^2}{z\sqrt{ab}}, \tag{5.30}$$

where a and b are given by (5.9). Let us expand the function $\frac{1}{2}R(z)f_1(z)$ near infinity and take the coefficient c_1 on z^{-1}. We have

$$c_1 = \frac{r^2}{2}\cos\frac{\psi}{2}, \tag{5.31}$$

where $\psi = \theta_2 - \theta_1$.

The second integral is calculated by Cauchy's integral formula applied to the function $\frac{t}{R(t)}$ analytically continued in the domain $|t| > r$ (Gakhov, 1966), see page 16,

$$f_2(z) = \frac{z}{R(z)} - 1. \tag{5.32}$$

The coefficient of the function $\frac{1}{2}R(z)f_2(z) = \frac{1}{2}[z - R(z)]$ on z^{-1} is equal to

$$c_2 = \frac{r^2}{16}\left[\exp(i\theta_1) - \exp(i\theta_2)\right]^2. \tag{5.33}$$

Then, (5.29) yields (up to $O(|z|^{-2})$)

$$\varphi(z) - \varphi(\infty) \simeq \frac{r^2}{z}\left\{\frac{1-\varrho}{2} - \frac{1+\varrho}{2}\left[\cos\frac{\psi}{2} + \frac{\left[\exp(i\theta_1) - \exp(i\theta_2)\right]^2}{8}\right]\right\}. \tag{5.34}$$

5.3   Maxwell's approach

Maxwell's approach (Maxwell, 1873) is based on the self-consistent method. The dipole of the potential of the homogenized medium bounded by the large circle $|z| = R$ is equated to the dipole of the local potential of the considered problem. Let $r_k^2 M_k$ denote the dipole of the kth component lying in the disk $|z| < R$, where r_k denotes the radius of inclusion.

Let σ_e denote the macroscopic conductivity tensor

$$\sigma_e = \begin{pmatrix} \sigma_e^{11} & \sigma_e^{12} \\ \sigma_e^{21} & \sigma_e^{22} \end{pmatrix}. \tag{5.35}$$

If the homogenized medium is isotropic,

$$\sigma_e = \sigma_e \mathbf{I},$$

where $\mathbf{I}$ stands for the identity matrix. The balance of dipoles yields for macroscopically isotropic composites the following equation (Maxwell, 1873)

$$\frac{\sigma_e - \sigma_0}{\sigma_e + \sigma_0} R^2 = \sum_{k=1}^{n} r_k^2 M_k. \tag{5.36}$$

If all the dipoles M_k identical, then (5.36) becomes

$$\frac{\sigma_e - \sigma_0}{\sigma_e + \sigma_0} = \frac{nr^2}{R^2} M. \tag{5.37}$$

In the anisotropic case, σ_e has to be replaced by the value $\sigma_e^{11} - i\sigma_e^{12}$, since the external pressure gradient is applied along the real axis (Gluzman et al., 2017). The value $\sigma_e^{22} + i\sigma_e^{12}$ can be calculated by analogous formula, considering the problem with the external flux applied along the imaginary axis.

The dipole $r^2 M$ is equal to the coefficient (with minus sign) of $\varphi(z)$ on $|z|^{-1}$ at infinity (Bigon et al., 1998, Rylko et al., 2013). It can be calculated in the considered case from (5.34). Simple trigonometric transformations yield

$$M = \rho \cos^2 \frac{\psi}{4} - \sin^2 \frac{\psi}{4} + \frac{1+\rho}{4} \sin^2 \frac{\psi}{2} [\cos(2\theta_1 + \psi) + i \sin(2\theta_1 + \psi)]. \tag{5.38}$$

Substituting (5.38) into (5.37) and taking the limit $n \to \infty$, we arrive at the formula

$$\sigma_e^{11} - i\sigma_e^{12} = \sigma_0 \frac{1 + fM}{1 - fM} + O(f^2), \tag{5.39}$$

where $f = \lim_{R\to\infty} \frac{nr^2}{R^2}$ is the concentration of inclusions. The estimation of the precision $O(f^2)$ can be found in Mityushev and Rylko (2013).

The value $\sigma_e^{22} + i\sigma_e^{12}$ can be obtained by rotation of the considered medium to $\frac{\pi}{2}$ radians. Then, the angles θ_1 and θ_2 transform onto $\frac{\pi}{2} - \theta_1$ and $\frac{\pi}{2} - \theta_2$ and we obtain

$$\sigma_e^{22} + i\sigma_e^{12} = \sigma_0 \frac{1 + fM'}{1 - fM'} + O(f^2), \tag{5.40}$$

where

$$M' = \rho \cos^2 \frac{\psi}{4} - \sin^2 \frac{\psi}{4} - \frac{1+\rho}{4} \sin^2 \frac{\psi}{2} [\cos(2\theta_1 + \psi) + i \sin(2\theta_1 + \psi)]. \tag{5.41}$$

In the case $\psi = 0 \Leftrightarrow \theta_1 = \theta_2$, we obtain $M = M' = \rho$. This case corresponds to the absence of cracks along the interface and leads to the Clausius–Mossotti approximation (5.2).

In the case $\varrho = -1$, we obtain the known value $M = -1$. This case corresponds to an impermeable inclusion described also by (5.2) with $\varrho = -1$.

In the case $\varrho = 1$, we have

$$M = \cos \frac{\psi}{2} + \frac{1}{2} \sin^2 \frac{\psi}{2} [\cos(2\theta_1 + \psi) + i \sin(2\theta_1 + \psi)]. \tag{5.42}$$

After trigonometric manipulations one can see that this result coincides with formula (4.7) from Rylko et al. (2013).

In the case $\varrho = 0$, we arrive at the homogeneous medium with impermeable cracks with

$$M = -\sin^2 \frac{\psi}{4} + \frac{1}{4} \sin^2 \frac{\psi}{2} [\cos(2\theta_1 + \psi) + i \sin(2\theta_1 + \psi)]. \tag{5.43}$$

The mathematical expectation $\langle \exp \theta_1 \rangle$ vanishes for uniformly distributed inclinations of cracks θ_1 on $(0, \pi]$. Then, (5.39) yields the scalar effective conductivity (5.4). It is worth noting that the inclusion becomes neutral, i.e. $\sigma_e = \sigma_0$, when $\varrho = \tan^2 \frac{\psi}{4}$. The typical dependence of $\frac{\sigma_e}{\sigma_0}$ on f is shown in Fig. 5.2.

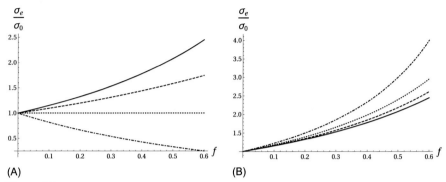

Figure 5.2 The conductivity calculated with (A) (5.4) and (B) (5.5) for the normalized $\sigma_0 = 1$ against the concentration f. Data are for $\varrho = 0.7$ and $\psi = 0$ (solid line), $\psi = \frac{\pi}{2}$ (dashed), $\psi = 2.79$ (the neutral inclusion displayed by dotted line in (A)), $\psi = 2\pi$ (dot–dashed).

Thence case of impermeable cracks is considered above. The case of highly conducting "anti-cracks" can be derived from the previous formulas by Keller's transformation (Keller, 1964). After Keller's transformation, conductivities σ_0 and σ_1 are interchanged, the contact along L becomes perfect and the pressure on L' becomes a constant, i.e. instead of (5.7) we get

$$u(z) = u_1(z) = constant, \qquad z \in L'. \tag{5.44}$$

For simplicity, consider macroscopically isotropic media. Then, the effective conductivity σ_e' of the medium with conductivity σ_1, with inclusions of conductivity σ_0, can be determined by means of the phase-interchange relation

$$\sigma_e' \sigma_e = \sigma_0 \sigma_1. \tag{5.45}$$

Substitution σ_e from (5.4) into (5.45) yields (5.5) after the corresponding replacements.

Following (Gluzman et al., 2017) we can extend the obtained results to 3D fibrous composites with de-bonding interface. The longitudinal conductivity does not depend on the distribution of de-bonding regions and is equal to the mean value $\sigma_\| = \sigma_0(1-f) + \sigma_1 f$. Let a geometric statistical distribution of de-bonding regions be fixed in a section perpendicular to fibers. This implies that there exists a cell in the section which represents the disks distribution and the span angles of cracks. This means that an arbitrarily chosen representative cell in the section obeys the same statistical distribution. The span angle distribution is independent on the disks distributions and is determined by the one-dimensional distributions of θ_1 and of ψ. Let this distribution be statistically homogeneous along fibers. Then, according to the theory of stochastic homogenization (Gluzman et al., 2017), the transverse effective conductivity of

3D fibers is equal to the 2D transverse effective conductivity. Therefore, formulas
(5.4)–(5.5) and (5.39)–(5.40) hold for random fibrous composites with de-bonding
interface.

The new analytical formulas derived above, can be applied to the macroscopic conductivity of media with voids and fractures. The significant influence of cracks onto
the macroscopic conductivity is demonstrated in Fig. 5.2 where the results based on
the classical formula (5.2) are shown by solid lines.

Moreover, measuring the conductivity can be applied to estimation of the effective fraction of cracks and voids. Consider, for instance, the case of macroscopically
isotropic media described by formula (5.4). Let σ_e be measured and the parameters
σ_0, σ_1, f are known. Substituting these values into (5.4) or (5.5) we arrive at a linear
equation on $f \langle \sin^2 \frac{\psi}{4} \rangle$. Solution to this equation yields a de-bonding coefficient which
can be used as the measure of effective fraction of highly conducting cracks

$$m_+ = \langle \sin^2 \frac{\psi}{4} \rangle f = \frac{\frac{\sigma_e'}{\sigma_0} - \frac{1+f\varrho}{1-f\varrho}}{(1-\varrho)(1-f\varrho)(\frac{\sigma_e'}{\sigma_0}+1)}. \tag{5.46}$$

Analogous value can be introduced for almost insulating cracks by using of (5.4)

$$m_- = \langle \sin^2 \frac{\psi}{4} \rangle f = \frac{\frac{1+f\varrho}{1-f\varrho} - \frac{\sigma_e}{\sigma_0}}{(1+\varrho)(1-f\varrho)(\frac{\sigma_e}{\sigma_0}+1)}. \tag{5.47}$$

These coefficients can be considered as the dimensionless de-bonding measure.

Effective conductivity of a random suspension of highly conducting spherical particles

6

6.1 Introduction

The effective properties of 3D random composites with non-overlapping spherical inclusions are of considerable interest in a number of fields. Basic method originally devised by Lord Rayleigh (1892) was applied to various regular arrays of spheres in McPhedran and McKenzie (1978), McKenzie et al. (1978), Poladian and McPhedran (1986), Poladian (1988a,b), where analytical approximate formulas for the effective conductivity were derived. Alternative methods based on the generalized functions and Fourier series were applied to the cubic lattice in Zuzovsky and Brenner (1977), Sangani and Acrivos (1983). Berdichevsky (1983) introduced triply periodic functions associated to an arbitrary 3D lattice and applied them to the considered problem. The effective conductivity tensor for such lattice was calculated up to $O(f^6)$ where f denotes the concentration of inclusions, see formula (6.90). Asymptotically equivalent expressions for the effective conductivity were discussed in Andrianov et al. (1999) and Gluzman et al. (2017, Chapter 8).

Various methods were developed to extend the results from regular to random composites. The example of such random composite with 1000 spheres per periodicity unit cell is displayed in Fig. 2.3. We do not discuss here some pure numerical methods useful for special geometries and concentrate attention on derivation of analytical approximate formulas. The conductivity problem is governed by the Laplace equation and the viscous flow of suspension by the Stokes equations. The Stokes equations are reduced to the bi-Laplace equation. Let equal spherical particles of conductivity σ_1 be embedded in a matrix of conductivity σ. Let $\beta = \frac{\sigma_1 - \sigma}{\sigma_1 + 2\sigma}$ denote the contrast parameter. The first order approximation in f for the effective conductivity is known as the Clausius–Mossotti approximation or Maxwell's formula (Markov, 1999, Milton, 2002, Torquato, 2002)

$$\frac{\sigma_e}{\sigma} = \frac{1 + 2\beta f}{1 - \beta f} + O(f^2). \tag{6.1}$$

In fluid mechanics, the effective viscosity of hard spherical particles embedded in a fluid can be found from the Einstein formula (Einstein, 1906), (Markov, 1999, Torquato, 2002)

$$\frac{\mu_e}{\mu} = 1 + \frac{5}{2}f + O(f^2), \tag{6.2}$$

where μ stands for the viscosity of fluid.

Applied Analysis of Composite Media. https://doi.org/10.1016/B978-0-08-102670-0.00015-9

Many attempts were made to extend these formulas to high concentrations of inclusions. Theoretical difficulties were embodied in a conditionally convergent integral (sum) arisen in the course of the spacial averaging. This integral for the effective conductivity of the regular cubic array was determined by Rayleigh (1892). Batchelor (Batchelor and Green, 1972) estimated the similar integral and the second order term in f for the effective viscosity μ_e of random suspensions by physical intuitive approach, rather than by a rigorous mathematical investigation. Jeffrey (1973) modified Batchelor's approach to conductivity problem having taken into account the two-sphere interactions. He derived the following formula for macroscopically isotropic composites

$$\frac{\sigma_e}{\sigma} \approx 1 + 3\beta f + 3\beta^2 f^2 + 3f^2\beta^3 \left(\frac{3}{4} + \frac{9}{16}\frac{\sigma_1+2}{2\sigma_1+3} + \cdots \right). \tag{6.3}$$

This formula was compared with others in Torquato (2002, p. 493).

In the present chapter, we reconsider Jeffrey's formula (6.3), and find that the f^2-term is equal to $3f^2$ in the case $\beta = 1$, corresponding to highly conducting inclusions. Jeffrey's formula (6.3) in this case gives different result $4.51 f^2$. We do not analyze where is the fallacy off Jeffrey (1973). We can say that it is certainly methodological, not a computational error, related to an intuitive physical treatment of the conditionally convergent integral discussed in Jeffrey (1973).

The critical review of the way of taking the limit and transitioning from finite to infinite number of inclusions in applications of self-consistent and cluster methods is presented in Appendix A.4 for 2D problems, see also (Rylko et al., 2013, Gluzman et al., 2017, Mityushev et al., 2018b, Mityushev, 2018).

The conductivity problem for a finite number of spheres in $\mathbb{R}^3$ was solved by Mityushev and Rylko (2013) in terms of the 3D Poincaré series by the method of functional equations. This method can be considered as the alternating method of Schwartz in the form of contrast or cluster expansions (Mityushev, 2015). The limit transition to infinite number of inclusions was performed in Gluzman et al. (2017, Chapter 8) in the special 3D case when inclusions form the regular cubic lattice.

In the present chapter in Section 6.2, we develop the method of functional equations to the problem with N spheres per unit cubic cell. First, the local field around a finite number n of highly conducting spheres arbitrarily located without mutual intersections is exactly written in Section 6.3. The averaged local field is calculated in Section 6.4. This first part of calculations actually repeats the scheme of (Gluzman et al., 2017, Chapter 8). However, the scheme was ultimately applied only to the regular cubic lattice. Here, we modify the method in order to apply it to an arbitrary positions of inclusions. The limit $n \to \infty$ of the averaged local field is calculated in Section 6.5 by means of the Eisenstein summation following (Rayleigh, 1892). The justification of this approach was given in McPhedran and McKenzie (1978, Section 2.4) by studying of the shape-dependent sums. It was shown that the Eisenstein summation yielded the vanishing polarization charge on the exterior surface having tended to infinity (for details see McPhedran and McKenzie (1978, Fig. 2)). Though this justification concerned a regular cubic lattice it can be applied verbatim to random composites.

Application of the Eisenstein summation to random composites yields an analytical formula for the effective conductivity tensor. Its components are explicitly written in the form (6.87)–(6.88) up to $O(f^{\frac{10}{3}})$.

Numerical examples with $N = 1000$ spheres per cell are presented in Section 6.6. Randomness is considered by the direct approach used in Gluzman et al. (2017). It can be shortly outlined as follows. First, a deterministic problem with an arbitrary location of non-overlapping spheres is solved and the effective conductivity tensor is explicitly written. Since centers of spheres $\mathbf{a}_k$ and their number per cell N are symbolically presented in the final formulas, we may consider them as random variable $\{\mathbf{a}_1, \mathbf{a}_2, \ldots, \mathbf{a}_N\}$ which obeys prescribed joint probabilistic distribution. We performed 10 numerical experiments with $N = 1000$ and obtained practically the same numerical values for coefficients in powers of f in the expansion of the effective conductivity tensor.

6.2 General formula for highly conducting spheres

6.2.1 Statement of the problem

Let vectors $\boldsymbol{\omega}_1 = (1, 0, 0)$, $\boldsymbol{\omega}_2 = (0, 1, 0)$ and $\boldsymbol{\omega}_3 = (0, 0, 1)$ form a cubic lattice. The fundamental periodicity cell (the $\mathbf{0}$-cell) is the cube $\mathcal{O} = \{\mathbf{x} \in \mathbb{R}^3 : -\frac{1}{2} < x_j < \frac{1}{2} \ (j = 1, 2, 3)\}$. Let centers $\mathbf{a}_k$ of mutually disjoint balls $D_k = \{\mathbf{x} \in \mathbb{R}^3 : |\mathbf{x} - \mathbf{a}_k| < r_k\}$ $(k = 1, 2, \ldots, N)$ lie in $\mathcal{O}$ and D denote the complement of all the balls $|\mathbf{x} - \mathbf{a}_k| \leq r_k$ to $\mathcal{O}$. One can consider the triply periodic set of balls $\{D_k + \sum_{i=1,2,3} k_i \boldsymbol{\omega}_i\}$ $(k_i \in \mathbb{Z})$.

Let $\mathbf{n} = (n_1, n_2, n_3)$ denote the unit outward normal vector to the sphere ∂D_k and $\frac{\partial}{\partial \mathbf{n}}$ the corresponding normal derivative. The normal vector has the form

$$\mathbf{n}(\mathbf{x}) = \frac{1}{r_k}(\mathbf{x} - \mathbf{a}_k), \quad \mathbf{x} \in \partial D_k \quad (k = 1, 2, \ldots, n). \tag{6.4}$$

We are looking for functions $u(\mathbf{x})$ harmonic in D and $u_k(\mathbf{x})$ harmonic in D_k $(k = 1, 2, \ldots, N)$, and continuously differentiable in closures of the considered domains with the conjugation (transmission) conditions

$$u = u_k, \quad \frac{\partial u}{\partial \mathbf{n}} = \sigma_1 \frac{\partial u_k}{\partial \mathbf{n}}, \quad |\mathbf{x} - \mathbf{a}_k| = r_k, \ k = 1, 2, \ldots, N. \tag{6.5}$$

Eqs. (6.5) express the condition of perfect contact between materials of conductivity σ_1 occupying the ball and the host of the normalized unit conductivity. It is assumed that functions $u(\mathbf{x})$ and $u_k(\mathbf{x})$ are quasi-periodic, namely,

$$[u]_1 = 1, \quad [u]_2 = 0, \quad [u]_3 = 0, \tag{6.6}$$

where $[u]_j := u(\mathbf{x} + \boldsymbol{\omega}_j) - u(\mathbf{x})$ stands for the jump of $u(\mathbf{x})$ per cell along the axis x_j.

Instead of the normal derivative in (6.5) we will also consider the derivative $\frac{\partial}{\partial r}$ where $r = |\mathbf{x} - \mathbf{a}_k|$ is the radial local coordinate near ∂D_k. For a fixed k, we have

$$\frac{\partial}{\partial \mathbf{n}} = \frac{x_1 - a_{k1}}{r_k} \frac{\partial}{\partial x_1} + \frac{x_2 - a_{k2}}{r_k} \frac{\partial}{\partial x_2} + \frac{x_3 - a_{k3}}{r_k} \frac{\partial}{\partial x_3} = \frac{\partial}{\partial r}, \quad r = r_k, \qquad (6.7)$$

where $\mathbf{a}_k = (a_{k1}, a_{k2}, a_{k3})$. Then, (6.5) becomes

$$u = u_k, \quad \frac{\partial u}{\partial r} = \sigma_1 \frac{\partial u_k}{\partial r}, \quad r = r_k, \ k = 1, 2, \ldots, N. \qquad (6.8)$$

We use below the Ostrogradsky–Gauss formula

$$\int_V \nabla \cdot \mathbf{F} \, d\mathbf{x} = \int_{\partial V} \mathbf{F} \cdot \mathbf{n} \, ds, \qquad (6.9)$$

where V is a domain with the smooth boundary ∂V, $d\mathbf{x} = dx_1 dx_2 dx_3$ is the volume and ds the surface differentials. The averaged flux $\langle \mathbf{q} \rangle$ can be calculated by application of this formula in $\mathcal{O}$. The flux component $\langle q_i \rangle$ along the axis x_i becomes

$$-\langle q_i \rangle = \int_D \frac{\partial u}{\partial x_i} d\mathbf{x} + \sigma_1 \sum_{k=1}^{N} \int_{D_k} \frac{\partial u_k}{\partial x_i} d\mathbf{x}$$

$$= \int_{\partial \mathcal{O}} u \, n_i ds + \sum_{k=1}^{N} \int_{\partial D_k} (\sigma_1 u_k - u) \, n_i ds, \qquad (6.10)$$

where $\mathbf{n} = (n_1, n_2, n_3)$. Formula (6.9) is applied three times. For instance, for $i = 1$, its application in D with $\mathbf{F} = (u, 0, 0)$ and in D_k with $\mathbf{F} = (u_k, 0, 0)$ yields the first equation (6.10). The integral over $\partial \mathcal{O} = \partial D - \sum_{k=1}^{n} \partial D_k$ in (6.10) is equal to Kronecker delta δ_{i1} because of (6.6). Using the first relation (6.5) and once again the Ostrogradsky–Gauss formula, we obtain

$$-\langle q_i \rangle = \delta_{i1} + (\sigma_1 - 1) \sum_{k=1}^{N} \int_{D_k} \frac{\partial u_k}{\partial x_i} d\mathbf{x}. \qquad (6.11)$$

The mean value theorem for harmonic functions (Tikhonov and Samarskii, 1977, p. 294) yields

$$-\langle q_i \rangle = \delta_{i1} + (\sigma_1 - 1) \sum_{k=1}^{N} \frac{4}{3} \pi r_k^3 \frac{\partial u_k}{\partial x_i}(\mathbf{a}_k). \qquad (6.12)$$

Let $\Sigma = \{\sigma_e^{ij}\}$ denote the effective conductivity tensor defined by the relation $\mathbf{q} = -\Sigma \langle \overline{\nabla u} \rangle$. Here, $\overline{\nabla u} = ([u]_1, [u]_2, [u]_3)$ stands for the jump vector of the potential u per cell. The normalized jump conditions (6.6) yield $\sigma_e^{ij} = -\langle q_i \rangle$, hence, (6.12) becomes

$$\sigma_e^{ij} = \delta_{i1} + (\sigma_1 - 1) \sum_{k=1}^{N} \frac{4}{3} \pi r_k^3 \frac{\partial u_k}{\partial x_i}(\mathbf{a}_k). \tag{6.13}$$

The formula (6.13) is derived for $j = 1$, i.e., for the external flux applied in the x_1-direction, see (6.6), and can be established analogously for $j = 2, 3$. Therefore, functions $u_k(\mathbf{x})$ from (6.14) depend also on j.

In the case of equal radii we arrive at the formula

$$\sigma_e^{ij} = \delta_{ij} + (\sigma_1 - 1) \frac{f}{N} \sum_{k=1}^{N} \frac{\partial u_k}{\partial x_i}(\mathbf{a}_k) \quad (i, j = 1, 2, 3), \tag{6.14}$$

where f denote the concentration of inclusions.

6.2.2 High conducting inclusions

Below, we study the case of highly conducting inclusions when $\sigma_1 \gg 1$ and use the ansatz

$$u = u^{(0)} + \frac{1}{\sigma_1} u^{(1)} + \frac{1}{\sigma_1^2} u^{(2)} + \dots, \quad u_k = u_k^{(0)} + \frac{1}{\sigma_1} u_k^{(1)} + \frac{1}{\sigma_1^2} u_k^{(2)} + \dots . \tag{6.15}$$

Substitute (6.15) into (6.5) and take the terms up to $O(\sigma_1^{-1})$. The zeroth coefficient yields the problem

$$u^{(0)} = u_k^{(0)}, \tag{6.16}$$

$$\frac{\partial u_k^{(0)}}{\partial \mathbf{n}} = 0, \quad |\mathbf{x} - \mathbf{a}_k| = r_k \ (k = 1, 2, \dots, N). \tag{6.17}$$

It follows from Eq. (6.17) that $u_k^{(0)}(\mathbf{x})$ is a constant for each k. Then Eq. (6.16) can be considered as the modified Dirichlet problem with undetermined constants $u_k^{(0)}$ on the boundary, to be discussed in the next section. The coefficient in σ_1^{-1} yields the problem

$$u^{(1)} = u_k^{(1)}, \tag{6.18}$$

$$\frac{\partial u^{(0)}}{\partial \mathbf{n}} = \frac{\partial u_k^{(1)}}{\partial \mathbf{n}}, \quad |\mathbf{x} - \mathbf{a}_k| = r_k \ (k = 1, 2, \dots, N). \tag{6.19}$$

The relation (6.19) can be considered as N independent Neumann problems on $u_k^{(1)}(\mathbf{x})$ for the balls D_k $(k = 1, 2, \dots, N)$ where $u^{(0)}(\mathbf{x})$ is taken after solving the modified Dirichlet problem (6.16). By solving the latter problem we arrive at the Dirichlet problem (6.18) for the domain D.

Using of (6.14) and (6.15) yields

$$\sigma_e^{ij} = \delta_{ij} + (\sigma_1 - 1) \sum_{k=1}^{N} \frac{4}{3}\pi r_k^3 \left[\frac{\partial u_k^{(0)}}{\partial x_i}(\mathbf{a}_k) + \frac{1}{\sigma_1} \frac{\partial u_k^{(1)}}{\partial x_i}(\mathbf{a}_k) \right] + O(\sigma_1^{-1}). \quad (6.20)$$

Taking into account that $u_k^{(0)}(\mathbf{x})$ is a constant we obtain

$$\sigma_e^{ij} = \delta_{ij} + \frac{\sigma_1 - 1}{\sigma_1} \sum_{k=1}^{N} \frac{4}{3}\pi r_k^3 \frac{\partial u_k^{(1)}}{\partial x_i}(\mathbf{a}_k) + O(\sigma_1^{-1}). \quad (6.21)$$

Formula (6.21) can be written in the form

$$\sigma_e^{ij} = \delta_{ij} + \sum_{k=1}^{N} \frac{4}{3}\pi r_k^3 \frac{\partial u_k^{(1)}}{\partial x_i}(\mathbf{a}_k) + O(\sigma_1^{-1}), \quad (6.22)$$

since $\sigma_1 \to \infty$.

In Eqs. (6.21) and (6.22), $u_k^{(1)}(\mathbf{x})$ is a solution of the Neumann problem (6.19) where $u_k^{(0)}(\mathbf{x})$ is a solution of the modified Dirichlet problem (6.16). In order to avoid solving the two problems we transform (6.22) as follows. First, application of the Green identity

$$\int_{D_k} (g\Delta f + \nabla f \cdot \nabla g) dx = \int_{\partial D_k} g \frac{\partial f}{\partial \mathbf{n}} ds, \quad (6.23)$$

for $f = u_k^{(1)}$, $g = x_i - a_{ki}$ and the mean value theorem yields

$$\frac{4}{3}\pi r_k^3 \frac{\partial u_k^{(1)}}{\partial x_i}(\mathbf{a}_k) = \int_{D_k} \frac{\partial u_k^{(1)}}{\partial x_i} dx = \int_{\partial D_k} (x_i - a_{ki}) \frac{\partial u_k^{(1)}}{\partial \mathbf{n}} ds. \quad (6.24)$$

Using (6.19) we can write (6.22) in the form

$$\sigma_e^{ij} = \delta_{ij} + \sum_{k=1}^{N} \frac{4}{3}\pi r_k^3 \int_{\partial D_k} (x_i - a_{ki}) \frac{\partial u^{(0)}}{\partial \mathbf{n}} ds + O(\sigma_1^{-1}), \quad (6.25)$$

where $u^{(0)}(\mathbf{x})$ is a solution of the modified Dirichlet problem (6.16). Using the radial derivative (6.7) we rewrite (6.25) in the equivalent form

$$\sigma_e^{ij} = \delta_{ij} + \sum_{k=1}^{N} \frac{4}{3}\pi r_k^3 \int_{\partial D_k} (x_i - a_{ki}) \frac{\partial u^{(0)}}{\partial r} ds + O(\sigma_1^{-1}). \quad (6.26)$$

6.3 Modified Dirichlet problem

6.3.1 Finite number of inclusions

In the present section we consider a modified Dirichlet problem for a finite domain $D(n)$ with n balls in $\mathbb{R}^3$ and construct its solution $u(\mathbf{x}, n)$. The sought triple periodic function can be constructed by the limit transition $u^{(0)}(\mathbf{x}) = \lim_{n \to \infty} u(\mathbf{x}, n)$. It will be convenient to introduce auxiliary functions $u_k(\mathbf{x}, n)$ harmonic in D_k and express the effective conductivity in terms of $\frac{\partial u_k}{\partial x_i}(\mathbf{a}_k, \infty)$. Formally, $u_k(\mathbf{x}, n)$ differ from the potential $u_k(\mathbf{x})$ in D_k for the general problem (6.5) with finite σ_1. The new functions $u_k(\mathbf{x}, n)$ can be considered as the potential in D_k satisfying the special interface condition (6.33) discussed below. We write D and $u(\mathbf{x})$ instead of $D(n)$ and $u(\mathbf{x}, n)$ for shortness.

Consider mutually disjoint balls $D_k = \{\mathbf{x} \in \mathbb{R}^3 : |\mathbf{x} - \mathbf{a}_k| < r_k\}$ $(k = 1, 2, \ldots, n)$ and the domain $\dot{D} = \mathbb{R}^3 \setminus \cup_{k=1}^n (D_k \cup \partial D_k)$. It is convenient to add the infinite point and introduce the domain $D = \dot{D} \cup \{\infty\}$ lying in the one-point compactification of $\mathbb{R}^3$. We find a function $u(\mathbf{x})$ harmonic in $\dot{D}$ and continuously differentiable in $\dot{D} \cup_{k=1}^n \partial D_k$ with the boundary conditions

$$u = c_k, \quad \mathbf{x} \in \partial D_k \quad (k = 1, 2, \ldots, n). \tag{6.27}$$

Here, c_k are undetermined constants which should be found during solution to the boundary value problem. Let the external flux at infinity be parallel to the x_j-axis, hence,

$$u(\mathbf{x}) - x_j \text{ tends to } 0, \text{ as } |\mathbf{x}| \to \infty. \tag{6.28}$$

We have

$$\int_{\partial D_k} \frac{\partial u}{\partial \mathbf{n}} ds = 0 \Longleftrightarrow \int_{\partial D_k} \frac{\partial u}{\partial r} ds = 0 \quad (k = 1, 2, \ldots, n). \tag{6.29}$$

The inversion with respect to a sphere ∂D_k is introduced by formula

$$\mathbf{x}^*_{(k)} = \frac{r_k^2}{r^2}(\mathbf{x} - \mathbf{a}_k) + \mathbf{a}_k, \tag{6.30}$$

where $r = |\mathbf{x} - \mathbf{a}_k|$ denotes the local spherical coordinate near the point $\mathbf{x} = \mathbf{a}_k$. The Kelvin transform with respect to ∂D_k has the form (Tikhonov and Samarskii, 1977)

$$\mathcal{K}_k w(\mathbf{x}) = \frac{r_k}{r} w(\mathbf{x}^*_{(k)}). \tag{6.31}$$

If a function $w(\mathbf{x})$ is harmonic in $|\mathbf{x} - \mathbf{a}_k| < r_k$, the function $\mathcal{K}_k w(\mathbf{x})$ is harmonic in $|\mathbf{x} - \mathbf{a}_k| > r_k$ and vanishes at infinity.

Let $v(\mathbf{x})$ be a given Hölder continuous function on a fixed sphere $|\mathbf{x} - \mathbf{a}_k| = r_k$. Consider an auxiliary Robin problem

$$\frac{1}{r_k} u_k + 2 \frac{\partial u_k}{\partial r} = v, \quad r = r_k, \tag{6.32}$$

on the function $u_k(\mathbf{x})$ harmonic in the ball $|\mathbf{x} - \mathbf{a}_k| < r_k$. This problem has a unique solution (Mityushev and Rogosin, 2000, Lanzani and Shen, 2004). The following assertion is evident from the physical point of view but in mathematics it requires a rigorous justification.

Let $u_k(\mathbf{x})$ be a solution to the problem (6.32) with $v = \frac{\partial u}{\partial r}$, i.e.,

$$\frac{1}{r_k}u_k + 2\frac{\partial u_k}{\partial r} = \frac{\partial u}{\partial r}, \quad r = r_k. \tag{6.33}$$

Introduce the function, piece-wise harmonic in $\cup_{k=1}^{n} D_k \cup \dot{D}$

$$U(\mathbf{x}) = \begin{cases} u_k(\mathbf{x}) + \sum_{m \neq k} \frac{r_m}{|\mathbf{x}-\mathbf{a}_m|} u_m(\mathbf{x}_{(m)}^*) + c_k, & |\mathbf{x} - \mathbf{a}_m| \leq r_k, \\ & k = 1, 2, \dots, n, \\ u(\mathbf{x}) + \sum_{m=1}^{n} \frac{r_m}{|\mathbf{x}-\mathbf{a}_m|} u_m(\mathbf{x}_{(m)}^*), & \mathbf{x} \in \dot{D}. \end{cases} \tag{6.34}$$

Let $U^+(\mathbf{x}) = \lim_{\cup_{k=1}^{n} D_k \ni \mathbf{y} \to \mathbf{x}} U(\mathbf{y})$ and $U^-(\mathbf{x}) = \lim_{D \ni \mathbf{y} \to \mathbf{x}} U(\mathbf{y})$. Calculate the jump $\Delta_k(\mathbf{x}) = U^+(\mathbf{x}) - U^-(\mathbf{x})$ across ∂D_k ($k = 1, 2, \dots, n$). Using the definition of $U(\mathbf{x})$ and the boundary condition (6.27) we get

$$\Delta_k = u_k + c_k - u - \frac{r_k}{|\mathbf{x}-\mathbf{a}_k|}u_k = 0. \tag{6.35}$$

Now, we calculate $\Delta_k' = \frac{\partial U^+}{\partial r} - \frac{\partial U^-}{\partial r}$ across ∂D_k. Using the definition of $U(\mathbf{x})$ we obtain for a fixed k

$$\Delta_k'(\mathbf{x}) = \frac{\partial u_k}{\partial r}(\mathbf{x}) - \frac{\partial u}{\partial r}(\mathbf{x}) - \frac{\partial}{\partial r}\left[\frac{r_k}{|\mathbf{x}-\mathbf{a}_k|}u_k(\mathbf{x}_{(k)}^*)\right], \quad \mathbf{x} \in \partial D_k. \tag{6.36}$$

In order to properly calculate the radial derivative in (6.36) we introduce the local spherical coordinates (r, θ, φ) near a fixed sphere ∂D_k and the functions

$$v(r) = u_k(\mathbf{x}) \equiv u_k(r, \theta, \varphi), \quad w(r) = u_k(\mathbf{x}_{(k)}^*) \equiv u_k\left(\frac{r_k^2}{r}, \theta, \varphi\right). \tag{6.37}$$

One can see that

$$\frac{\partial w}{\partial r} = -\frac{\partial v}{\partial r}, \quad r = r_k. \tag{6.38}$$

Therefore, (6.36) implies that

$$\Delta_k' = \left[\frac{\partial u_k}{\partial r} - \frac{\partial u}{\partial r} + \frac{r_k}{r}\frac{\partial u_k}{\partial r} - u_k\frac{\partial}{\partial r}\left(\frac{r_k}{r}\right)\right]\Bigg|_{r=r_k} = 2\frac{\partial u_k}{\partial r} + \frac{1}{r_k}u_k - \frac{\partial u}{\partial r}. \tag{6.39}$$

Application of (6.33) yields $\Delta_k' = 0$. Then, the harmonic continuation principle implies that the function $U(\mathbf{x})$ is harmonic in $\dot{D}$. It follows from the definition of $U(\mathbf{x})$ near infinity and (6.28) that $U(\mathbf{x}) - x_j$ is harmonic in $\mathbb{R}^3 \cup \{\infty\}$. Therefore, $U(\mathbf{x}) - x_j$

is a constant by Liouville's theorem. Moreover, $U(\mathbf{x}) - x_j$ vanishes at infinity, hence, this constant is equal to zero and $U(\mathbf{x}) = x_j$ for all $\mathbf{x} \in \mathbb{R}^3 \cup \{\infty\}$. Eq. (6.34) written in the considered domains gives the system of functional equations

$$u_k(\mathbf{x}) = -\sum_{m \neq k} \frac{r_m}{|\mathbf{x} - \mathbf{a}_m|} u_m(\mathbf{x}^*_{(m)}) + x_j - c_k, \quad |\mathbf{x} - \mathbf{a}_m| \leq r_k \ (k = 1, 2, \ldots, n), \quad (6.40)$$

and

$$u(\mathbf{x}) = -\sum_{m=1}^{n} \frac{r_m}{|\mathbf{x} - \mathbf{a}_m|} u_m(\mathbf{x}^*_{(m)}) + x_j, \quad \mathbf{x} \in \dot{D}. \tag{6.41}$$

After solving the system (6.40) the auxiliary functions $u_k(\mathbf{x})$ are substituted into (6.41). This gives the solution of the problem (6.28), the function $u(\mathbf{x})$.

Introduce the space $C(D_k)$ of functions harmonic in all D_k and continuous in their closures endowed with the norm $\|h\|_k = \max_{\mathbf{x} \in \partial D_k} |h(\mathbf{x})|$. Let $C = C\left(\cup_{k=1}^n D_k\right)$ denote the space of functions harmonic in all D_k and continuous in their closures. The norm in C has the form $\|h\| = \max_k \|h\|_k = \max_k \max_{\mathbf{x} \in \partial D_k} |h(\mathbf{x})|$. The system of functional equations (6.40) can be considered as an equation in the space C on a function equal to $u_k(\mathbf{x})$ in each closed ball $D_k \cup \partial D_k$.

Theorem 6.1 (Mityushev and Rylko, 2013). *Let a given function $h(\mathbf{x})$ belong to C. The system of functional equations*

$$u_k(\mathbf{x}) = -\sum_{m \neq k} \frac{r_m}{|\mathbf{x} - \mathbf{a}_m|} u_m(\mathbf{x}^*_{(m)}) + h(\mathbf{x}), \quad |\mathbf{x} - \mathbf{a}_m| \leq r_k \ (k = 1, 2, \ldots, n) \quad (6.42)$$

has a unique solution in C. This solution can be found by successive approximations converging in C, i.e., uniformly in $\cup_{k=1}^n (D_k \cup \partial D_k)$.

Let k_s run over $1, 2, \ldots, n$. Consider the sequence of inversions with respect to the spheres $\partial D_{k_1}, \partial D_{k_2}, \ldots, \partial D_{k_m}$ determined by the recurrence formula

$$x^*_{(k_m k_{m-1} \ldots k_1)} := \left(x^*_{(k_{m-1} k_{m-2} \ldots k_1)}\right)^*_{k_m}. \tag{6.43}$$

It is supposed that no equal neighbor numbers exist in the sequence $k_1, k_2, \ldots, k_m$. The transformations (6.43) for $m = 1, 2, \ldots$ with the identity map form the Schottky group S of maps acting in $\mathbb{R}^3$, see the 2D theory (Mityushev, 2011).

Straightforward application of the successive approximations described in Theorem 6.1 gives the exact formula

$$u_k(\mathbf{x}) = (Ph)(\mathbf{x}), \quad \mathbf{x} \in D_k \cup \partial D_k, \tag{6.44}$$

where the operator P acts in the space C and has the form $P = P_k$ in $D_k \cup \partial D_k$; the operator P_k is defined in terms of the series

$$(P_k h)(\mathbf{x}) = h(\mathbf{x})$$

$$-\sum_{m \neq k} \frac{r_m}{|\mathbf{x} - \mathbf{a}_m|} h(\mathbf{x}^*_{(m)}) + \sum_{m \neq k} \sum_{k_1 \neq k} \frac{r_m}{|\mathbf{x} - \mathbf{a}_m|} \frac{r_{k_1}}{|\mathbf{x}^*_{(m)} - \mathbf{a}_{k_1}|} h(\mathbf{x}^*_{(k_1 m)})$$

$$-\sum_{\substack{m \neq k \\ k_2 \neq k_1}} \sum_{k_1 \neq m} \frac{r_m}{|\mathbf{x} - \mathbf{a}_m|} \frac{r_{k_1}}{|\mathbf{x}^*_{(m)} - \mathbf{a}_{k_1}|} \frac{r_{k_2}}{|\mathbf{x}^*_{(k_1 m)} - \mathbf{a}_{k_2}|} h(\mathbf{x}^*_{(k_2 k_1 m)}) + \cdots,$$

$$(6.45)$$

where for instance $\sum_{\substack{k_1 \neq m \\ k_2 \neq k_1}} := \sum_{k_1 \neq m} \sum_{k_2 \neq k_1}$. Every sum $\sum_{k_s \neq k_{s-1}}$ contains terms
with $k_s = 1, 2, \ldots, n$ except $k_s = k_{s-1}$. Following Mityushev and Rogosin (2000, Chapter 4) and Mityushev and Rylko (2013) one can prove compactness of P in C. The uniqueness of solution established in Theorem 6.1 for the linear system (6.42) implies that the successive approximations can be applied separately to $h_1(\mathbf{x})$ and to $h_2(\mathbf{x})$ when $h(\mathbf{x}) = h_1(\mathbf{x}) + h_2(\mathbf{x})$. The unique result will be the same. Thus, P can be applied separately to x_j and to the piece-wise constant function $c(\mathbf{x}) = c_k$ where $\mathbf{x} \in D_k$ ($k = 1, 2, \ldots, n$). Then, the solution of the system (6.40) can be written in the form

$$u_k(\mathbf{x}) = (P x_j)(\mathbf{x}) - (P c)(\mathbf{x}), \quad \mathbf{x} \in D_k \cup \partial D_k \ (k = 1, 2, \ldots, n). \tag{6.46}$$

Substitution of (6.46) into (6.41) yields

$$u(\mathbf{x}) = x_j + (P_0 x_j)(\mathbf{x}) - (P_0 c)(\mathbf{x}), \quad \mathbf{x} \in \dot{D}, \tag{6.47}$$

where the operator P_0 is introduced as follows

$$(P_0 h)(\mathbf{x}) = -\sum_{k=1}^{n} \frac{r_k}{|\mathbf{x} - \mathbf{a}_k|} h(\mathbf{x}^*_{(k)})$$

$$+ \sum_{k=1}^{n} \sum_{k_1 \neq k} \frac{r_k}{|\mathbf{x} - \mathbf{a}_k|} \frac{r_{k_1}}{|\mathbf{x}^*_{(k)} - \mathbf{a}_{k_1}|} h(\mathbf{x}^*_{(k_1 k)})$$

$$- \sum_{k=1}^{n} \sum_{\substack{k_1 \neq k \\ k_2 \neq k_1}} \frac{r_k}{|\mathbf{x} - \mathbf{a}_k|} \frac{r_{k_1}}{|\mathbf{x}^*_{(k)} - \mathbf{a}_{k_1}|} \frac{r_{k_2}}{|\mathbf{x}^*_{(k_1 k)} - \mathbf{a}_{k_2}|} h(\mathbf{x}^*_{(k_2 k_1 k)})$$

$$+ \sum_{k=1}^{n} \sum_{\substack{k_1 \neq k \\ k_2 \neq k_1 \\ k_3 \neq k_2}} \frac{r_k}{|\mathbf{x} - \mathbf{a}_k|} \frac{r_{k_1}}{|\mathbf{x}^*_{(k)} - \mathbf{a}_{k_1}|} \frac{r_{k_2}}{|\mathbf{x}^*_{(k_1 k)} - \mathbf{a}_{k_2}|} \frac{r_{k_3}}{|\mathbf{x}^*_{(k_2 k_1 k)} - \mathbf{a}_{k_3}|} h(\mathbf{x}^*_{(k_3 k_2 k_1 k)})$$

$$+ \cdots, \quad \mathbf{x} \in \dot{D}. \tag{6.48}$$

Consider a class of functions $\mathcal{R}$ harmonic in $\mathbb{R}^3$ except a finite set of isolated points located in D where at most polynomial growth can take place. It follows from (6.45) that the operator $P : \mathcal{R} \to \mathcal{R}$ is properly defined. The series $(P_0 h)(\mathbf{x})$ for $h \in \mathcal{R}$ converges uniformly in every compact subset of $\mathbf{x} \in \dot{D} \cup \partial D$. We call it by the 3D Poincaré θ-series associated to the 3D Schottky group $\mathcal{S}$ for the function $h(\mathbf{x})$.

One can see from (6.47) that $u(\mathbf{x})$ contains the undetermined constants c_k only in the part

$$
\begin{aligned}
(P_0 c)(\mathbf{x}) = & -\sum_{k=1}^{n} \frac{r_k}{|\mathbf{x} - \mathbf{a}_k|} c_k + \sum_{k=1}^{n} \sum_{k_1 \neq k} \frac{r_k}{|\mathbf{x} - \mathbf{a}_k|} \frac{r_{k_1}}{|\mathbf{x}_{(k)}^* - \mathbf{a}_{k_1}|} c_{k_1} \\
& -\sum_{k=1}^{n} \sum_{\substack{k_1 \neq k \\ k_2 \neq k_1}} \frac{r_k}{|\mathbf{x} - \mathbf{a}_k|} \frac{r_{k_1}}{|\mathbf{x}_{(k)}^* - \mathbf{a}_{k_1}|} \frac{r_{k_2}}{|\mathbf{x}_{(k_1 k)}^* - \mathbf{a}_{k_2}|} c_{k_2} \\
& +\sum_{k=1}^{n} \sum_{\substack{k_1 \neq k \\ k_2 \neq k_1 \\ k_3 \neq k_2}} \frac{r_k}{|\mathbf{x} - \mathbf{a}_k|} \frac{r_{k_1}}{|\mathbf{x}_{(k)}^* - \mathbf{a}_{k_1}|} \frac{r_{k_2}}{|\mathbf{x}_{(k_1 k)}^* - \mathbf{a}_{k_2}|} \frac{r_{k_3}}{|\mathbf{x}_{(k_2 k_1 k)}^* - \mathbf{a}_{k_3}|} c_{k_3} \\
& +\ldots
\end{aligned}
\tag{6.49}
$$

The function $(P_0 c)(\mathbf{x})$ is represented in the form $(P_0 c)(\mathbf{x}) = \sum_{m=1}^{n} c_m p_m(\mathbf{x})$. Let δ_{km} denote the Kronecker symbol. Functions $p_m(\mathbf{x})$ can be obtained by Lemma 4.3 from Mityushev and Rogosin (2000, p. 152) or by substitution $c(\mathbf{x}) = \delta_{km}$, $\mathbf{x} \in D_k$ ($k = 1, 2, \ldots, n$) into (6.49)

$$
\begin{aligned}
p_m(\mathbf{x}) = & -\frac{r_m}{|\mathbf{x} - \mathbf{a}_m|} + \sum_{k \neq m} \frac{r_k}{|\mathbf{x} - \mathbf{a}_k|} \frac{r_m}{|\mathbf{x}_{(k)}^* - \mathbf{a}_m|} \\
& -\sum_{k=1}^{n} \sum_{k_1 \neq k, m} \frac{r_k}{|\mathbf{x} - \mathbf{a}_k|} \frac{r_{k_1}}{|\mathbf{x}_{(k)}^* - \mathbf{a}_{k_1}|} \frac{r_m}{|\mathbf{x}_{(k_1 k)}^* - \mathbf{a}_m|} \\
& +\sum_{k=1}^{n} \sum_{\substack{k_1 \neq k \\ k_2 \neq k_1, m}} \frac{r_k}{|\mathbf{x} - \mathbf{a}_k|} \frac{r_{k_1}}{|\mathbf{x}_{(k)}^* - \mathbf{a}_{k_1}|} \frac{r_{k_2}}{|\mathbf{x}_{(k_1 k)}^* - \mathbf{a}_{k_2}|} \frac{r_m}{|\mathbf{x}_{(k_2 k_1 k)}^* - \mathbf{a}_m|} + \ldots
\end{aligned}
\tag{6.50}
$$

Then, (6.29) yields the system of linear algebraic equations on constants c_m ($m = 1, 2, \ldots, n$).

$$
\sum_{m=1}^{n} c_m \int_{\partial D_k} p_m(\mathbf{x}) ds = \int_{\partial D_k} (P_0 x_j)(\mathbf{x}) ds, \quad k = 1, 2, \ldots, n.
\tag{6.51}
$$

It follows from the general theory of boundary value problems (Mikhlin, 1964) that the system (6.51) has a unique solution. Because in the opposite case we arrive at the non-uniqueness of solution to the modified Dirichlet problem (6.27) that contradicts to Mikhlin (1964). It is worth noting that constants c_m depend on the external potential x_j, i.e., on j.

6.3.2 Computation of undetermined constants

Though the undetermined constants c_k $(k = 1, 2, \ldots, n)$ can be found from the system (6.51), we will find c_k by another explicit method. First, (6.29) by application of (6.33) is reduced to

$$\int_{\partial D_k} u_k ds = 0, \quad k = 1, 2, \ldots, n. \tag{6.52}$$

Here, the relation $\int_{\partial D_k} \frac{\partial u_k}{\partial \mathbf{n}} ds = 0$ for u_k harmonic in D_k is used. The integral from (6.52) is calculated from the mean value theorem for harmonic functions (Tikhonov and Samarskii, 1977, p. 294). Then, (6.52) becomes

$$u_k(\mathbf{a}_k) = 0, \quad k = 1, 2, \ldots, n. \tag{6.53}$$

The function u_k is expressed exactly by (6.46)–(6.48). Consider the case of equal radii $r_k = r_0$ and write $u_k(\mathbf{x})$ explicitly up to $O(r_0^7)$

$$
\begin{aligned}
u_k(\mathbf{x}) = & \, x_j - c_k - \sum_m \frac{r_0(a_{mj} - c_m)}{|\mathbf{x} - \mathbf{a}_m|} + \sum_{m,l} \frac{r_0^2(a_{lj} - c_l)}{|\mathbf{x} - \mathbf{a}_m||\mathbf{a}_m - \mathbf{a}_l|} \\
& - \sum_m \frac{r_0^3(x_j - a_{mj})}{|\mathbf{x} - \mathbf{a}_m|^3} - \sum_{m,k,l,s} \frac{r_0^3(a_{sj} - c_s)}{|\mathbf{x} - \mathbf{a}_m||\mathbf{a}_m - \mathbf{a}_l||\mathbf{a}_l - \mathbf{a}_s|} \\
& - \sum_{m,l} \frac{r_0^4(a_{lj} - c_l)\sum_{i=1}^{3}(x_i - a_{mi})(a_{mi} - a_{li})}{|\mathbf{x} - \mathbf{a}_m|^3|\mathbf{a}_m - \mathbf{a}_l|^3} \\
& + \sum_{m,l} \frac{r_0^4(a_{mj} - a_{lj})}{|\mathbf{x} - \mathbf{a}_m||\mathbf{a}_m - \mathbf{a}_l|^3} + \sum_{m,l,s,t} \frac{r_0^4(a_{tj} - c_t)}{|\mathbf{x} - \mathbf{a}_m||\mathbf{a}_m - \mathbf{a}_l||\mathbf{a}_l - \mathbf{a}_s||\mathbf{a}_s - \mathbf{a}_t|} \\
& - \sum_{m,l,s,t,w} \frac{r_0^5(a_{wj} - c_w)}{|\mathbf{x} - \mathbf{a}_m||\mathbf{a}_m - \mathbf{a}_l||\mathbf{a}_l - \mathbf{a}_s||\mathbf{a}_s - \mathbf{a}_t||\mathbf{a}_t - \mathbf{a}_w|} \\
& - \sum_{m,l,s} \frac{r_0^5(a_{lj} - a_{sj})}{|\mathbf{x} - \mathbf{a}_m||\mathbf{a}_m - \mathbf{a}_l||\mathbf{a}_l - \mathbf{a}_s|^3} \\
& + \sum_{m,l,s} \frac{r_0^5(a_{sj} - c_s)\sum_{i=1}^{3}(x_i - a_{mi})(a_{mi} - a_{li})}{|\mathbf{x} - \mathbf{a}_m|^3|\mathbf{a}_m - \mathbf{a}_l|^3|\mathbf{a}_l - \mathbf{a}_s|} \\
& + \sum_{m,l,s} \frac{r_0^5(a_{sj} - c_s)\sum_{i=1}^{3}(a_{mi} - a_{li})(a_{li} - a_{si})}{|\mathbf{x} - \mathbf{a}_m||\mathbf{a}_m - \mathbf{a}_l|^3|\mathbf{a}_l - \mathbf{a}_s|^3} \\
& - \sum_{m,l} \frac{3r_0^6(a_{lj} - c_l)\sum_{i=1}^{3}(x_i - a_{mi})(a_{mi} - a_{li})}{2|\mathbf{x} - \mathbf{a}_m|^5|\mathbf{a}_m - \mathbf{a}_l|^5} \\
& + \sum_{m,l} \frac{r_0^6(x_j - a_{mj})}{|\mathbf{x} - \mathbf{a}_m|^3|\mathbf{a}_m - \mathbf{a}_l|^3}
\end{aligned}
$$

$$-\sum_{m,l} \frac{r_0^6(a_{lj} - c_l) \sum_{i=1}^{3}(x_i - a_{mi})^2}{2|\mathbf{x} - \mathbf{a}_m|^5 |\mathbf{a}_m - \mathbf{a}_l|^3}$$

$$-\sum_{m,l} \frac{3r_0^6(a_{mj} - a_{lj}) \sum_{i=1}^{3}(x_i - a_{mi})(a_{mi} - a_{li})}{|\mathbf{x} - \mathbf{a}_m|^3 |\mathbf{a}_m - \mathbf{a}_l|^5}$$

$$-\sum_{m,l,s,t} \frac{r_0^6(a_{tj} - c_t) \sum_{i=1}^{3}(x_i - a_{mi})(a_{mi} - a_{li})}{|\mathbf{x} - \mathbf{a}_m|^3 |\mathbf{a}_m - \mathbf{a}_l|^3 |\mathbf{a}_l - \mathbf{a}_s| |\mathbf{a}_s - \mathbf{a}_t|}$$

$$-\sum_{m,l,s,t} \frac{r_0^6(a_{tj} - c_t) \sum_{i=1}^{3}(a_{mi} - a_{li})(a_{li} - a_{si})}{|\mathbf{x} - \mathbf{a}_m| |\mathbf{a}_m - \mathbf{a}_l|^3 |\mathbf{a}_l - \mathbf{a}_s|^3 |\mathbf{a}_s - \mathbf{a}_t|}$$

$$+\sum_{m,l,s,t} \frac{r_0^6(a_{sj} - a_{tj})}{|\mathbf{x} - \mathbf{a}_m| |\mathbf{a}_m - \mathbf{a}_l| |\mathbf{a}_l - \mathbf{a}_s| |\mathbf{a}_s - \mathbf{a}_t|^3}$$

$$-\sum_{m,l,s,t} \frac{r_0^6(a_{tj} - c_t) \sum_{i=1}^{3}(a_{li} - a_{si})(a_{si} - a_{ti})}{|\mathbf{x} - \mathbf{a}_m| |\mathbf{a}_m - \mathbf{a}_l| |\mathbf{a}_l - \mathbf{a}_s|^3 |\mathbf{a}_s - \mathbf{a}_t|^3}$$

$$+\sum_{m,l,s,t,w,v} \frac{r_0^6(a_{vj} - c_v)}{|\mathbf{x} - \mathbf{a}_m| |\mathbf{a}_m - \mathbf{a}_l| |\mathbf{a}_l - \mathbf{a}_s| |\mathbf{a}_s - \mathbf{a}_t| |\mathbf{a}_t - \mathbf{a}_w| |\mathbf{a}_w - \mathbf{a}_v|}$$

$$+ O(r_0^7). \tag{6.54}$$

Introduce the designation $z_k := a_{kj} - c_k$ and substitute $\mathbf{x} = \mathbf{a}_k$ into (6.54). Then, (6.53) becomes

$$z_k - r_0^3 \sum_{m \neq k} \frac{a_{kj} - a_{mj}}{|\mathbf{a}_k - \mathbf{a}_m|^3} - \sum_{m \neq k} \frac{r_0 z_m}{|\mathbf{a}_k - \mathbf{a}_k|} + \sum_{\substack{m \neq k \\ m \neq l}} \frac{r_0^2 z_l}{|\mathbf{a}_k - \mathbf{a}_m| |\mathbf{a}_m - \mathbf{a}_l|}$$

$$-\sum_{\substack{m \neq k \\ l \neq m \\ s \neq l}} \frac{r_0^3 z_s}{|\mathbf{a}_k - \mathbf{a}_m| |\mathbf{a}_m - \mathbf{a}_l| |\mathbf{a}_l - \mathbf{a}_s|} = O(r_0^4), \quad k = 1, 2, \cdots, n. \tag{6.55}$$

It can be easily obtained from (6.55) that

$$c_k = a_{kj} - r_0^3 \sum_{m \neq k} \frac{a_{kj} - a_{mj}}{|\mathbf{a}_k - \mathbf{a}_m|^3} + O(r_0^4), \quad k = 1, 2, \cdots, n. \tag{6.56}$$

Substitute (6.56), computed up to $O(r_0^7)$, into (6.54)

$$u_k(\mathbf{x}) = x_j - a_{kj} + r_0^3 \sum_{m \neq k} \frac{a_{kj} - a_{mj}}{|\mathbf{a}_k - \mathbf{a}_m|^3} - \sum_{m \neq k} \frac{r_0^3(x_j - a_{mj})}{|\mathbf{x} - \mathbf{a}_m|^3}$$

$$-\sum_{\substack{m \neq k \\ l \neq m}} \frac{r_0^6 (a_{kj} - a_{mj})}{|\mathbf{a}_k - \mathbf{a}_m|^3 |\mathbf{a}_m - \mathbf{a}_l|^3} + \sum_{\substack{m \neq k \\ l \neq m}} \frac{r_0^6 (x_j - a_{mj})}{|\mathbf{x} - \mathbf{a}_m|^3 |\mathbf{a}_m - \mathbf{a}_l|^3}$$

$$-\sum_{\substack{m \neq k \\ l \neq m}} \frac{3 r_0^6 (a_{mj} - a_{lj}) \sum_{i=1}^{3} (x_i - a_{mi})(a_{mi} - a_{li})}{|\mathbf{x} - \mathbf{a}_m|^3 |\mathbf{a}_m - \mathbf{a}_l|^5}$$

$$+\sum_{\substack{m \neq k \\ l \neq m}} \frac{3 r_0^6 (a_{mj} - a_{lj}) \sum_{i=1}^{3} (a_{ki} - a_{mi})(a_{mi} - a_{li})}{|\mathbf{a}_k - \mathbf{a}_m|^3 |\mathbf{a}_m - \mathbf{a}_l|^5} + O(r_0^7). \qquad (6.57)$$

This is the main analytical approximate formula which to be used below to compute the effective conductivity tensor.

6.4 Averaged conductivity

The countable set of centers forms the lattice described in Section 6.1 and can be ordered in the following way $\{\mathbf{a}_k + \sum_{s=1,2,3} m_s \boldsymbol{\omega}_s\}$. Here, $(m_1, m_2, m_3) \in \mathbb{Z}^3$ is the number of the cell; points $\mathbf{a}_k$ $(k = 1, 2, \ldots, N)$ belong to the $\mathbf{0}$-cell.

First, we consider the finite number of inclusions $n = NM$ in the space $\mathbb{R}^3$ using the linear order numeration $\mathbf{a}_k$ $(k = 1, 2, \ldots, n)$, i.e., M is the number of cells and the fixed number N is the number of inclusions per cell. We consider equal spheres, i.e., $r_k = r_0$ for simplicity. Introduce the quantity

$$\sigma_e^{ij}(n) = \delta_{ij} + \frac{1}{M} \sum_{k=1}^{n} \int_{\partial D_k} (x_i - a_{ki}) \frac{\partial u}{\partial \mathbf{n}} ds. \qquad (6.58)$$

It corresponds to σ_e^{ij} from (6.25), more precisely,

$$\sigma_e^{ij} = \lim_{n \to \infty} \sigma_e^{ij}(n) + O(\sigma_1^{-1}). \qquad (6.59)$$

We use the Eisenstein summation in the limit $n \to \infty \iff \{M \to \infty, N \text{ is fixed}\}$. The subscript $j = 1, 2, 3$ denotes the potential jump per cell along the axis x_j in the triple periodic problem.

It follows from (6.33) that for every fixed k

$$\int_{\partial D_k} (x_i - a_{k1}) \frac{\partial u}{\partial \mathbf{n}} ds = \int_{\partial D_k} \frac{x_i - a_{k1}}{r_0} u_k ds + 2 \int_{\partial D_k} (x_i - a_{ki}) \frac{\partial u_k}{\partial \mathbf{n}} ds. \qquad (6.60)$$

The first integral can be calculated by the Ostrogradsky–Gauss formula (6.9) with the non-zero jth coordinate of $\mathbf{F}$ equal to u_k. We have

$$\int_{\partial D_k} \frac{x_i - a_{ki}}{r_0} u_k ds = \int_{D_k} \frac{\partial u_k}{\partial x_i} dx, \qquad (6.61)$$

since the ith component of the unit outward normal vector to the sphere ∂D_k is equal to $\frac{x_i - a_{ki}}{r_k}$. Let $|D_k|$ denote the volume of the ball D_k. Application of the mean value theorem for harmonic functions to (6.61) yields

$$\int_{D_k} \frac{\partial u_k}{\partial x_i} d\mathbf{x} = |D_k| \frac{\partial u_k}{\partial x_i} (\mathbf{a}_k, n). \tag{6.62}$$

Green's identity (6.23) for $f = u_k - F_k$, $g = x_i - a_{ki}$ and the mean value theorem imply that

$$\int_{\partial D_k} (x_i - a_{ki}) \frac{\partial u_k}{\partial \mathbf{n}} ds = |D_k| \frac{\partial u_k}{\partial x_i} (\mathbf{a}_k, n). \tag{6.63}$$

Substitution of (6.61)–(6.63) into (6.58), (6.60) yields

$$\sigma_e^{ij}(n) = \delta_{ij} + \frac{3}{M} \sum_{k=1}^{n} |D_k| \frac{\partial u_k}{\partial x_i} (\mathbf{a}_k, n), \tag{6.64}$$

where the value $\frac{\partial u_k}{\partial x_1}(\mathbf{a}_k)$ can be calculated by the approximation (6.57). When the external flux is applied in the x_1-direction, i.e., $j = 1$, we have

$$\begin{aligned}
\frac{\partial u_k}{\partial x_1}(\mathbf{a}_k, n) &= 1 + r_0^3 \sum_{m \neq k} \frac{2(a_{k1} - a_{m1})^2 - (a_{k2} - a_{m2})^2 - (a_{k3} - a_{m3})^2}{|\mathbf{a}_k - \mathbf{a}_m|^5} \\
&+ r_0^6 \sum_{m,s} \frac{[2(a_{k_1} - a_{m_1})^2 - (a_{k_2} - a_{m_2})^2 - (a_{k_3} - a_{m_3})^2]}{|\mathbf{a}_k - \mathbf{a}_m|^5 |\mathbf{a}_m - \mathbf{a}_s|^5} \\
&\quad \times [2(a_{m_1} - a_{s_1})^2 - (a_{m_2} - a_{s_2})^2 - (a_{m_3} - a_{s_3})^2] \\
&+ 9r_0^6 \sum_{m,s} \frac{(a_{k_1} - a_{m_1})(a_{m_1} - a_{s_1})(a_{k_2} - a_{m_2})(a_{m_2} - a_{s_2})}{|\mathbf{a}_k - \mathbf{a}_m|^5 |\mathbf{a}_m - \mathbf{a}_s|^5} \\
&+ 9r_0^6 \sum_{m,s} \frac{(a_{k_1} - a_{m_1})(a_{m_1} - a_{s_1})(a_{k_3} - a_{m_3})(a_{m_3} - a_{s_3})}{|\mathbf{a}_k - \mathbf{a}_m|^5 |\mathbf{a}_m - \mathbf{a}_s|^5} \\
&+ O(r_0^7),
\end{aligned} \tag{6.65}$$

where the double sum $\sum_{m,s} := \sum_{m \neq k} \sum_{s \neq m}$ is used, and

$$\begin{aligned}
\frac{\partial u_k}{\partial x_2}(\mathbf{a}_k, n) &= 3r_0^3 \sum_{m,s} \frac{(a_{k_1} - a_{m_1})(a_{k_2} - a_{m_2})}{|\mathbf{a}_k - \mathbf{a}_m|^5} \\
&+ 3r_0^6 \sum_{m,s} \frac{(a_{k_1} - a_{m_1})(a_{k_2} - a_{m_2})[2(a_{m_1} - a_{s_1})^2 - (a_{m_2} - a_{s_2})^2 - (a_{m_3} - a_{s_3})^2]}{|\mathbf{a}_k - \mathbf{a}_m|^5 |\mathbf{a}_m - \mathbf{a}_s|^5} \\
&+ 3r_0^6 \sum_{m,s} \frac{(a_{m_1} - a_{s_1})(a_{m_2} - a_{s_2})[-(a_{k_1} - a_{m_1})^2 + 2(a_{k_2} - a_{m_2})^2 - (a_{k_3} - a_{m_3})^2]}{|\mathbf{a}_k - \mathbf{a}_m|^5 |\mathbf{a}_m - \mathbf{a}_s|^5} \\
&+ 9r_0^6 \sum_{m,s} \frac{(a_{k_2} - a_{m_2})(a_{k_3} - a_{m_3})(a_{m_1} - a_{s_1})(a_{m_3} - a_{s_3})}{|\mathbf{a}_k - \mathbf{a}_m|^5 |\mathbf{a}_m - \mathbf{a}_s|^5} + O(r_0^7).
\end{aligned} \tag{6.66}$$

$$\frac{\partial u_k}{\partial x_3}(\mathbf{a}_k, n) = 3r_0^3 \sum_{m,s} \frac{(a_{k_1} - a_{m_1})(a_{k_3} - a_{m_3})}{|\mathbf{a}_k - \mathbf{a}_m|^5}$$

$$+ 3r_0^6 \sum_{m,s} \frac{(a_{k_1} - a_{m_1})(a_{k_3} - a_{m_3})}{|\mathbf{a}_k - \mathbf{a}_m|^5 |\mathbf{a}_m - \mathbf{a}_s|^5}[2(a_{m_1} - a_{s_1})^2 - (a_{m_2} - a_{s_2})^2 - (a_{m_3} - a_{s_3})^2]$$

$$+ 3r_0^6 \sum_{m,s} \frac{(a_{m_1} - a_{s_1})(a_{m_3} - a_{s_3})[-(a_{k_1} - a_{m_1})^2 - (a_{k_2} - a_{m_2})^2 + 2(a_{k_3} - a_{m_3})^2]}{|\mathbf{a}_k - \mathbf{a}_m|^5 |\mathbf{a}_m - \mathbf{a}_s|^5}$$

$$+ 9r_0^6 \sum_{m,s} \frac{(a_{k_2} - a_{m_2})(a_{k_3} - a_{m_3})(a_{m_1} - a_{s_1})(a_{m_2} - a_{s_2})}{|\mathbf{a}_k - \mathbf{a}_m|^5 |\mathbf{a}_m - \mathbf{a}_s|^5} + O(r_0^7). \tag{6.67}$$

6.5 Effective conductivity

Consider the regular infinite array of spheres described in Section 6.2.1 with $N = 1$ and $M = n$, i.e., one sphere per periodicity unit cell. Rayleigh (1892) calculated the limit (see formula after (64) in the paper) using the Eisenstein summation

$$\lim_{M \to \infty} \frac{1}{M} \sum_{\substack{m=1 \\ m \neq k}}^{M} \frac{2(a_{k_1} - a_{m_1})^2 - (a_{k_2} - a_{m_2})^2 - (a_{k_3} - a_{m_3})^2}{|\mathbf{a}_k - \mathbf{a}_m|^5} = \frac{4}{3}\pi. \tag{6.68}$$

Substitution of (6.68) into (6.65) and further into the limit expression (6.64) yields the effective conductivity of the simple cubic array of spheres

$$\sigma_{11} = 1 + 3f + 3f^2 + O(r_0^7), \tag{6.69}$$

where $f = \frac{4}{3}\pi r_0^3$ denotes the concentration of spheres (one sphere per unit cubic cell). For general N the formula (6.69) is replaced by the following relation

$$\sigma_{11} = 1 + 3f + 3f^2 \frac{3}{4\pi} e_{11} + O(r_0^7), \tag{6.70}$$

where

$$e_{11} = \frac{1}{N^2} \sum_{k,m=1}^{N} E_{11}(\mathbf{a}_k - \mathbf{a}_m), \tag{6.71}$$

$$E_{11}(\mathbf{x}) = \sum_{\mathbf{R}} \frac{2(x_1 - R_1)^2 - (x_2 - R_2)^2 - (x_3 - R_3)^2}{|\mathbf{x} - \mathbf{R}|^5}. \tag{6.72}$$

Vectors $\mathbf{R} = (R_1, R_2, R_3) = \sum_{i=1,2,3} m_i \omega_i$ ($m_i \in \mathbb{Z}$) generate the simple cubic lattice. The Eisenstein summation is used in (6.72), and it is assumed for shortness that $E_{11}(\mathbf{0}) := \frac{4}{3}\pi$, when $\mathbf{a}_k = \mathbf{a}_m$ is taken in (6.71).

The following relation holds $E_{11} = -\frac{\partial E_1}{\partial x_1}$, where $E_1(\mathbf{x})$ is the first coordinate of the function (8.4.77) introduced in (Gluzman et al., 2017)

$$E_1(\mathbf{x}) = \sum_{\mathbf{R}} \frac{x_1 - R_1}{|\mathbf{x} - \mathbf{R}|^3}. \tag{6.73}$$

This function is related to Berdichevsky's function $\wp_{11}(\mathbf{x})$ considered in Berdichevsky (1983)[1] (see also Gluzman et al. (2017, Section 3, Chapter 8))

$$E_{11}(\mathbf{x}) = \frac{4}{3}\pi + \wp_{11}(\mathbf{x}). \tag{6.74}$$

The function $\wp_{11}(\mathbf{x})$ was introduced in Berdichevsky (1983) through the absolutely convergent series

$$\wp_{11}(\mathbf{x}) = \frac{2x_1^2 - x_2^2 - x_3^2}{|\mathbf{x}|^5}$$
$$+ \sum_{\mathbf{R}}' \left[-\frac{1}{|\mathbf{x} - \mathbf{R}|^3} + \frac{1}{|\mathbf{R}|^3} + 3\left(\frac{(x_1 - R_1)^2}{|\mathbf{x} - \mathbf{R}|^5} - \frac{R_1^2}{|\mathbf{R}|^5}\right) \right], \tag{6.75}$$

where $\mathbf{R}$ runs over $\mathbb{Z}^3$ in the infinite sum $\sum_{\mathbf{R}}'$ except $\mathbf{R} = \mathbf{0}$. A power series of $E_{11}(\mathbf{x})$ on the coordinates of $\mathbf{x}$ can be applied to effectively compute the values $E_{11}(\mathbf{a}_k - \mathbf{a}_m)$. Using the standard multidimensional Taylor expansion in the coordinates of $\mathbf{x}$ we obtain

$$E_{11}(\mathbf{x}) = \frac{4}{3}\pi + \frac{2x_1^2 - x_2^2 - x_3^2}{|\mathbf{x}|^5} + \sum_{\mathbf{R}}' \frac{3R_1\left(2R_1^2 - 3\left(R_2^2 + R_3^2\right)\right)}{|\mathbf{R}|^7} x_1$$
$$+ \sum_{\mathbf{R}}' \frac{3R_2\left(4R_1^2 - R_2^2 - R_3^2\right)}{|\mathbf{R}|^7} x_2 + \sum_{\mathbf{R}}' \frac{3R_3\left(4R_1^2 - R_2^2 - R_3^2\right)}{|\mathbf{R}|^7} x_3$$
$$+ \dots, \tag{6.76}$$

where, for instance,

$$\sum_{\mathbf{R}}' \left[\frac{1}{|\mathbf{R}|^3} - \frac{1}{|\mathbf{x} - \mathbf{R}|^3} \right] = 3 \sum_{\mathbf{R}}' \frac{R_1 x_1 + R_2 x_2 + R_3 x_3}{|\mathbf{R}|^7} + \dots. \tag{6.77}$$

All the sums $\sum_{\mathbf{R}}'$ in (6.76) are at least of order $|\mathbf{R}|^{-4}$, hence, they converge absolutely (Berdichevsky, 1983). In this case we can change the order of summation and use their symmetry in R_i. In particular, this implies that the coefficients in x_j vanish.

The function $E_{12} = -\frac{\partial E_1}{\partial x_2}$ coincides with Berdichevsky's function $\wp_{12}(\mathbf{x})$ (Berdichevsky, 1983) and can be presented as the Eisenstein series

$$E_{12}(\mathbf{x}) = 3 \sum_{\mathbf{R}} \frac{(x_1 - R_1)(x_2 - R_2)}{|\mathbf{x} - \mathbf{R}|^5}. \tag{6.78}$$

---

[1] The 3D $\wp$-functions are discussed only in the Russian edition (Berdichevsky, 1983).

The function $\wp_{12}(\mathbf{x})$ was expanded by Berdichevsky (1983) into the absolutely convergent series

$$\wp_{12}(\mathbf{x}) = 3\left\{\frac{x_1 x_2}{|\mathbf{x}|^5} + {\sum_{\mathbf{R}}}'\left[\frac{(x_1 - R_1)(x_2 - R_2)}{|\mathbf{x} - \mathbf{R}|^5} - \frac{R_1 R_2)}{|\mathbf{R}|^5}\right]\right\}. \tag{6.79}$$

The functions $E_{21}(\mathbf{x})$ and $\wp_{21}(\mathbf{x})$ are introduced analogously to (6.78) and (6.79) by interchanging subscripts 1 and 2. It follows from the absolute convergence of (6.79) that

$$\wp_{12}(\mathbf{x}) = \wp_{21}(\mathbf{x}), \text{ hence, } E_{12}(\mathbf{x}) = E_{21}(\mathbf{x}). \tag{6.80}$$

The function $E_{13} = -\frac{\partial E_1}{\partial x_3}$ is similar to $E_{12}(\mathbf{x})$ and can be calculated analogously by replacing the subscript 2 by 3.

Introduce the absolutely convergent sums $(i_1 + i_2 + \ldots + i_m \leq \ell - 4)$ symmetric with respect to R_i

$$e_\ell^{(i_1, i_2, i_3)} = {\sum_{\mathbf{R}}}'\frac{R_1^{i_1} R_2^{i_1} R_3^{i_3}}{|\mathbf{R}|^\ell}. \tag{6.81}$$

Note that $e_\ell^{(i_1, i_2, i_3)} = 0$ when at least one superscript i_k is odd. Then, (6.76) is reduced to

$$\begin{aligned}
E_{11}(\mathbf{x}) =& \frac{4}{3}\pi + \frac{2x_1^2 - x_2^2 - x_3^2}{|\mathbf{x}|^5} - \frac{21}{2}\left(2x_1^2 - x_2^2 - x_3^2\right)\left(3e_9^{2,2,0} - e_9^{4,0,0}\right) \\
&+ \frac{45}{8}\left(2x_1^4 - 6\left(x_2^2 + x_3^2\right)x_1^2 - x_2^4 - x_3^4 + 12x_2^2 x_3^2\right) \\
&\times \left(30e_{13}^{2,2,2} - 15e_{13}^{4,2,0} + e_{13}^{6,0,0}\right) \\
&+ \frac{693}{16}\left(2x_1^6 - 15\left(x_2^2 + x_3^2\right)x_1^4 + 15\left(x_2^4 + x_3^4\right)x_1^2 - x_2^6 - x_3^6\right) \\
&\times \left(35e_{17}^{4,4,0} - 28e_{17}^{6,2,0} + e_{17}^{8,0,0}\right) \\
&- \frac{585}{128}\left(10x_1^8 - 140\left(x_2^2 + x_3^2\right)x_1^6 + 70\left(x_2^4 + 24x_3^2 x_2^2 + x_3^4\right)x_1^4\right. \\
&+ 28\left(x_2^6 - 30x_3^2 x_2^4 - 30x_3^4 x_2^2 + x_3^6\right)x_1^2 - 5x_2^8 - 5x_3^8 + 112x_2^2 x_3^6 \\
&\left. - 140x_2^4 x_3^4 + 112x_2^6 x_3^2\right)\left(630e_{21}^{4,4,2} - 504e_{21}^{6,2,2}\right. \\
&\left. - 42e_{21}^{6,4,0} + 45e_{21}^{8,2,0} - e_{21}^{10,0,0}\right) + \ldots. \tag{6.82}
\end{aligned}$$

The function $E_{12}(\mathbf{x})$ is calculated from the similar formula

$$\begin{aligned}
E_{12}(\mathbf{x}) =& \frac{3x_1 x_2}{|\mathbf{x}|^5} + 21x_1 x_2\left(3e_9^{2,2,0} - e_9^{4,0,0}\right) \\
&- \frac{45}{2}x_1 x_2\left(x_1^2 + x_2^2 - 6x_3^2\right)\left(30e_{13}^{2,2,2} - 15e_{13}^{4,2,0} + e_{13}^{6,0,0}\right)
\end{aligned}$$

$$-\frac{693}{8}x_1x_2\left(3x_1^4-10x_2^2x_1^2+3x_2^4\right)\left(35e_{17}^{4,4,0}-28e_{17}^{6,2,0}+e_{17}^{8,0,0}\right)$$

$$+\frac{585}{16}x_1x_2\left(5x_1^6-7\left(x_2^2+12x_3^2\right)x_1^4-7\left(x_2^4-20x_3^2x_2^2-10x_3^4\right)x_1^2\right.$$

$$\left.+5x_2^6-28x_3^6+70x_2^2x_3^4-84x_2^4x_3^2\right)\left(630e_{21}^{4,4,2}-504e_{21}^{6,2,2}\right.$$

$$\left.-42e_{21}^{6,4,0}+45e_{21}^{8,2,0}-e_{21}^{10,0,0}\right)+\ldots. \tag{6.83}$$

It is convenient for computations to replace combinations of $e_\ell^{(i_1,i_2,i_3)}$ with Coulomb lattice sums (Huang, 1999)

$$\mathcal{L}_n^m = \sum_{\mathbf{R}}' A_n^m \left(\frac{\partial}{\partial x}+i\frac{\partial}{\partial y}\right)^m \left(\frac{\partial}{\partial z}\right)^{n-m}\left(\frac{1}{|\mathbf{R}|}\right),$$

where

$$A_n^m = \frac{(-1)^n}{\sqrt{(n-m)!(n+m)!}}.$$

As an example let us expand the sum $\mathcal{L}_4^0$

$$\mathcal{L}_4^0 = \frac{1}{8}\sum_{\mathbf{R}}' \frac{3R_1^4+6R_2^2R_1^2-24R_3^2R_1^2+3R_2^4+8R_3^4-24R_2^2R_3^2}{|\mathbf{R}|^9}.$$

Hence, by (6.81) we have

$$\mathcal{L}_4^0 = \frac{1}{8}\left(3e_9^{(4,0,0)}+3e_9^{(0,4,0)}+8e_9^{(0,0,4)}+6e_9^{(2,2,0)}-24e_9^{(2,0,2)}-24e_9^{(0,2,2)}\right).$$

Since sums (6.81) are absolutely convergent, the permutation of indexes i_k do not change their values. Therefore, the considered lattice sum simplifies to

$$\mathcal{L}_4^0 = -\frac{7}{4}\left(3e_9^{2,2,0}-e_9^{4,0,0}\right).$$

Other lattice sums can be rewritten along similar lines

$$\mathcal{L}_6^0 = \frac{3}{8}\left(30e_{13}^{2,2,2}-15e_{13}^{4,2,0}+e_{13}^{6,0,0}\right)$$

$$\mathcal{L}_8^0 = \frac{99}{64}\left(35e_{17}^{4,4,0}-28e_{17}^{6,2,0}+e_{17}^{8,0,0}\right)$$

$$\mathcal{L}_{10}^0 = -\frac{65}{128}\left(630e_{21}^{4,4,2}-504e_{21}^{6,2,2}-42e_{21}^{6,4,0}+45e_{21}^{8,2,0}-e_{21}^{10,0,0}\right).$$

Functions $E_{11}(\mathbf{x})$ and $E_{12}(\mathbf{x})$ take the form

$$E_{11}(\mathbf{x})=\frac{4}{3}\pi+\frac{2x_1^2-x_2^2-x_3^2}{|\mathbf{x}|^5}+6\mathcal{L}_4^0\left(2x_1^2-\left(x_2^2+x_3^2\right)\right)$$

$$+ 15\mathcal{L}_6^0 \left(2x_1^4 - 6\left(x_2^2 + x_3^2\right)x_1^2 - x_2^4 - x_3^4 + 12x_2^2 x_3^2\right)$$

$$+ 28\mathcal{L}_8^0 \left(2x_1^6 - 15\left(x_2^2 + x_3^2\right)x_1^4 + 15\left(x_2^4 + x_3^4\right)x_1^2 - x_2^6 - x_3^6\right)$$

$$+ 9\mathcal{L}_{10}^0 \left(10x_1^8 - 140\left(x_2^2 + x_3^2\right)x_1^6 + 70\left(x_2^4 + 24x_3^2 x_2^2 + x_3^4\right)x_1^4\right.$$

$$+ 28\left(x_2^6 - 30x_3^2 x_2^4 - 30x_3^4 x_2^2 + x_3^6\right)x_1^2 - 5(x_2^8 - x_3^8)$$

$$+ 112(x_2^2 x_3^6 + x_2^6 x_3^2) - 140x_2^4 x_3^4\right) + \dots, \tag{6.84}$$

$$E_{12}(\mathbf{x}) = \frac{3x_1 x_2}{|\mathbf{x}|^5} - 12\mathcal{L}_4^0 x_1 x_2 - 60\mathcal{L}_6^0 x_1 \left(x_1^2 + x_2^2 - 6x_3^2\right)x_2$$

$$- 56\mathcal{L}_8^0 x_1 \left(3x_1^4 - 10x_2^2 x_1^2 + 3x_2^4\right)x_2$$

$$- 72\mathcal{L}_{10}^0 x_1 \left(5x_1^6 - 7\left(x_2^2 + 12x_3^2\right)x_1^4\right.$$

$$-7\left(x_2^4 - 20x_3^2 x_2^2 - 10x_3^4\right)x_1^2 + 5x_2^6 - 28x_3^6 + 70x_2^2 x_3^4 - 84x_2^4 x_3^2\right)x_2$$

$$+ \dots. \tag{6.85}$$

The function $E_{13}(\mathbf{x})$ can be obtained from (6.85) by interchanging x_2 and x_3.

Introduce the discrete spatial convolution sums constructed from sums (6.65)–(6.67)

$$e_{ij*pl} = \frac{1}{N^3} \sum_{k,m,s=1}^{N} E_{ij}(\mathbf{a}_k - \mathbf{a}_m) E_{pl}(\mathbf{a}_m - \mathbf{a}_s). \tag{6.86}$$

The values e_{ij} and e_{ij*pl} will be called the *structural sums*. They generalize the classic lattice sums constructed for regular arrays. The approximation (6.70) is extended by application of (6.65)

$$\sigma_{11} = 1 + 3f + 3f^2 \frac{3}{4\pi} e_{11} + 3f^3 \left(\frac{3}{4\pi}\right)^2 [e_{11*11} + 3(e_{12*12} + e_{13*13})] + O(f^{\frac{10}{3}}). \tag{6.87}$$

It follows from (6.66) that

$$\sigma_{12} = 9f^2 \left[\frac{3}{4\pi} e_{12} + f\left(\frac{3}{4\pi}\right)^2 (e_{12*11} + e_{12*22} + e_{13*13} + e_{23*13})\right] + O(f^{\frac{10}{3}}). \tag{6.88}$$

The component σ_{13} can be derived from (6.67). Ultimately, it is written by the interchange of subscripts 2 and 3 in (6.88).

Remark 6.1. In the case of simple cubic array when $N = 1$, the convolution (6.86) is simplified to $e_{11*11} = \left(\frac{4\pi}{3}\right)^2$ and $e_{12*12} = e_{13*13} = 0$. The effective conductivity of the simple cubic array can be calculated from the Clausius–Mossotti approximation (6.1) for $\beta = 1$

$$\sigma_{11} = \frac{1 + 2f}{1 - f} + O(f^4), \tag{6.89}$$

in accord with the general formula (6.87). The most advanced formula for the simple cubic array was established by Berdichevsky (1983, formula (11.80))

$$\sigma_e = \frac{1 + 2f}{1 - f} + \frac{3.913 f^{\frac{13}{3}}}{(1 - f)^2} + \frac{1.469 f^{\frac{17}{3}}}{(1 - f)^2} + O(f^6), \tag{6.90}$$

where $\sigma_e = \sigma_{11}$ is the scalar effective conductivity of the simple cubic array.

An important lesson can be learned. Derivation of analytical (approximate) formulas for the effective properties of composites requires a subtle mathematical study of the conditionally convergent sums arisen in the course of spatial averaging. The first order approximations (6.1) and (6.2) in f were obtained as solutions to the single-inclusion problem.

It would be interesting to compare our analytical formulas (6.87)–(6.88) with others. But we have to leave aside some known numerical results for fixed locations of inclusions, since they do not contain the symbolic dependencies on the locations. Although such a dependence can be investigated by Monte Carlo methods, it requires a tedious, time-consuming computer simulations. It is rather impossible at the present time to get accurate results by not following our methodology. It is worth noting that our numerical results presented in Section 6.6 are obtained for 10 generated locations with $N = 1000$ balls each, by using a standard laptop computer.

The present section contains an algorithm for computation of the effective conductivity tensor with an arbitrary precision in f. The symbolic computations are performed up to $O(f^{\frac{10}{3}})$. As a result, the analytical approximate formulas (6.87)–(6.88) are derived for arbitrary locations of non-overlapping spheres, and (6.98) for a class of macroscopically homogeneous, isotropic composites. These formulas explicitly demonstrate the dependence of the effective conductivity tensor on the deterministic and probabilistic distributions of inclusions.

6.6   Numerical results for random composite

We now proceed with application of the (6.86) to the numerical estimation of the effective conductivity of random macroscopically isotropic composites. For samples generation we applied Random Sequential Adsorption (RSA) protocol, where consecutive objects are placed randomly in the cell, rejecting those that overlap with previously absorbed ones. We generated 10 samples from the RSA distribution with $N = 1000$ balls each. The concentration $f = 0.3$ is fixed during the generation. Such a sample is displayed in Fig. 2.3.

Table 6.1 Numerical results for the structural sums for the 10 samples (top) and the corresponding mean values (bottom). The bar stands for the mean values.

e_{11}	e_{11*11}	e_{12*12}	e_{13*13}
4.13295	19.5071	1.41568	1.41133
4.19783	19.6056	1.39797	1.55950
4.17890	19.2992	1.50897	1.49774
4.22268	19.5005	1.40287	1.38913
4.19561	19.6061	1.41058	1.42387
4.16198	19.0934	1.41535	1.52156
4.21459	19.6623	1.47737	1.42095
4.18675	19.5398	1.45704	1.44189
4.22531	19.5500	1.40073	1.29346
4.19565	19.3031	1.39020	1.58077

$\overline{e_{11}}$	$\overline{e_{11*11}}$	$\overline{e_{12*12}}$	$\overline{e_{13*13}}$
4.19122	19.4667	1.42768	1.45402

In computations of (6.87) we used the following approximated values of convergent lattice sums

$$\mathcal{L}_4^0 = 3.10822, \quad \mathcal{L}_6^0 = 0.573329, \quad \mathcal{L}_8^0 = 3.25929, \quad \mathcal{L}_{10}^0 = 1.00922,$$

computed as partial sums of (6.81) for $-250 \le R_k \le 250$ ($k = 1, 2, 3$). Obtained approximations agree to at least five significant digits with known numerical values (Huang, 1999). Numerical results are presented in Table 6.1.

The structural sums (6.71) and (6.86) are special case of *discrete multidimensional convolution of functions* defined in Nawalaniec (2017). In our computations we applied special, efficient algorithms developed therein. The total time of calculations of Table 6.1, ran on a standard notebook equipped with Intel Core i7 Processor of 4th generation, was about 30 minutes.

One can see that $\overline{e_{11}}$ is close to $\frac{4\pi}{3} \approx 4.18879$. We conjecture that $e_{11} = \frac{4\pi}{3}$ for ideally macroscopically isotropic composites. The conjecture can be confirmed by the following observations. The expansion (6.84) of $E_{11}(\mathbf{x})$ is a linear combination of the even, homogeneous polynomials $P_l(x_1, x_2, x_3)$. One can directly check for a few l that

$$P_l(x_1, x_2, x_3) + P_l(x_2, x_1, x_3) + P_l(x_3, x_2, x_1) = 0. \tag{6.91}$$

We suggest that (6.91) holds for all $P_l(x_1, x_2, x_3)$ in the expansion (6.84). Consider 24 points $(\pm a_1, \pm a_2, \pm a_3)$, $(\pm a_2, \pm a_1, \pm a_3)$, $(\pm a_3, \pm a_2, \pm a_1)$ in the **0**-cell generating a macroscopically isotropic structure. If (6.91) is true, than $e_{11} = \frac{4\pi}{3}$ for these 24 points. Therefore, (6.87) for the considered macroscopically isotropic composites yields the scalar effective conductivity

$$\sigma_e = 1 + 3f + 3f^2 + 4.80654 f^3 + O(f^{\frac{10}{3}}). \tag{6.92}$$

Is it possible to extrapolate (6.92) to all f, and find the critical index and amplitude from the truncated series? Assume that the threshold is known and corresponds to RCP, $f_c = 0.637$. Then in the vicinity of RCP it is widely assumed that

$$\sigma_e \simeq A(f_c - f)^{-s}, \tag{6.93}$$

where the superconductivity critical index s is expected to have the value of 0.73 ± 0.01 (Clerc et al., 1990). There is also a slightly larger estimate, $s \approx 0.76$ (Bergman and Stroud, 1992).

Let us estimate the value of s based on asymptotic information encapsulated in (6.92). There is a possibility to obtain for σ_e, the simplest factor approximant with fixed position of singularity and floating critical index, from the requirement of asymptotic equivalence with (6.92). We obtain the following result

$$\sigma^*(f) = \mathcal{F}_4^*(f) = \frac{(2.48123 f + 1)^{0.766996}}{(1 - 1.56986 f)^{0.698732}}. \tag{6.94}$$

The (6.94) suggests the value of 0.7 for the superconductivity critical index.

Assume that in the vicinity of threshold, $\sigma(f) \sim (0.637 - f)^{-(1+s')}$, with unity to be expected from the usual contribution from the radial distribution function at the particles contact $G(2, f)$, (Brady, 1993, Losert et al., 2000), and the value of s' coming from the particle interactions in the composite. One can suggest a simple root approximant

$$r(z) = 1 + b_1 z(1 + b_2 z)^{s'}, \quad z = \frac{f}{f_c - f},$$

which explicitly takes the unity-contribution into account. After imposing the asymptotic equivalence with (6.92), we find all three parameters with the final result

$$\sigma^*(f) = 1 + \frac{1.911 f}{\left(\frac{0.709259 f + 0.637}{0.637 - f}\right)^{0.212373} (0.637 - f)}. \tag{6.95}$$

The expression (6.95) allows to estimate $s' \approx -0.212$. Total value of the critical index, $s \approx 0.788$ is still close enough to the expected values.

Let us employ the systematic methodology used to construct the table of indices. The method of construction was explained in great detail on page 45. It is based on considering iterated root approximants as functions of the critical index by itself. The index is to be found by imposing optimization condition in the form of minimal difference on critical amplitudes (Gluzman and Yukalov, 2017b). The series (6.92) allows to get only three estimates for the critical index s, see Table 6.2. All three are fairly close to 0.72 and to the expected value of 0.73.

Table 6.2 Critical indices s_k for the superconductivity obtained from the optimization
conditions $\Delta_{kn}(s_k) = 0$.

s_k	$\Delta_{k\,k+1}(s_k) = 0$	$\Delta_{k\,3}(s_k) = 0$
s_1	0.725	0.721
s_2	0.715	0.715

For possible applications, one can simply adjust the iterated root approximants to
the most plausible value 0.73 for the critical exponent,

$$\mathcal{R}_2^*(f) = \left(\frac{2.56706 f^2 + 2.06109 f + 0.405769}{(0.637 - f)^2} \right)^{0.365},$$

$$\mathcal{R}_3^*(f) = \left(\left(\frac{2.56706 f^2 + 2.06109 f + 0.405769}{(0.637 - f)^2} \right)^{3/2} - \frac{0.372504 f^3}{(0.637 - f)^3} \right)^{0.243333}. \qquad (6.96)$$

The two expressions are very close numerically, and the critical amplitude can be
found from (6.96). From the former approximant we obtain $A \approx 1.449$, and from the
latter $A \approx 1.456$, giving practically the same result. The 4th order coefficient found
from $\mathcal{R}_3^*(f)$ equals 7.48.

Various techniques bring very close results, especially for the critical amplitude.
For instance, one can simply extract the singularity first, and then apply the Padé
technique, obtaining the following approximant

$$Cor^*(f) = \frac{22.1332(f + 0.439227)}{(1 - 1.56986 f)^{0.73} \left(-f^2 + 4.1095 f + 9.72153 \right)}, \qquad (6.97)$$

which brings the value of $A \approx 1.44$ close to other estimates, but the 4th order coefficient is different and equal to 6.59^2. The critical index for conductivity t is considered
in Subsubsection 1.3.3.2.

Remark 6.1 concerning the simple cubic array can be used to improve the polynomial formulas (6.87)–(6.88). These formulas are derived by the method of successive
approximations applied to the system of functional equations (6.40) for $r_m = r_0$. We
write exactly all the terms up to $O(r_0^7)$ taking into account interactions among all the
spheres. We shall increase this precision in future work. But now we can exactly obtain
some high order terms in the considered case of macroscopically isotropic composites. If we keep in the pth iteration applied to (6.40) only the term with r_0^p we obtain a
sequence of terms which leads to the Clausius–Mossotti approximation for the effective conductivity $1 + 3f + 3f^2 + 3f^3 + \ldots = \frac{1+2f}{1-f}$. Berdichevsky's formula (6.90)
takes into account interactions between spheres located in different periodicity cells.
The similar terms have to be in the formula for random case. The above argumentation

---

[2] The error in Jeffrey's series estimate of c_2 manifests itself in the critical index. In terms of the variable
$z = \frac{f}{f_c - f}$, after standard calculations from Section 1.3.3, page 37, we obtain the two estimates for the
critical index, $s_1 = 0.96$, $s_2 = 1.12$, and $s \approx 1.04 \pm 0.08$. The estimate is close to the effective medium
result $s = 1$.

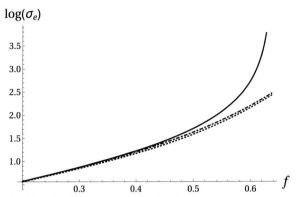

Figure 6.1 The derived formula (6.97) is shown with solid line. Our asymptotic suggestion (6.98) is shown with dot–dashed line. Berdichevsky formula for simple cubic array given by (6.90) is shown with dotted line.

makes possible to rewrite (6.92) in the asymptotically equivalent form

$$\sigma_e = \frac{1+2f}{1-f} + 1.80654\,f^3 + \frac{3.913\,f^{\frac{13}{3}}}{(1-f)^2} + \frac{1.469\,f^{\frac{17}{3}}}{(1-f)^2} + O(f^{\frac{10}{3}}). \qquad (6.98)$$

The scheme, due to its simplicity, can lead to the analytical expression. See Fig. 6.1 for comparison of the different formulas.

We conclude that derivation of analytical (approximate) formulas for the effective conductivity of composites requires a subtle mathematical study of the conditionally convergent sums arisen in the spatial averaging. The first-order approximations (6.1) and (6.2) in f were obtained by solution to the single-inclusion problem.

For many years it was thought that Maxwell's and Clausius–Mossotti approximations can be improved to higher order terms in f by taking into account interactions between pairs of spheres, triplets of spheres, and so on. However, it was recently demonstrated (Gluzman et al., 2017, Mityushev et al., 2018b) that the field around a finite cluster of inclusions yieldS a formula for the effective conductivity only for a dilute cluster. The higher-order term can be properly found by the subtle study of the conditionally convergent series. In the present chapter, the Eisenstein summation is used following (Rayleigh, 1892) for a regular cubic lattice. The justification of the Eisenstein summation was given in McPhedran and McKenzie (1978, Section 2.4) by studying the shape-dependent sums.

The present chapter contains an algorithm to compute the effective conductivity tensor with an arbitrary precision in f. The symbolic computations are performed up to $O(f^{\frac{10}{3}})$. As a result, the analytical approximate formulas (6.87)–(6.88) are derived for an arbitrary location of non-overlapping spheres and (6.98) for a class of macroscopically isotropic composites.

6.7 Foams

Maxwell's formula is still useful for applications, in particular when some appropriate asymptotic conditions are satisfied. Foams are often used in their solidified form as electrical (thermal) insulators. Because of the modest proportion of liquid in a foam and the large fraction of gas which has a much lower (thermal) conductivity the effective conductivity of the foam is much less than that of a liquid body made of the same material. The gas bubbles are pressed together to form the foam and are separated by thin films. Where films meet, there is a liquid-filled interstitial channel called a Plateau border. In a real foam some liquid will collect within the edges at which the films meet. The amount of liquid available for these borders depends on the total amount of liquid left in the foam.

For a dispersion of bubbles of a non-conducting gas in a continuous liquid phase of very high volume fraction (very-wet regime) $\epsilon \to 1$ ($f = 1 - \epsilon \to 0$ gives fraction of bubbles), Maxwell's expression could be adapted (Jeffrey, 1973),

$$\sigma(\epsilon) = \frac{2\epsilon}{3 - \epsilon} \simeq 1 - \frac{3(1 - \epsilon)}{2} + \frac{3}{4}(1 - \epsilon)^2. \tag{6.99}$$

Here $\sigma = \frac{\sigma_{sample}}{\sigma_{liquid}}$, and ϵ stands for the volume fraction of the continuous liquid phase.

On the other end of very-dry regime, as $\epsilon \to 0$, the foam structure is polyhedral, and the entire condensed phase comprises a network of slender randomly oriented channels (Plateau borders). The model network is one of straight borders of uniform cross-section, isotropic, i.e., uniformly distributed in orientation, and meeting at symmetric tetrahedral vertices (Lemlich, 1978, Phelan et al., 1996). Width of the Plateau borders is small compared with their length, and the contribution of the films is ignored. The conductivity in this limit of $\epsilon \to 0$ is given as follows,

$$\sigma(\epsilon) = \frac{\epsilon}{3} + O(\sqrt{\epsilon}), \tag{6.100}$$

as explained in Lemlich (1978, 1985), Phelan et al. (1996), Feitosa et al. (2005). The leading term in (6.100) is shown to give an upper bound for the conductivity (Durand et al., 2004).

Between the two extremes the bubble shape varies from spherical to polyhedral as ϵ decreases. The electrical conductivity in the intermediate regime can be deduced from the two asymptotic expressions. In fact, the data in the wet and dry regimes match smoothly, and can be described by a simple empirical formula (Feitosa et al., 2005), which also respects (6.99) and (6.100),

$$\sigma(\epsilon) = \frac{3.8\epsilon^{3/2} + \epsilon}{-2.8\epsilon + 4.6\sqrt{\epsilon} + 3}. \tag{6.101}$$

It expands at small ϵ as

$$\sigma(\epsilon) \simeq \frac{\epsilon}{3} + 0.76\epsilon^{3/2}.$$

The sign of the leading correction appears to be positive, while another estimate based on formulas from Phelan et al. (1996), Feitosa et al. (2005) gives negative sign.

The leading correction may be calculated more systematically. To this end let us construct and check all possible two-point Padé approximants of the type $\frac{\epsilon}{3}P_{n,m}(\sqrt{\epsilon})$. It turns out that all of them give positive leading correction to the linear term. The following approximant is closest to (6.101), with maximal percentage error around 2%,

$$\sigma(\epsilon) = \frac{\left(-4\epsilon + 3\sqrt{\epsilon} + 3\right)\epsilon}{9 - 7\sqrt{\epsilon}}, \tag{6.102}$$

and it behaves at small ϵ as

$$\sigma(\epsilon) \simeq \frac{\epsilon}{3} + \frac{16\epsilon^{3/2}}{27}.$$

Formula (6.102) is also in good agreement with yet different fit from Feitosa et al. (2005), which includes only integer powers of ϵ. Thus, we constructed a good approximation to various experimental data relying only on asymptotic expressions (6.104) and (6.99), without invoking fitting.

Consider also the two-dimensional foam, or monolayer of foam bubbles. As $\epsilon \to 1$, Maxwell's expression could be written as follows (see, for example, Andrianov et al. (2010)),

$$\sigma(\epsilon) = \frac{\epsilon}{2 - \epsilon} \simeq 1 - 2(1 - \epsilon) + 2(1 - \epsilon)^2. \tag{6.103}$$

At the other end of very-dry regime, as liquid fraction $\epsilon \to 0$, the foam structure is two-dimensionally polygonal, and in the entire condensed phase comprises a network of slender randomly oriented channels. The conductivity as $\epsilon \to 0$, is given as follows

$$\sigma(\epsilon) = \frac{\epsilon}{2} + O(\sqrt{\epsilon}), \tag{6.104}$$

as explained in Durand (2007).

Just as in the 3D case we construct the Padé approximant $P_{3,2}(\sqrt{\epsilon})$, with the following result for the conductivity

$$\sigma(\epsilon) = \frac{\epsilon\left(\epsilon - 3\sqrt{\epsilon} + 5\right)}{10 - 7\sqrt{\epsilon}} \tag{6.105}$$

which behaves at small ϵ as

$$\sigma(\epsilon) \simeq \frac{\epsilon}{2} + \frac{\epsilon^{3/2}}{20}.$$

Remarkably, the conductivity formula corresponding to the approximant $P_{2,3}(\sqrt{\epsilon})$ appears to be almost identical with Maxwell's formula (6.103)! The difference with (6.105) is minuscule. We conclude that for the 2D foams the dilute regime extends to the whole region of ϵ, due to a quasi-linear structure of the conducting network.

6.8   Slits, platelets

We estimate the effective conductivity of a two-dimensional medium consisting of a random distribution of thin, needle-like, non-conducting inclusions (slits, cracks) in an otherwise homogeneous matrix. The effective medium theory due to Maxwell is qualitatively correct in application to such objects. To get the case of slits (needles) one has to consider them as the particular limit-case of elliptic inclusions, with infinite aspect ratio and minimal excluded area (Li and Ostling, 2015). Maxwell's formula also estimates the onset of criticality with surprising accuracy (Zimmerman, 1996). Maxwell's method does not explicitly prevent particles from overlapping, while methodology of Pesetskaya et al. (2018) does. Accordingly, the threshold value estimated in Pesetskaya et al. (2018) is smaller.

The problem of calculating the effective conductivity (electrical, thermal, etc.) of random fractured media is of considerable interest in a number of fields. We restrict our attention to analytical formulas for 2D random macroscopically isotropic media, where insulating cracks are modeled by N equal slits of the length l randomly located in the area Q. Instead of the volume fraction, for thin needle-like objects one can use the crack density introduced in Zimmerman (1996), Pesetskaya et al. (2018), $\rho = \frac{Nl^2}{4Q}$, where N is the number of slits (cracks) of length l, per unit of the area Q. In the continuum percolation theory some very similar definition for the density of objects, $n = 4\rho$, is used. The percolation threshold for sticks is well known numerically, $n_c \approx 5.637$ (Mertens and Moore, 2012, Li and Ostling, 2015). Then, $\rho_c = \frac{n_c}{4} \approx 1.4093$. The objects are allowed to overlap, and to mimic intersection of slits/cracks. By allowing the needles to intersect, deformation and branching of realistic structures are modeled by means of rigid objects of simple shape.

For small ρ in a dilute regime only the linear formula for conductivity is known,

$$\sigma_e = 1 + c_1 \rho + O(\rho^2), \tag{6.106}$$

with $c_1 = -\frac{\pi}{2}$. Its mean-filed extension

$$\sigma_e = \frac{1}{1 + \frac{\pi}{2}\rho} + O(\rho^2),$$

sometimes is used, as written in Davis and Knopoff (2008), Pesetskaya et al. (2018).

As $\rho \to \rho_c$, the following critical behavior is considered (Garboczi et al., 1991, Davis and Knopoff, 2008),

$$\sigma_e \sim (\rho_c - \rho)^{1.3}, \tag{6.107}$$

with the 2D conductivity critical index, the same for different objects. Thus, percolation theory is assumed in the vicinity of ρ_c, and the critical index is the same for continuum and lattice percolation.

The effective medium theory due to Maxwell,

$$\sigma_e(\rho) = \frac{1 - \frac{\pi\rho}{4}}{1 + \frac{\pi\rho}{4}} \tag{6.108}$$

is shown with dot–dashed line in Fig. 6.2. Although qualitatively correct, it underestimates the onset of criticality as noted in Zimmerman (1996).

Let us reconstruct the expression for all ρ based on the two asymptotic forms. We proceed by analogy with the case studied in Eqs. (3.3)–(3.4) from (Gluzman et al., 2013), literally following it to obtain the following approximant

$$\sigma_e^*(\rho) = e^{-0.64832\rho}(1 - 0.709597\rho)^{1.3}. \tag{6.109}$$

Following Eqs. (3.5)–(3.8) from (Gluzman et al., 2013), one can obtain at least two more approximants based on the same input data, but they appear to be very close to (6.109) and will not be shown here. Formula (6.109) appears to be a reasonable upper bound when various formulas are compared.

The lower bound, on the other hand is given by another formula, which simply rewrites the linear expression with account to the critical point. It is also given in Gluzman et al. (2013),

$$\sigma_e(\rho) = (1 - 0.709597\rho)^{1.3}(1 - 0.64832\rho). \tag{6.110}$$

Note that formula (6.109) may be considered as the result of (6.110), being re-summed in the strongest way allowed by algebraic self-similar renormalization. No wonder that the two expressions give rather tight bounds.

Their simple arithmetic average can serve as our final approximation for the conductivity

$$\sigma_e(\rho) = e^{-0.64832\rho}(1 - 0.709597\rho)^{1.3}\left(e^{0.64832\rho}(0.5 - 0.32416\rho) + 0.5\right). \tag{6.111}$$

It is shown with solid line in Fig. 6.2. Spline-interpolation composed from the mean-filed (6.106) and (6.107) as in Davis and Knopoff (2008), is shown with dotted line in Fig. 6.2. It is clearly inferior to other formula, because its exaggerates the extent of critical region.

The empirical formula from Garboczi et al. (1991) is convenient for comparison, except that a higher threshold of 1.475 is employed to better fit the numerical data,

$$\sigma_e(\rho) = \frac{(1 - 0.677966\rho)(-0.640769\rho^2 + 0.677966\rho + 1)}{1.20846\rho + 1}, \tag{6.112}$$

and it shown with dashed line in Fig. 6.2. There are three distinct regimes captured best of all by (6.111). For small ρ slits do not interact or overlap, while at the intermediate ρ they are allowed to interact, but not to intersect. While in the vicinity of ρ_c they do intersect, forming percolation network.

The problem considered above was discussed by Davis and Knopoff (2008) for the effective share modulus and anti-plane cracks, but no analytical solution to describe the whole curve was deduced. Using phase-interchange symmetry, one can readily solve the dual problem of effective conductivity (shear modulus) for ideally conducting (perfectly rigid) needles embedded at random into the 2D matrix. The model is

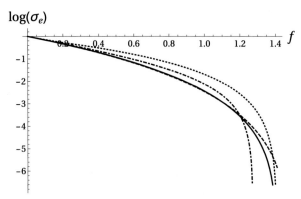

Figure 6.2 The empirical formula (6.112) is shown with dashed line. Our suggestion (6.111) is shown with solid line. Naive spline interpolation composed from the mean-filed (6.106) and (6.107) is shown with dotted line. Maxwell's formula (6.108) is shown with dot–dashed line.

widely used to simulate nano-composites. One simply has to invert the formula (6.111) to get the answer.

As was noted above for the problem of slits, the spline, as well as a naive extrapolation of the critical law (6.107) to the whole region,

$$\sigma_e = \left(1 - \frac{\rho}{\rho_c}\right)^{1.3},$$

appear to be worse than other, more complex approximations. But this is not true for the effective conductivity of the random distribution of overlapping, non-conducting disks. In this example, for convenience, we can go back to the familiar volume fraction, with known parameter values (Novak et al., 2009), $f_c = 0.6763$, $c_1 = -2$, and same critical index of 1.3.

Then, a naive "critical" formula

$$\sigma_e = \left(1 - \frac{f}{f_c}\right)^{1.3} = (1 - 1.47863 f)^{1.3}, \tag{6.113}$$

as well as spline, both have the same accuracy as the more sophisticated analogs of (6.109) and (6.110),

$$\sigma_e^* = e^{-0.0777761 f}(1 - 1.47863 f)^{1.3};$$
$$\sigma_e = (1 - 1.47863 f)^{1.3}(1 - 0.0777761 f), \tag{6.114}$$

in comparison with simulation data of Novak et al. (2009).

The naive formula predicts the linear coefficient $c_1 \approx -1.93$ with good accuracy. In this sense the diluted regime is derived from the critical behavior, in contrast with the case of needles.

In (Novak et al., 2009) it is also suggested to model the internal membranes pertinent to cellular structures, by disks with small aspect ratios, also called platelets

(Nagy and Duxbury, 2002). More precisely, the inclusion can be made in the form of penny-shaped platelet of radius a and thickness $2b$. Random 3D composite is made consisting of aligned non-conducting platelets with large aspect ratio $\frac{a}{b} \gg 1$ immersed in a matrix. As in the case of slits, one can think about penny-shaped cracks. The effective properties such as normal to the platelet surface conductivity, permeability or diffusion constant are expressed as the function of non-dimensional $x = \frac{af}{b}$, and f is the platelet volume fraction (Nagy and Duxbury, 2002). In a case of small x, the leading contribution to the conductivity from isolated penny-shaped platelets is known,

$$\sigma_e \simeq 1 - \frac{4}{3\pi}x, \tag{6.115}$$

and corresponds to the single inclusion approximation.

The relevant parameter x is typically not small, since for a large aspect ratio even a small inclusion volume fraction leads to large values of x. Then one has to calculate the effective property, e.g., the diffusion constant, in regimes where x is not small and derive the formula which represents the data well. Authors of Nagy and Duxbury (2002) suggested, and found numerically that for large x there is a simple power-law dependence,

$$\sigma_e \sim x^{-2}. \tag{6.116}$$

The quadratic term dominates for large x, due to a narrow necks between platelets not included in single inclusion theories. The critical point is located at infinity, but the critical index equals 2, just to the value accepted for the critical index t for 3D conductivity. The typical mean-field formula

$$\sigma_e = \frac{1}{1 + \frac{4}{3\pi}x},$$

can be suggested, but is not "aware" of the latter regime.

In Nagy and Duxbury (2002) the two regimes were matched through numerical simulations for the diffusion coefficient. The simple quadratic fit was found leading to the form

$$\sigma_e \sim \frac{1}{1 + 0.44x + 0.05x^2}, \tag{6.117}$$

for the effective conductivity. It is in a good agreement with theoretical formula (6.115) valid for small x, and gives $\sigma_e \sim 20x^{-2}$, at large x. The fit works well for non-overlapping, well-dispersed platelets.

We attempt to connect smoothly the two regimes for conductivity, based on asymptotic equivalence with (6.115), (6.116). To this end only the simplest root approximant

$$\sigma_e^* = \frac{1}{\left(\frac{2x}{3\pi} + 1\right)^2} \approx \frac{1}{1 + 0.424413x + 0.0450316x^2}, \tag{6.118}$$

is suited. It appears to be close to the fit (6.117), well within the error margins for the fitting coefficients given in Nagy and Duxbury (2002). Similar considerations can be formally applied in 2D for the case of rectangular platelets modeling cracks/slits, but with much less convincing agreement between approximant and fit.

Permeability of porous media

7

7.1 Models of critical permeability

Quantitative understanding of processes that govern flow and transport in porous and discontinuous media is of utmost importance in many geophysical processes. For instance, the presence of fluid is one of the leading candidates to solve a variety of problems in the physics of earthquakes such as the apparent weakness of mature faults. Understanding the transport properties of fluids in the fragmented crust in the presence of discontinuities occurring at many length scales is an essential component in the ultimate goal of understanding earthquakes. Understanding the properties of transport of fluids in complex porous and cracked media at many scales is also of fundamental importance from an environmental point of view as well as for a good stewardship of the storage of contaminants or pollutants or to remedy sites from contamination that arose during nuclear weapons production. The theory of porous and fractured porous media is widely applicable to engineering problems of oil industry.

Critical behavior of permeability of certain, power-series-treatable models of Darcy flow in porous media is studied in this chapter. Permeability of spatially periodic arrays of cylinders is found in an analytical form. Transverse flow past hexagonal and square arrays of cylinders is studied as well, based on expansions for small concentrations and lubrication approximation for high concentrations of cylinders. 3D periodic arrays of spherical obstacles are discussed as well. Formulas for the drag force exerted by various lattices of obstacles are derived from the low-concentration expansions.

Also, the Stokes flow through a 2D and 3D channels enclosed by two wavy walls is studied by means of the analytical-numerical algorithm. Efficient formulas for the permeability are derived in the form of series for small values of wave amplitude. Various power-laws are found in the regime of large amplitudes, based only on expansions at small amplitudes. Lubrication approximation is shown to break down, but accurate formulas for the effective permeability for arbitrary values of the wave amplitude are derived from the expansions.

This chapter deals exclusively with constructive analytical solutions. In other words with approximate analytical solutions, when the resulting formulas contain the main physical and geometrical parameters. The available truncated series are considered as polynomials. They "remember" their infinite expansions, so that with a help of some additional resummation procedure one can extrapolate to the whole series by means of special constructive forms called approximants, that are asymptotically equivalent to the truncated series. The approximants are richer than original polynomials in a sense that they also suggest an infinite additional number of the coefficients. The quality of approximants significantly improves when more than one asymptotic regime can be studied and incorporated into the approximant.

For low Reynolds numbers $\mathcal{R}$, the flow of a viscous fluid through a channel is described by the well-known Darcy's law which corresponds to a linear relation be-

Applied Analysis of Composite Media. https://doi.org/10.1016/B978-0-08-102670-0.00016-0

tween the pressure gradient $\overline{\nabla p}$ and the average velocity $\overline{u}$ along the pressure gradient (Adler, 1992)

$$|\overline{\nabla p}| = \frac{\eta}{K}\overline{u}, \tag{7.1}$$

where K is the permeability and η is the dynamic viscosity of the fluid. The permeability characterizes the amount of viscous fluid flow through a porous medium per unit time and unit area when a unit macroscopic pressure gradient is applied to the system. The classical Poiseuille flow in the channel bounded by two parallel planes separated by a distance $2b$, generated by an average pressure gradient $\overline{\nabla p}$ is a classic example which yields the Darcy's law. The flow profile is parabolic when the Reynolds number $\mathcal{R}$ is small.

When the channel is not straight and when the Reynolds number is not negligible, additional terms appear in this relation (Malevich et al., 2006, Adler et al., 2013). Yet Darcy law holds for the Stokes flow through a channel with three-dimensional wavy walls enclosed by two wavy walls whose amplitude is proportional to the mean clearance of the channel multiplied by the small dimensionless parameter ε. Mind that special technology used for stimulation of petroleum wells leads to the formation of highly conductive channels, so-called wormholes. The channel can be just a single conduit with a minimum of branching. It seems feasible to represent the wormhole by a wavy-walled channel. Channels can be considered jointly as a network of one-dimensional channels. The permeability of a single channel models the permeability of the corresponding edge in the network. Application of the standard methods of flows in networks yields the macroscopic permeability of the considered media. The ultimate result essentially depends on the graph, modeling the network.

Another problem when Darcy low does hold, corresponds to the longitudinal or transverse permeability of a spatially periodic square or hexagonal array of circular cylinders, when a Newtonian fluid is flowing at low Reynolds number along and perpendicular to the cylinders (Sangani and Acrivos, 1982, Drummond and Tahir, 1984, Sangani and Yao, 1988, Gebart, 1992, Mityushev and Adler, 2002a,b, Scholle et al., 2009, Tamayol and Bahrami, 2010b, Yazdchi et al., 2011, Hale et al., 2014, Scholle and Marner, 2018). Non of the currently available expressions for the transverse permeability can be considered satisfactory simultaneously in both, high-and-low porosity limits. Although the problems of permeability and conductivity belong to two different classes (Torquato, 2002), for small and moderate volume fractions mathematical structure of the permeability problem is Laplacian, and can be addressed by the same methodology as in the chapter dedicated to conductivity. But for high volume fractions, such reduction does not work, since it appears that problems of permeability and conductivity are characterized by different critical exponents.

Darcy low also is prerequisite for lattice and continuum percolation models, used to predict critical behavior of the fluid permeability. A random pipe network is a generic model of percolation. It is equivalent to a random resistor network, and is used to simulate fluid flow in the vital case of a sea ice, with profound implications for Climate (Golden et al., 1998, Golden, 2009). Even such proposal is being considered to repair the Ice Sheet Instability by constructing either a continuous artificial sill, or finding isolated artificial pinning points to counter a collapse (Wolovick and Moore, 2018).

The fluid permeability $K(f)$ of random network, where f stands for the pipe/bond fraction, is the ratio between the macroscopic fluid current density and the applied pressure gradient. The bonds could be interpreted as open or closed pipes, and the permeability follows the power-law in the vicinity of a percolation threshold f_c,

$$K(f) \sim (f - f_c)^{\varkappa},$$

where $\varkappa$ is the fluid permeability exponent (Torquato, 2002).

Consider a discrete lattice percolation. Usually, it is tacitly assumed that the target random structure is obtained from the uniform identical distribution of bonds. This corresponds to the "complete stirring" in the James Bond paradigm (Gluzman et al., 2017). Though different authors use theoretically same uniform identical distribution, it is known that the ultimate result can depend on the simulation protocol. We suppose that the discussed results were obtained on the basis of the same theoretical model and similar protocols, i.e., we assume that the same "game of chance" was played. Under acceptance of such rules concerning the structure of media, one believes that $\varkappa = t$, where t is the conductivity critical exponent. For lattices the index t is believed to be universal, depending only on space dimensions. In 2D, $t \approx 1.3$, and in 3D, $t \approx 2.0$. Thus, the critical index for permeability is determined by analogy and from numerical experimentation. Although continuum models can exhibit a non-universal behavior, with the critical exponent values different from the lattice case. A rough estimate for the fluid permeability critical exponent $\varkappa$ for sea ice, is about 2.5 (Golden et al., 1998). Strikingly, the two classical models mentioned above based on direct hydrodynamic solutions, also demonstrate power-law behavior around their corresponding thresholds, with the very same value of the critical exponent $5/2$! Mind that the permeability exponent for the Swiss cheese model of continuum percolation was determined numerically to be close with the value of 2.53 (Murat et al., 1986).

Thus, reality places obvious limitations on application of pure lattice percolation models, but stresses the necessity to study critical phenomena directly from the hydrodynamic equations to deduce transport coefficients via homogenization procedure. If the transport medium is taken as empty space where viscous fluid can flow through, and the cylinders are taken as solid phase blocking the fluid, one still encounters the critical exponent describing the behavior of fluid permeability K near the threshold for the flow. But in situations considered below in this chapter, the critical exponent can be calculated directly from the solution of Stokes problem.

7.2 Permeability of regular arrays

7.2.1 Transverse permeability. Square array

A porous media is considered as consisting of parallel circular cylinders /fibers of radius r suspended in to a continuous matrix involved into a Stokes flow. The fibers are arranged into periodic pattern very much like conducting inclusions within the matrix. Corresponding square array is shown in Fig. 2.11. The flow perpendicular to fibers is constricted while the flow along the fibers is not. Square arrays are important, both theoretically and practically. In particular, they are used to model a micro-pillar arrays which has numerous practical applications (Yazdchi et al., 2011, Hale et al., 2014). The flow can be decomposed onto longitudinal and transverse components. This implies decomposition of the 3D permeability tensor onto 2D transverse tensor $K_\perp$ and the scalar longitudinal permeability $K_{||}$. Below, we consider only macroscopically isotropic 2D structure. Hence, $K_\perp$ is a scalar denoted for shortness as K. Analytical solutions of Sangani and Acrivos (1982) for the non-dimensional permeability of the transverse flow are valid for high porosity $\epsilon \equiv 1 - f$, where $f = \pi r^2$ is the area fraction of the cylinders, and the designation $K' = \frac{K}{(2r)^2}$ used below

$$K'_{SA}(f) = \frac{K(f)}{(2r)^2} = \frac{4.076 f^3 - 1.774 f^2 + 2f - \log(f) - 1.476}{32 f}. \tag{7.2}$$

The lubrication approximation for the non-dimensional permeability of transverse flow through square cylinder arrays is valid at low porosity (Gebart, 1992),

$$K'_l(f) = \frac{K(f)}{(2r)^2} = \frac{4}{9\sqrt{2\pi}} \left(\sqrt{\frac{f_c}{f}} - 1 \right)^{5/2}. \tag{7.3}$$

It is assumed that the resistance to transverse flow comes from the narrow gaps formed between the fibers. As $f \to f_c = \frac{\pi}{4}$, one would simply have the following expression,

$$K'_l(f) \simeq \frac{16 \left(\frac{\pi}{4} - f \right)^{5/2}}{9\pi^{7/2}}, \tag{7.4}$$

which matches the general critical form with the critical index for permeability $\varkappa = 5/2$, and the critical amplitude $A = \frac{16}{9\pi^{7/2}}$.

The problem of interest can be formulated mathematically as follows. Given the approximation (7.2) for the permeability and, assuming that the critical behavior (7.4) is also known, to solve an interpolation problem and derive the formula for permeability for all f. The main technical difficulty with the (7.2) originates from the logarithmic term, which should be attended to in order to prevent unwanted contributions at high porosity. To allow for the standard techniques to be applied, let us take the derivative and work on the expression

$$V(f) = f \frac{dK'_{SA}}{df}(f),$$

which is obviously a conventional power series. It can be transformed into expression which satisfy the correct limits for the differentiated quantities. In fact, for convenience, we are going to work with the quantity $32 f K'$. After integration $\int \frac{V^*_{sq}(f)}{f} df$,

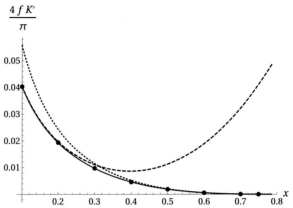

Figure 7.1 Comparison of the formula (7.5) with numerical data of Sangani and Acrivos (1982), for several values of area fraction f. The expansion (7.2) (dashed) and critical form (7.3) (dotted), are shown as well. The non-dimensional quantity $\frac{4fK'(f)}{\pi}$ is shown.

we obtain the expression,

$$
\begin{aligned}
32fK'_{sq}(f) &= 1.10801(f - 3.2344)\sqrt{f_c - f} + 2\tanh^{-1}\left(1.12838\sqrt{f_c - f}\right) \\
&\quad - (0.029876 + 0.223703i)\tan^{-1}\left((0.134275 + 1.03886i)\sqrt{f_c - f}\right) \\
&\quad + (0.223703 + 0.029876i)\tanh^{-1}\left((1.03886 + 0.134275i)\sqrt{f_c - f}\right).
\end{aligned}
\tag{7.5}
$$

Comparison of the formula (7.5) with numerical data of Sangani and Acrivos (1982), for several values of f, is presented in Fig. 7.1. Excellent agreement is achieved for all f.

7.2.2   Transverse permeability. Hexagonal array

The same method can be applied to the hexagonal arrays, with fibers arranged into the periodic pattern very much like conducting inclusions within the matrix. In the case of hexagonal array of cylinders, the following analytical solution (Drummond and Tahir, 1984) is available for high porosity,

$$
K'_{TD}(f) = \frac{K}{(2r)^2} = \frac{-\log(f) - 1.497 + 2f - \frac{f^2}{2} - 0.739f^4}{32f},
\tag{7.6}
$$

for the non-dimensional permeability of the transverse flow. Lubrication approximation for the non-dimensional permeability of transverse flow through hexagonal cylinder arrays is valid at low porosity (Gebart, 1992),

$$
K'_l(f) = \frac{K}{(2r)^2} = \frac{4}{9\sqrt{6\pi}}\left(\sqrt{\frac{f_c}{f}} - 1\right)^{5/2},
\tag{7.7}
$$

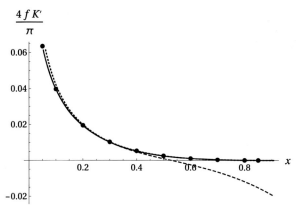

Figure 7.2 Comparison of the formula (7.9) with numerical data of Sangani and Acrivos (1982). The expansion (7.6) (dashed) and critical form (7.7) (dotted), are shown as well. The non-dimensional quantity $\frac{4fK'(f)}{\pi}$ is shown.

where $f_c = \frac{\pi}{2\sqrt{3}}$. As $f \to f_c$ one would simply have the following expression,

$$K'_l(f) \simeq \frac{2\sqrt{2}}{3\sqrt[4]{3}\pi^{7/2}} (f_c - f)^{5/2},$$
(7.8)

which matches the general critical form with the critical index for permeability $\bar{e} = 5/2$, with the critical amplitude $A = \frac{2\sqrt{2}}{3\sqrt[4]{3}\pi^{7/2}}$. Following literally the same approach as presented above in the case of square lattice, we obtain the following formula for the permeability of the hexagonal array of cylinders,

$$\begin{aligned}
32f K'_{hex}(f) &= 2.2194(f + 5.662)\sqrt{0.9069 - f} - 31.9869 \tan^{-1}\left(0.5274\sqrt{0.9069 - f}\right) \\
&\quad + 0.1395 \tan^{-1}\left(1.3573\sqrt{0.9069 - f}\right) + 2\tanh^{-1}\left(1.05008\sqrt{0.9069 - f}\right).
\end{aligned}$$
(7.9)

Comparison of the formula (7.9) with numerical data of Sangani and Acrivos (1982), for several values of f, is given in Fig. 7.2. The formula faithfully represents the numerical data.

7.2.3 Longitudinal permeability. Square array

We study the longitudinal permeability of a spatially periodic rectangular array of circular cylinders, when a Newtonian fluid is flowing at low Reynolds number along the cylinders. Longitudinal laminar flow between unidirectional cylinders is governed by the two-dimensional Poisson equation (Adler, 1992)

$$\nabla^2 w = 1,$$
(7.10)

where w is the component of velocity which is parallel to cylinders; viscosity μ and the pressure gradient are taken equal to 1. In general c and w satisfy a classical boundary condition (Dirichlet, Neumann and their generalizations); velocity vanishes on ∂D, the boundary of a domain D, where Eqs. (7.10) are fulfilled. Poisson equation can be transformed into a functional equation. This equation can be solved by the method of successive approximations. The major advantage of this technique is that the permeability of the array can be expressed analytically in terms of the radius of the cylinders and of the aspect ratio of the unit cell. The unit cell is a rectangle which contains a single circular disc, which is the cross-section of a cylinder. The effective permeability $K_{||}$ (Adler, 1992) is defined as the double integral of the flow velocity over the unit cell

$$-K_{||} = \int_D w(x_1, x_2) dx_1 dx_2. \tag{7.11}$$

Series for the longitudinal permeability of the regular square lattice array of cylinders was calculated by Mityushev and Adler (2002a)

$$\begin{aligned}
K_{||}(f) = \tfrac{1}{4\pi}[&-\log(f) - 1.47644 + 2f - 0.5f^2 - 0.0509713 f^4 + 0.077465 f^8 \\
&- 0.109757 f^{12} + 0.122794 f^{16} - 0.146135 f^{20} + 0.244536 f^{24} \\
&- 0.322667 f^{28} + 0.310566 f^{32} - 0.541237 f^{36} + 0.820399 f^{40}] \\
&+ O(f^{41}),
\end{aligned} \tag{7.12}$$

where $f = \pi r^2$ is the area fraction of the cylinders. In the case being studied, application of the Padé approximants to (7.12), (while leaving the log-term outside), does not give any significant improvement. Such behavior is to be expected for the convergent series within the region of their convergence and with sufficiently large number of terms preserved after truncation. Since the parallel flow solutions are idealized solutions for the flow through cigarette filters, the series (7.12) has certain practical value.

In order to explain the methodological difference between our analytical approach and various simplifications, consider the following expression for the non-dimensional permeability for parallel flow through square array of cylinders from Tamayol and Bahrami (2010a),

$$K'_{||}(f) = \frac{K_{||}}{(2r)^2} = \frac{-0.0186 f^4 - \frac{f^2}{2} + 2f - \log(f) - 1.479}{16 f}. \tag{7.13}$$

The series (7.12) by itself provides better accuracy than (7.13), compared with numerical results of Sangani and Yao (1988). The percentage error given by the series (7.12) equals to 0.193% as $f = 0.7$, while formula (7.13) gives the error of 7.177%. Longitudinal permeability remains finite at f_c. The seepage at $f = f_c$ predicted by the series (7.12) is significantly, by 17.9% smaller than prediction of (7.13). The paper (Tamayol and Bahrami, 2010a) is based on the intuitive approximation of the considered array by a 1D array. Unfortunately, a disagreement between (7.12) and (7.13) is frequently explained by engineers as a difference in basic modeling. Due to the principle of democracy, both models have equal rights. However, here we are in the framework of the properly stated mathematical model when the permeability is uniquely determined. The model is already fixed and is the same in both cases. Actually we discuss not a model but a method of solution. The comparison of (7.12) and (7.13) leads us to conclusion that the term f^4 in the formula (7.13) is wrong.

7.3   3D periodic arrays of spherical obstacles

Hasimoto obtained a low-concentration formula for nondimensional permeability of three different lattices of obstacles, characterizing slow fluid flow past spheres located on the sites of the lattice. He obtained the four terms in powers of $f^{\frac{1}{3}}$ in the series for the dimensionless fluid permeability (Hasimoto, 1959, Torquato, 2002), and found that for all lattices the diluted regime is described by rather close expressions.

Much more terms in the series for the inverse permeability up to the terms f^{10} can be found in Sangani and Acrivos (1983). After this order, the series behave highly irregularly, forbidding to reach the region of $f > 0.85 f_c$. Numerical results for the whole region of concentrations were obtained by different numerical method (Zick and Homsy, 1982, Sangani and Acrivos, 1983).

For the region of higher f, Carman suggested the following formula for inverse $K(f)$ (Zick and Homsy, 1982),

$$F(f) = (K(f))^{-1} = \frac{10f}{(1-f)^3}, \qquad (7.14)$$

which involves a non-physical singularity, where the critical index equals 3, as $f \to 1$. $F(f)$ is nothing else but a non-dimensional drag force exerted by the fluid moving with some average speed on a representative sphere in the array, given as a function of the volume fraction f of the spheres (Sangani and Acrivos, 1983). Primarily we are interested in the value of drag force $[K(f_c)]^{-1}$ at the maximum concentration.

As remarked by Zick and Homsy (1982), the Carman formula falls within the 15% of their results for at least one of the three of packing for concentrations greater than 0.5. This suggests similarity of the flow through randomly packed beds of spheres and periodic arrays of spheres. Carman formula is going to serve as a guide for constructing the initial approximation, which is going to be corrected by accounting for the information from the long series (Sangani and Acrivos, 1983), extracted by means of the diagonal Padé approximants. We obtain below an analytical formulas for the drag force $F(f)$ exerted by three most studied regular arrays.

7.3.1   BCC and SC lattices of spherical obstacles

For the BCC (body-centered cubic) lattice of obstacles Hasimoto series is given as follows,

$$K(f) = 1 - 1.79186 \sqrt[3]{f} + f - 0.329 f^2, \qquad (7.15)$$

and $f_c = \frac{\sqrt{3}\pi}{8}$. Let us follow the strategy for resummation developed by Gluzman and Yukalov (2016), Gluzman et al. (2017), leading to the corrected Padé approximants. With the simplest control function with fixed position of singularity and estimated critical index,

$$\mathcal{F}^*_{2,bcc}(f) = \frac{1}{\left(1 - \sqrt[3]{f}\right)^{1.79186}} \qquad (7.16)$$

one can ensure the numerical convergence of the sequence of Padé approximants.

We find that it is possible to correct (7.16) with the Padé approximant of high order,

$$F_{bcc}(f) = Cor^*_{bcc}(f) = \mathcal{F}^*_{2,bcc}(f)P_{bcc}(f), \qquad (7.17)$$

where

$P_{bcc}(f) = \frac{Q(f)}{W(f)},$

$Q(f) =$
$\quad 1 - 1.43733\sqrt[3]{f} + 2.90434 f^{2/3} - 2.82422 f + 3.42749 f^{4/3} - 0.64898 f^{5/3}$
$\quad + 2.8144 f^2 - 3.4264 f^{7/3} + 5.80474 f^{8/3} - 1.51059 f^3 + 5.91231 f^{10/3}$
$\quad - 2.87197 f^{11/3} + 4.68723 f^4 - 9.80947 f^{13/3} - 0.280041 f^{14/3} + 7.36241 f^5;$

$W(f) =$
$\quad 1 - 1.43733\sqrt[3]{f} + 2.19489 f^{2/3} - 2.12643 f + 1.7466 f^{4/3} + 0.64224 f^{5/3}$
$\quad + 0.836806 f^2 - 3.61494 f^{7/3} + 3.62097 f^{8/3} + 0.27101 f^3 + 2.32196 f^{10/3}$
$\quad - 3.13063 f^{11/3} + 0.27979 f^4 - 9.4974 f^{13/3} - 1.538 f^{14/3} + 11.221 f^5.$

$$(7.18)$$

From (7.22) one can evaluate the value of drag force at $f = f_c$, as $Cor^*_{BCC}(f_c) = 162.54$. It is in a good agreement with the numerical result of 163 (Zick and Homsy, 1982). Maximum deviation of the formula $Cor^*_{BCC}(f)$ from the results of Zick and Homsy (1982) equals 0.905%. The approximation (7.17) agrees well with the series from Sangani and Acrivos (1983), and less so with Carman expression (7.14), as shown in Fig. 7.3.

For the SC (simple cubic) lattice Hasimoto formula is given as follows,

$$K(f) = 1 - 1.76011 f^{1/3} + f - 1.5593 f^2, \qquad (7.19)$$

while $f_c = \frac{\pi}{6}$. Let us follow the same approach as for the BCC lattice and define the control function

$$\mathcal{F}^*_{2,sc}(f) = \frac{1}{\left(1 - \sqrt[3]{f}\right)^{1.76012}}, \qquad (7.20)$$

which is close to the control function of the BCC. The approximant (7.20) can be vastly improved with the high-order Padé approximant, exploiting high-order coefficients from Sangani and Acrivos (1983).

$$F_{sc}(f) = Cor^*_{sc}(f) = \mathcal{F}^*_{2,sc}(f)P_{sc}(f), \qquad (7.21)$$

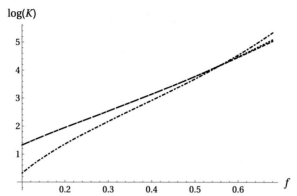

Figure 7.3 Formula (7.17) (dotted) for the drag force $F_{bcc}(f)$, is compared with the series from Sangani and Acrivos (1983) (dashed), and Carman approximation (7.14) (dot–dashed).

where

$$P_{sc}(f) = \frac{Q(f)}{W(f)}, \qquad (7.22)$$

with

$$
\begin{aligned}
Q(f) = &\; 1 - 0.933278\sqrt[3]{f} + 4.1886 f^{2/3} + 7.4086 f - 8.9357 f^{4/3} \\
&+ 29.4615 f^{5/3} - 27.4915 f^2 + 36.1936 f^{7/3} - 16.08 f^{8/3} \\
&+ 27.4624 f^3 - 14.237 f^{10/3} + 50.3806 f^{11/3};
\end{aligned}
$$

$$
\begin{aligned}
W(f) = &\; 1 - 0.933278\sqrt[3]{f} + 3.51965 f^{2/3} + 7.802 f - 11.4977 f^{4/3} \\
&+ 23.7119 f^{5/3} - 24.4125 f^2 + 18.4205 f^{7/3} - 5.02996 f^{8/3} \\
&- 6.40273 f^3 - 10.3503 f^{10/3} + 37.0912 f^{11/3}.
\end{aligned}
$$

$$(7.23)$$

The value of the drag force at $f = f_c$, found from (7.21), equals $\mathrm{Cor}^*_{sc}(f_c) = 41.418$, and is in a good agreement with the numerical result 42.1 (Zick and Homsy, 1982). Maximum deviation of the $\mathrm{Cor}^*_{sc}(f)$ from the numerical results from Zick and Homsy (1982) is equal to 2.03%. Formula (7.21) is shown in Fig. 7.4 and compared with the series from Sangani and Acrivos (1983) and formula (7.14). Interesting that an accurate Padé approximant for the nondimensional drag force can be found for the SC lattice,

$$F_{sc}(f) = \frac{P(f)}{Q(f)},$$

$$
\begin{aligned}
P(f) = &\; 3.03506 f^{2/3} + 6.48454 f^{4/3} + 6.95235 f^{5/3} - 8.10749 f^{7/3} \\
&+ 30.1647 f^{8/3} + 47.8278 f^{10/3} + 0.983938 f^{11/3} + 45.4896 f^4 \\
&- 4.90258 f^3 + 15.6135 f^2 + 0.0365858\sqrt[3]{f} + 0.472782 f + 1, \quad (7.24)
\end{aligned}
$$

$$
\begin{aligned}
Q(f) = &\; 2.97067 f^{2/3} + 5.68898 f^{4/3} - 1.45556 f^{5/3} - 29.3381 f^{7/3} \\
&+ 50.872 f^{8/3} + 47.4423 f^{10/3} - 61.6716 f^{11/3} + 31.928 f^4 \\
&- 43.291 f^3 + 2.39245 f^2 - 1.72353\sqrt[3]{f} - 3.86929 f + 1,
\end{aligned}
$$

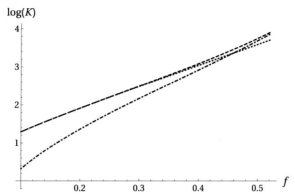

Figure 7.4 Formula (7.21) (dotted) for the drag force $F_{sc}(f)$ is compared with the series from Sangani and Acrivos (1983) (dashed), and Carman approximation (7.14) (dot–dashed).

with maximum deviation 1.026%. It also turns out that although for the BCC lattice one can similarly construct the Padé formula, but its maximum deviation from the numerical data 2.187%, gets worse than corresponding quantity for the formula (7.17).

Most importantly, the direct application of the Padé approximants completely fails in the case of FCC (face-centered cubic) lattice. The accuracy and behavior of the approximants get close to the original series from Sangani and Acrivos (1983), and the error is equal to 35.33% at $f = f_c$. If for SC lattice the influence of Carman singularity on the behavior in the physical region can be ignored, in the case of FCC it becomes crucial. BCC lattice represents an intermediate case.

7.3.2 Formula for FCC lattice

Hasimoto also obtained a low-concentration formula for the FCC lattice of obstacles, with $f_c = \frac{\sqrt{2}\pi}{6}$, characterizing slow fluid flow past spheres located on the sites of the lattice. For the dimensionless fluid permeability (Hasimoto, 1959, Torquato, 2002), he obtained

$$K(f) = 1 - 1.79175\sqrt[3]{f} + f - 0.302f^2, \tag{7.25}$$

while much more terms in the series can be found in Sangani and Acrivos (1983). The simple control function employed for BCC and SC lattices does not work for the FCC. Let us employ same idea as above, but take as the control function the higher-order factor approximant. It is a convenient tool when one wants to match an inverse for (7.25), and also estimate parameters of singularity at $f = 1$. The factor approximant can be obtained readily,

$$\mathcal{F}^*_{4,fcc}(f) = \frac{1}{\left(1 - \sqrt[3]{f}\right)^{2.86402} \left(0.322995\sqrt[3]{f} + 1\right)^{3.31977}}. \tag{7.26}$$

It uses the terms up to f^1 inclusively, and predicts the critical index close to that of Carman. The approximant (7.26) is remarkable, since its expansion can produce four more coefficients in the small f expansions with rather good accuracy, in average better than 1%. Indeed, the following expansion of the factor approximant

$$\mathcal{F}^*_{4,fcc}(f) \simeq 1 + 1.79175\sqrt[3]{f} + 3.21036 f^{2/3} + 4.75215 f$$
$$+ 6.66309 f^{4/3} + 8.81575 f^{5/3} + 11.259 f^2 + 13.9647 f^{7/3} + \ldots, \tag{7.27}$$

compares well with the series for the drag force from Sangani and Acrivos (1983),

$$\mathcal{F}(f) \simeq 1 + 1.79175 f^{1/3} + 3.21036 f^{2/3} + 4.75215 f$$
$$+ 6.72291 f^{4/3} + 8.83539 f^{5/3} + 11.3807 f^2 + 14.2098 f^{7/3} + \ldots. \tag{7.28}$$

Such approach is analogous to selection of the best equation of state, which is often based on the idea of comparison of the predicted coefficients in the expansion with the known coefficients, as discussed in Gluzman and Yukalov (2016). Note that Carman formula (7.14) reminds the popular Carnahan–Starling equation of state (see e.g. Gluzman and Yukalov (2016) and Section 8.1 below).

The factor approximant (7.26) can be further corrected with high-order Padé approximant, exploiting high-order coefficients from Sangani and Acrivos (1983),

$$F_{fcc}(f) = \mathbf{Cor}^*_{fcc}(f) = \mathcal{F}^*_{4,fcc}(f) P_{fcc}(f), \tag{7.29}$$

where

$$P_{fcc}(f) = \frac{Q(f)}{W(f)}, \tag{7.30}$$

$$Q(f) = 1 -$$
$$8.59556\sqrt[3]{f} + 6.11157 f^{2/3} - 17.5659 f + 22.0293 f^{4/3} - 39.4082 f^{5/3}$$
$$+ 41.0955 f^2 - 45.5243 f^{7/3} + 29.1888 f^{8/3} - 22.1831 f^3 - 3.41307 f^{10/3}$$
$$+ 8.19781 f^{11/3} + 26.781 f^4 - 74.5148 f^{13/3},$$

$$W(f) = 1 -$$
$$8.59556\sqrt[3]{f} + 6.11157 f^{2/3} - 17.5659 f + 21.9695 f^{4/3} - 38.8065 f^{5/3}$$
$$+ 39.891 f^2 - 43.2816 f^{7/3} + 26.576 f^{8/3} - 17.008 f^3 - 11.0216 f^{10/3}$$
$$+ 19.1284 f^{11/3} + 26.6879 f^4 - 62.5965 f^{13/3}.$$

$$\tag{7.31}$$

The estimated value at the critical point, $\mathbf{Cor}^*_{fcc}(f_c) = 441.334$, is in reasonable agreement with the numerical result 435 (Zick and Homsy, 1982). Maximum deviation of the $F^*_{fcc}(f)$ from the numerical results is equal to 2.53%. Note that numerical value 435, is larger than available experimental value of 398, and another available estimate of 412 (Zick and Homsy, 1982). Let us bring up one more approximant which is in the same mold as (7.29), but turns out to agree better with all three estimates,

$$F_{fcc}(f) = \mathbf{Cor}^*_{fcc}(f) = \mathcal{F}^*_{4,fcc}(f) P_{fcc}(f) \tag{7.32}$$

where

$$P_{fcc}(f) = \frac{Q_1(f)}{W_1(f)} \tag{7.33}$$

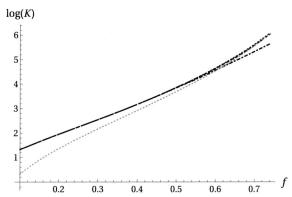

Figure 7.5 Formulas (7.29) (dotted) and (7.32) (dashed) for the drag force $F_{fcc}(f)$, are compared with the series from Sangani and Acrivos (1983) (dot–dashed) and Carman approximation (7.14) (gray–dotted).

and

$$Q_1(f) = 1 +$$
$$4.15034\sqrt[3]{f} + 0.37461\,f^{2/3} + 1.56139\,f - 0.0805799\,f^{4/3} - 2.42314\,f^{5/3}$$
$$+ 15.5682\,f^2 - 9.31077\,f^{7/3} + 40.7622\,f^{8/3} - 44.9789\,f^3 + 79.1364\,f^{10/3}$$
$$- 95.7832\,f^{11/3} + 116.049\,f^4 - 67.774\,f^{13/3} + 89.7841\,f^{14/3} - 48.2428\,f^5$$
$$- 68.3451\,f^{16/3} - 1.51239\,f^{17/3} + 266.478\,f^6,$$

$$W_1(f) = 1 +$$
$$4.15034\sqrt[3]{f} + 0.37461\,f^{2/3} + 1.56139\,f - 0.140395\,f^{4/3} - 2.58386\,f^{5/3}$$
$$+ 15.8226\,f^2 - 9.81739\,f^{7/3} + 40.536\,f^{8/3} - 44.7884\,f^3 + 77.8792\,f^{10/3}$$
$$- 96.7872\,f^{11/3} + 105.929\,f^4 - 65.0472\,f^{13/3} + 79.7924\,f^{14/3} - 28.0273\,f^5$$
$$- 84.5372\,f^{16/3} + 6.32732\,f^{17/3} + 228.662\,f^6.$$

$$(7.34)$$

Formula (7.32) maximally deviates from the numerical results of Zick and Homsy (1982) by 5.48%, while $\mathbf{Cor}^*_{fcc}(f_c) = 420.465$ is in between the numerical and experimental results. On logarithmic scale the two formulas are practically indistinguishable, as shown in Fig. 7.5. They do respect the series from Sangani and Acrivos (1983) at smaller f and tend to the Carman formula for larger f.

7.4 Permeability in wavy-walled channels

In Ref. (Malevich et al., 2006) a general asymptotic analysis was applied to a Stokes flow in curvilinear three-dimensional channels bounded by walls of the form (see Fig. 7.6)

$$z = S^+(x_1, x_2) \equiv b\big(1 + \varepsilon T(x_1, x_2)\big), \tag{7.35}$$

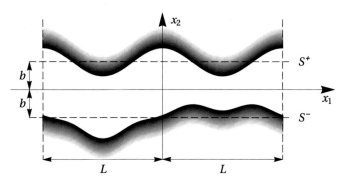

Figure 7.6 Curvilinear three-dimensional channels bounded by walls of the form (7.35), (7.36).

$$z = S^-(x_1, x_2) \equiv -b\big(1 + \varepsilon B(x_1, x_2)\big). \tag{7.36}$$

In what follows, the formally small dimensionless parameter $\varepsilon \geq 0$ is employed to present the form (7.35) as the perturbation around the straight channel.

Such approach generalizes an original Pozrikidis work on two-dimensional channels (Pozrikidis, 1987). Recently, the approach was extended to the gravity driven Stokesian flow past the wavy bottom (Wojnar and Bielski, 2015, 2018), to wavy tubes (Bielski and Wojnar, 2018) and to the Navier–Stokes equations (Adler et al., 2013, Rasoulzadeh and Panfilov, 2018), to the stationary heat conduction (Czapla et al., 2010), to electrokinetic phenomena in two-dimensional channels (Mityushev and Rylko, 2010). An arbitrary profiles $S^\pm(x_1, x_2)$ were considered, satisfying some natural conditions. For definiteness, it was assumed that

$$|T(x_1, x_2)| \leqslant 1 \quad \text{and} \quad |B(x_1, x_2)| \leqslant 1. \tag{7.37}$$

For an infinitely differentiable $T(x_1, x_2)$ and $B(x_1, x_2)$, in order to calculate velocities and permeability, one has to solve a cascade of boundary value problems for the Stokes equations for a straight channel (Malevich et al., 2006). Influence of curvilinear edges on the flow is of fundamental interest, since it illustrates the mechanism of viscous flow under different geometrical conditions. Apart from its theoretical importance, the flow through curvilinear channels has application in porous media (Adler, 1992, Adler and Thovert, 1999).

General expression for permeability as a function of ε can be deduced and the limiting value when eddies arise can be estimated as well. Application of the method of perturbation on ε turned out to be an efficient way to solve the problem, i.e. to calculate velocities and permeability in the form of an ε-expansion.

Let $\mathbf{u} = \mathbf{u}(x_1, x_2, x_3)$ be the velocity vector, and $p = p(x_1, x_2, x_3)$ the pressure. The flow of a viscous fluid through a channel is considered under the assumption that the Reynolds number is small enough for Stokes flow approximation to be made. Thus the fluid is governed by the Stokes equations

$$\begin{aligned} \mu\nabla^2\mathbf{u} &= \nabla p, \\ \nabla \cdot \mathbf{u} &= \mathbf{0} \end{aligned} \tag{7.38}$$

with the boundary conditions

$$\mathbf{u} = \mathbf{0} \quad \text{on} \quad S^{\pm}. \tag{7.39}$$

The solution $\mathbf{u}$ of (7.38)–(7.39) belongs to the class of periodic functions with period $2L$ in x_1 and x_2.

Let also u be the x-component of $\mathbf{u}$. And let us apply an overall external gradient pressure $\overline{\nabla p}$ along the x_1-direction. It can be described by a constant jump $2L\overline{\nabla p}$ along the x_1-axis of the periodic cell. The permeability of the channel in the x_1-direction is defined as follows

$$K_{x_1}(\varepsilon) = -\frac{\mu}{\overline{\nabla p}\,|\tau|} \int\limits_{-L}^{L} \int\limits_{-L}^{L} dx_1\, dx_2 \int\limits_{S^-(x_1,x_2)}^{S^+(x_1,x_2)} u(x_1, x_2, x_3)\, dx_3, \tag{7.40}$$

where $|\tau|$ is the volume of the unit cell Q of the channel,

$$|\tau| = \int\limits_{-L}^{L} \int\limits_{-L}^{L} dx_1\, dx_2 \int\limits_{S^-(x_1,x_2)}^{S^+(x_1,x_2)} dx_3. \tag{7.41}$$

$K_{x_1}(\varepsilon)$ in (7.40) is considered as a function in ε. For convenience the ratio $K = K(\varepsilon)$ of the dimensional permeability for the curvilinear channel and of the Poiseuille flow

$$K(\varepsilon) = \frac{K_{x_1}(\varepsilon)}{K_{x_1}(0)}, \tag{7.42}$$

is considered. The case $\varepsilon = 0$ corresponds to the well-known Poiseuille flow, for which (7.40) yields the permeability

$$K_{x_1}(0) = \frac{b^2}{3}, \tag{7.43}$$

as the flow profile in this limit-case obeys the parabolic law.

More precisely, the formula derived by Malevich et al. (2006) determines the coefficients of a Taylor expansion for the permeability

$$K(\varepsilon) = \sum_{m=0}^{\infty} c_m \varepsilon^m$$

and normalization (7.42) is used for K. In practical computations $K(\varepsilon)$ is approximated by the Taylor polynomial

$$K_N(\varepsilon) = \sum_{m=0}^{N} c_m \varepsilon^m.$$

The domain of application of this formula is restricted, since the corresponding Taylor series can be divergent for larger ε.

7.4.1 Symmetric sinusoidal two-dimensional channel. Breakdown of lubrication approximation

Let us consider the two-dimensional channel bounded by the surfaces

$$z = b(1 + \varepsilon \cos x), \quad z = -b(1 + \varepsilon \cos x). \tag{7.44}$$

The permeability was calculated up to $O(\varepsilon^{32})$, for $b = 0.5$. This example is typical among the references cited in Malevich et al. (2006). The following series for permeability as the function of "waviness" parameter ε was obtained,

$$
\begin{aligned}
K_{30}(\varepsilon) = \\
1 &- 3.14963\varepsilon^2 + 4.08109\varepsilon^4 - 3.48479\varepsilon^6 + 2.93797\varepsilon^8 - 2.56771\varepsilon^{10} \\
&+ 2.21983\varepsilon^{12} - 1.93018\varepsilon^{14} + 1.67294\varepsilon^{16} - 1.45302\varepsilon^{18} + 1.26017\varepsilon^{20} \\
&- 1.09411\varepsilon^{22} + 0.949113\varepsilon^{24} - 0.823912\varepsilon^{26} + 0.714804\varepsilon^{28} - 0.620463\varepsilon^{30} \\
&+ O(\varepsilon^{32}).
\end{aligned}
\tag{7.45}
$$

Also, a lubrication approximation in the case of two cylinders of different radii that are almost in contact with one another along a line was discussed by Adler (1992). For equal radii of cylinders a, the flow rate q per unit length is proportional to the pressure variation Δp

$$q = -\frac{K_l}{\mu}\Delta p, \tag{7.46}$$

where K_l is given by

$$K_l = \frac{2}{9\pi}\sqrt{\frac{\delta^5}{a}} \tag{7.47}$$

and δ is the gap between the cylinders.

For the channel (7.44), if ε is close to unity, the aperture at $x = -\pi$ is close to zero. Hence, one can apply (7.47) to this local channel with $\delta = 2b(1 - \varepsilon)$ and $a = b\varepsilon$. As $\varepsilon \to \varepsilon_c = 1$, one would simply have the following expression,

$$K_l \simeq \frac{8\sqrt{2}\sqrt{b^4}(1 - \varepsilon)^{5/2}}{9\pi}, \tag{7.48}$$

which matches the general critical form with the critical index for permeability $\varkappa = 5/2$, and the critical amplitude $A = \frac{8\sqrt{2}b^2}{9\pi}$. Thus, in the case under consideration $A = 0.100035$.

The lubrication approximation (Adler, 1992) considers situations when the amplitude of the wall oscillations is smaller than the channel width (Scholle, 2004). It also turns out that the channel width should be small when compared to a characteristic length of the channel, i.e., $\varepsilon \ll b \ll 2\pi$. The main assumption of the lubrication approximation is that the velocity has a parabolic profile. It is demonstrated for the plane channels (Malevich et al., 2006), that lubrication approximation gives correct results only for channels in which the mean surface

$$S(x_1, x_2) = b + \frac{\varepsilon}{2}\big(T(x_1, x_2) - B(x_1, x_2)\big)$$

is sufficiently close to a plane and for small value of ε.

Our main goal is to avoid using the lubrication approximation and find the way to approach the critical region (walls nearly touch) only based on (7.45).

The problem of interest can be formulated mathematically as follows. Given the polynomial approximation (7.45) of the function $K(\varepsilon)$, to determine critical index and amplitude(s) of the asymptotically equivalent approximation near $\varepsilon = \varepsilon_c$ for the permeability

$$K(\varepsilon) \simeq A(1 - \varepsilon)^{\varkappa}.$$

When such extrapolation problem is solved, we proceed to solve an interpolation problem, assuming that the critical behavior is known in advance and derive the compact formula for all ε.

Standard way to proceed with critical index calculations when the value of the threshold is known (Gluzman and Yukalov, 2014). One would first apply the following transformation widely used throughout the book,

$$z = \frac{\varepsilon}{\varepsilon_c - \varepsilon} \Leftrightarrow \varepsilon = \frac{z\varepsilon_c}{z + 1},$$

to the series (7.45) in order to make application of the different approximants more convenient.

Then, to such transformed series $M_1(z)$ one has to apply the $DLog$ transformation (differentiate Log of $M_1(z)$), and let us call the transformed series $M(z)$. In terms of $M(z)$ we can readily obtain the sequence of approximations $\varkappa_n$ for the critical index $\varkappa$,

$$\varkappa_n = -\lim_{z \to \infty} (zPadeApproximant[M[z], n, n + 1]). \tag{7.49}$$

There is an excellent convergence within the approximations for the critical index generated by the sequence of Padé approximants with their order increasing, to the value $5/2$:

$$\varkappa_1 = 2.57972, \quad \varkappa_2 = 2.30995, \quad \varkappa_3 = 2.47451, \quad \varkappa_4 = 2.49689,$$

$$\varkappa_5 = 2.4959, \quad \varkappa_6 = 2.49791, \quad \varkappa_7 = 2.49923, \quad \varkappa_8 = 2.50113,$$

$$\varkappa_9 = 2.50028, \quad \varkappa_{10} = 2.49783, \quad \varkappa_{11} = 2.49778,$$

$$\varkappa_{12} = 2.49829, \quad \varkappa_{13} = 2.49836.$$

If $B_n(z) = Pade\,Approximant[M[z], n, n+1]$, then

$$K_n^*(\varepsilon) = \exp\left(\int_0^{\frac{\varepsilon}{\varepsilon_c - \varepsilon}} B_n(z)\,dz\right), \tag{7.50}$$

and one can compute the corresponding amplitude numerically

$$A_n = \lim_{\varepsilon \to \varepsilon_c} (\varepsilon_c - \varepsilon)^{-\varkappa_n} K_n^*(\varepsilon), \tag{7.51}$$

with $A_9 = 3.7758$, by order of magnitude larger than the value anticipated from the lubrication approximation. With critical index fixed to 5/2, one can calculate A using standard Padé technique (Bender and Boettcher, 1994), leading to the very close value of 3.77188.

In the compact form permeability can be expressed in terms of factor approximant (Gluzman and Yukalov, 2014), which is asymptotically equivalent to (7.45) up to 16th order inclusively,

$$K_{1/2}^*(\varepsilon) = \frac{\left(1 - \varepsilon^2\right)^{2.5} \left(0.239311\varepsilon^2 + 1\right)^{0.591597}}{\left(1 - 0.722851\varepsilon^2\right)^{0.00840612} \left(1 - 0.260764\varepsilon^2\right)^{0.270545} \left(0.867799\varepsilon^2 + 1\right)^{1.00004}}, \tag{7.52}$$

with amplitude $A = 3.77177$. Mind that there are "spare", higher order coefficients, not employed in construction of (7.52). The maximal error in reproducing these higher-order coefficients defines the accuracy of the studied crossover formula. From the formula (7.52) one can readily obtain the higher-order coefficients (7.45), not employed in the final formula,

$$c_{18} = -1.453, \quad c_{20} = 1.26014, \quad c_{22} = -1.09408, \quad c_{24} = 0.949078,$$

$$c_{26} = -0.823874, \quad c_{28} = 0.714764, \quad c_{30} = -0.620422.$$

Formula (7.52) appears to be exceptionally accurate in reproducing the coefficients in the expansion (7.45) not employed in its construction. The maximal error appears in reproducing the 30th order, and it equals just 0.0066%. We conclude that the lubrication approximation breaks down even in a close vicinity of ε_c, as anticipated in Malevich et al. (2006). Note that lubrication theory gives correct estimate for the critical index. In Fig. 7.7 formula (7.52) is compared with the asymptotic regimes.

7.4.2 Symmetric sinusoidal two-dimensional channel. Breakdown continued

Let us consider now the channel bounded by the surfaces (7.44) with $b = 0.25$. $K(\varepsilon)$ was obtained in Malevich et al. (2006),

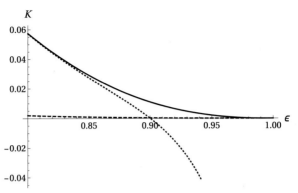

Figure 7.7 Formula (7.52) (solid) is compared with the power series (dotted) and lubrication approximation (dashed).

$$
\begin{aligned}
K(\varepsilon) = {}& 1 - 3.03748\varepsilon^2 + 3.54570\varepsilon^4 - 2.33505\varepsilon^6 + 1.35447\varepsilon^8 - 0.83303\varepsilon^{10} \\
& + 0.49762\varepsilon^{12} - 0.30350\varepsilon^{14} + 0.18185\varepsilon^{16} - 0.11083\varepsilon^{18} \\
& + 0.06636\varepsilon^{20} - 0.04051\varepsilon^{22} + 0.02419\varepsilon^{26}0.00880\varepsilon^{28} - 0.00544\varepsilon^{30} \\
& + O(\varepsilon^{32}).
\end{aligned}
\tag{7.53}
$$

There is an excellent convergence within the approximations for the critical index generated by the sequence of Padé approximants, corresponding to their order increasing:

$$\varkappa_1 = 2.64456, \quad \varkappa_2 = 2.41346, \quad \varkappa_3 = 2.49488, \quad \varkappa_4 = 2.49992,$$

$$\varkappa_5 = 2.49991, \quad \varkappa_6 = 2.50026, \quad \varkappa_7 = 2.50068, \quad \varkappa_8 = 2.50087,$$

$$\varkappa_9 = 2.50086, \quad \varkappa_{10} = 2.50063, \quad \varkappa_{11} = 2.50063, \quad \varkappa_{12} = 2.50086,$$

$$\varkappa_{13} = 2.50087, \quad \varkappa_{14} = 2.50068, \quad \varkappa_{15} = 2.50026.$$

Evidently this sequence implies the same value for the index, $\varkappa = 5/2$. The value of amplitude is estimated by $A_{15} = 3.77362$. Both amplitude and index are in complete agreement with the results for different b obtained above.

With critical index fixed to $5/2$, one can calculate the amplitude A, using standard Padé technique (Bender and Boettcher, 1994), leading to the very close value of $A \approx 3.77316$. In the compact form permeability can be expressed through the following factor approximant

$$
K_{1/4}^* = \frac{\left(1 - \varepsilon^2\right)^{2.5}}{\left(1 - 0.0437141\varepsilon^2\right)^{1.37166} \left(0.606745\varepsilon^2 + 1\right)^{0.984665}},
\tag{7.54}
$$

with amplitude calculated as $A = 3.77062$. This value is by orders of magnitude different from the value 0.02501, suggested by the lubrication approximation. From the crossover formula (7.54) one can estimate the higher-order coefficients from (7.53),

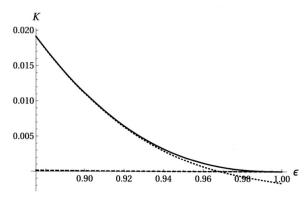

Figure 7.8 Formula (7.54) (solid) is compared with the power series (dotted) and lubrication approximation (dashed).

not employed in the derivation,

$$c_{10} = -0.833087, \quad c_{12} = 0.497914, \quad c_{14} = -0.304065, \quad c_{16} = 0.1825,$$

$$c_{18} = -0.111442, \quad c_{20} = 0.0668887, \quad c_{22} = -0.0409299, \quad c_{24} = 0.0245076,$$

$$c_{26} = -0.0150614, \quad c_{28} = 0.00896906, \quad c_{30} = -0.00555677.$$

Formula (7.52) appears to be accurate in reproducing the coefficients in the expansion (7.53), even not employed in its construction. The maximal error is in reproducing c_{30}, and it is equal to 2.147%. In Fig. 7.8 formula (7.54) is compared with the asymptotic regimes.

The amplitude and overall behavior of permeability in the vicinity of ε_c, practically does not depend on the parameter b. One can think that some universal (not dependent on b) mechanism is at work here, the chief suspect being celebrated similarity-solutions with complex exponent, known as viscous Moffat eddies (Moffatt, 1964), obviously not covered by the lubrication theory (Scholle, 2004). Eddies manifest themselves as reversed-flow regions near the walls. Onset of eddies is expected in the vicinity of ε_e, corresponding to zero of polynomials (7.45), (7.53), and would lead to a total disappearance of permeability, an artifact being corrected by a detailed consideration of the region $\varepsilon \sim 1$ with a power-law ansatz. With b decreasing, the value of ε_e moves closer to ε_c.

Due to Moffat it is understood that steady two-dimensional low-Reynolds number sharp corner flow between two fixed plane rigid boundaries has a remarkably universal form. The motion is being driven by some arbitrary stirring mechanism far from the corner. We largely follow below Moffat's brilliant, even poetic description of the eddies. The stream function ought to satisfy the biharmonic equation which has only complex solutions. The velocity components appear to oscillate infinitely albeit quite strongly damped, implying the existence of an infinitely reversing sequence of eddies as the corner is approached. The solution is truly universal, providing the asymptotic form of the generic two-dimensional flow near any sharp corner, irrespective of the nature of the remote forcing. Even if the remote flow conditions are in the high-Reynolds-number regime, the local Reynolds number near any perfectly sharp corner is always small, so that the low-Reynolds-number eddies are always present, but on an extremely small scale in the immediate vicinity of the corner! It is, however, quite clear that, although the mathematics implies an infinite geometrical sequence of eddies, not more than two or three of these eddies will ever be observed in practice. In reality the corner may be not quite sharp, in which case the infinite sequence of eddies is simply replaced by a finite sequence, the number depending in an obvious way on the degree of rounding of the corner. It appears that the tiny corner eddies, too weak to be observed under steady conditions, do play a role when the remote conditions are time-periodic: they emerge successively from the corner, one in each half-period, growing in stature like Kabuki actors on a stage, and ultimately taking the lead role (Moffatt, 1964)!

7.4.3 Parallel sinusoidal two-dimensional channel. Novel critical index

Consider yet different channel bounded by the surfaces

$$z = b(1 + \varepsilon \cos x), \quad z = -b(1 - \varepsilon \cos x), \tag{7.55}$$

with $b = 0.5$. There is no possibility of the walls touching and permeability remains finite, but is expected to decay as a power-law as ε becomes large. Instead of a critical transition from permeable to non-permeable phase, we have a non-critical transition, or crossover from high-to low permeability with increasing parameter ε. The crossover can be still characterized by the power-law, with corresponding index at large ε. There are no eddies in such channel even for very large ε, and lubrication approximation does not work at all (Malevich et al., 2006). However, for large b eddies are not excluded.

The permeability is calculated up to $O(\varepsilon^{32})$,

$$
\begin{aligned}
K_{30}(\varepsilon) = {} & \\
& 1 - 2.53686 \times 10^{-1}\varepsilon^2 + 4.28907 \times 10^{-2}\varepsilon^4 - 5.46188 \times 10^{-3}\varepsilon^6 \\
& + 4.54695 \times 10^{-4}\varepsilon^8 + 9.0656 \times 10^{-6}\varepsilon^{10} - 1.41572 \times 10^{-5}\varepsilon^{12} \\
& + 3.76584 \times 10^{-6}\varepsilon^{14} - 6.72021 \times 10^{-7}\varepsilon^{16} + 7.58331 \times 10^{-8}\varepsilon^{18} \\
& + 2.34495 \times 10^{-9}\varepsilon^{20} - 4.59993 \times 10^{-9}\varepsilon^{22} + 1.88446 \times 10^{-9}\varepsilon^{24} \\
& - 8.6005 \times 10^{-11}\varepsilon^{26} + 3.34156 \times 10^{-9}\varepsilon^{28} + 1.63748 \times 10^{-9}\varepsilon^{30}.
\end{aligned}
\tag{7.56}
$$

The velocity is analytic in ε in the disk $|\varepsilon| < \varepsilon_0$. Therefore, (7.56) is valid for $\varepsilon < \varepsilon_0$, where ε_0 is of order $\frac{1}{b\chi}$, with χ being the maximal wave number of $T(x_1, x_2)$ and $B(x_1, x_2)$. In order to calculate $K(\varepsilon)$ for $\varepsilon \geqslant \varepsilon_0$, one can apply Padé approximation to the polynomial (7.56) which agrees up to $O(\varepsilon^{32})$. The Padé approximant of the order $(10, 20)$ was written down by Malevich et al. (2006),

$$K_{10,20}(\varepsilon) = \frac{P_{10}(\varepsilon)}{Q_{20}(\varepsilon)}, \tag{7.57}$$

where

$$P_{10}(\varepsilon) = 1 - 3.14215\varepsilon^2 + 6.59346\varepsilon^4 + 34.7591\varepsilon^6 + 13.3065\varepsilon^8 + 1.53446\varepsilon^{10},$$

$$Q_{20}(\varepsilon) = 1 - 2.88846\varepsilon^2 + 5.81781\varepsilon^4 + 36.3643\varepsilon^6 + 22.2659\varepsilon^8 + 5.65641\varepsilon^{10}$$
$$+ 0.675967\varepsilon^{12} + 0.033858\varepsilon^{14} + 0.000131\varepsilon^{16} - 0.000010\varepsilon^{18} + 0.000001\varepsilon^{20}.$$
$$\tag{7.58}$$

This approximant yields $K_{10,20}(\varepsilon) \sim \varepsilon^{-10}$, as $\varepsilon \to \infty$. More generally one can conjecture that permeability decays as $K(\varepsilon) \sim \varepsilon^{\varkappa}$, as $\varepsilon \to \infty$, with some unknown negative exponent (index) $\varkappa$. When a function $K(\varepsilon)$ at asymptotically large variable behaves as

$$K(\varepsilon) \simeq B\varepsilon^{\varkappa}, \quad \text{as } \varepsilon \to \infty, \tag{7.59}$$

then the index can be represented by the limit

$$\varkappa = \lim_{\varepsilon \to \infty} \varepsilon \frac{d}{d\varepsilon} \log K(\varepsilon). \tag{7.60}$$

Assuming that the small-variable expansion for the function is given by the truncated sum $K_{tr}(\varepsilon)$, as in (7.56), we have the corresponding small-variable expression for the effective critical exponent

$$N(\varepsilon) \equiv \varepsilon \frac{d}{d\varepsilon} \log K_{tr}(\varepsilon). \tag{7.61}$$

Applying to the obtained series $N(\varepsilon)$ the method of Padé approximants, as has been discussed above in Subsection 1.3.2, the sought approximate expression for the critical exponent arises,

$$\varkappa_k = \lim_{\varepsilon \to \infty} \varepsilon P_{k,k+1}(\varepsilon), \tag{7.62}$$

which naturally depends on the approximation order k. Note that the value of the critical amplitude B does not enter the consideration at all. Application of the method to series of interest (7.56), is pretty much straightforward and suggests strongly the value of $\varkappa = -4$, as can be seen from Fig. 7.9. The amplitude B, corresponding to $k = 14$, is equal to 44.5872. Assuming that $\varkappa = -4$, we construct for the original series the sequence of Padé approximants $P_{n,n+4}$. There is a good convergence in approximation sequence for the amplitude B, to the value of 43.2 as shown in Fig. 7.10.

As the expression for permeability we suggest the Padé approximation of the order (12, 16).

$$K_{12,16}(\varepsilon) = \frac{P_{12}(\varepsilon)}{Q_{16}(\varepsilon)}, \tag{7.63}$$

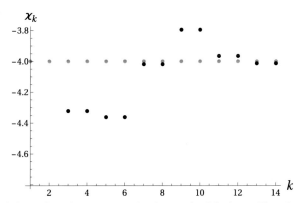

Figure 7.9 The index $\varkappa$ dependence on approximation number k is shown. The values calculated from (7.62), are shown with circles, and compared with the most plausible value of -4 (shown with squares).

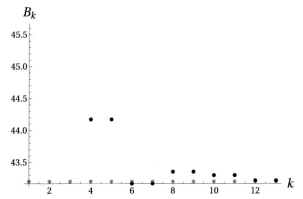

Figure 7.10 The amplitude B dependence on approximation number k is shown with the convergence to the value of 43.2.

where

$$P_{12}(\varepsilon) = 1 - 5.86404\varepsilon^2 - 3.84897\varepsilon^4 + 1.12295\varepsilon^6 + 0.867771\varepsilon^8 \\ + 0.151922\varepsilon^{10} + 0.00735283\varepsilon^{12},$$

$$Q_{16}(\varepsilon) = 1 - 5.61035\varepsilon^2 - 5.31512\varepsilon^4 + 0.0206681\varepsilon^6 + 1.06989\varepsilon^8 \\ + 0.395962\varepsilon^{10} + 0.0645092\varepsilon^{12} + 0.0051812\varepsilon^{14} + 0.000170141\varepsilon^{16}.$$

$$(7.64)$$

Comparison of the two Padé approximants is presented in Fig. 7.11. Remarkably, the permeability can be expressed also through the factor approximant of relatively low order, which uses terms up to the 14th order,

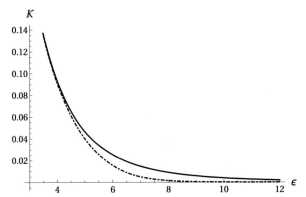

Figure 7.11 The two Padé approximants for permeability, $K_{10,20}$ (dot–dashed) and $K_{12,16}$ (solid) are compared.

$$\mathcal{F}_{14}^*(\varepsilon) =$$
$$\left(1 + (0.0925028 - 0.0501527i)\varepsilon^2\right)^{-0.957209 - 0.738327i}$$
$$\times \left(1 + (0.0925028 + 0.0501527i)\varepsilon^2\right)^{-0.957209 + 0.738327i} \tag{7.65}$$
$$\times \left(1 + (0.218267 - 0.101021i)\varepsilon^2\right)^{-0.042791 + 0.0798889i}$$
$$\times \left(1 + (0.218267 + 0.101021i)\varepsilon^2\right)^{-0.042791 - 0.0798889i},$$

which is very close to $K_{12,16}$, thus testifying for a high-quality of the original series. The factor approximant (7.65) is "smart", since it can predict three more coefficients, c_{16}, c_{18}, c_{20}, with average accuracy better than 1%. From (7.65) it also follows the value of amplitude $B = 43.3$, in agreement with calculations above. Note that the condition $\varkappa = -4$ was imposed in (7.65). Alternatively, we can use instead one more term from the expansion and calculate the index from the corresponding factor approximant with practically unchanged value, $\varkappa \approx -4.02$.

7.4.4 Symmetric sinusoidal three-dimensional channel. Two-fluid model

Following (Malevich et al., 2006) let us consider the channel restricted by the surfaces

$$z = \pm b\left(1 + \frac{1}{2}\varepsilon\big(\cos(x + y) + \cos(x - y)\big)\right), \tag{7.66}$$

with $b = 0.3$. The permeability is calculated up to $O(\varepsilon^{14})$

$$K_{14}(\varepsilon) = 1 - 0.465674\varepsilon^2 + 0.329218\varepsilon^4 - 0.261666\varepsilon^6 - 0.004467\varepsilon^8$$
$$- 0.0386987\varepsilon^{10} - 0.0177808\varepsilon^{12} - 0.0239319\varepsilon^{14}. \tag{7.67}$$

For $\varepsilon = \varepsilon_c = 1$, the surfaces (7.66) start touching, though the permeability remains finite at ε_c. The permeability series (7.67) is obtained with numerical precision of 10^{-3}

for values of ε up to 0.61. The "seepage" value of permeability at ε_c is considerable, $K_{14}(\varepsilon_c) = 0.517$, as is simply estimated from the series (7.67).

In the case being studied, application of the diagonal Padé approximants to (7.67), brings the following results $P_{6,6}(\varepsilon_c) = 0.51277$, $P_{8,8}(\varepsilon_c) = 0.490636$. The full form of the higher order Padé approximant is given below,

$$P_{6,6}(\varepsilon) = \frac{-0.272534\varepsilon^6 + 0.22825\varepsilon^4 - 0.657553\varepsilon^2 + 1}{-0.0363255\varepsilon^6 - 0.190321\varepsilon^4 - 0.191879\varepsilon^2 + 1};$$
$$P_{8,8}(\varepsilon) = \frac{-0.266547\varepsilon^8 - 0.131478\varepsilon^6 - 0.363105\varepsilon^4 + 0.256413\varepsilon^2 + 1}{-0.0832011\varepsilon^8 - 0.273346\varepsilon^6 - 0.356065\varepsilon^4 + 0.722087\varepsilon^2 + 1}.$$

(7.68)

Very important is to establish bounds for the solution, and the upper and lower Padé bounds for $P_{6,6}(\varepsilon)$ are given as follows (Baker and Graves-Moris, 1996):

$$P_{6,4}(\varepsilon) = \frac{-0.25985\varepsilon^6 + 0.27733\varepsilon^4 - 0.664548\varepsilon^2 + 1}{-0.144498\varepsilon^4 - 0.198874\varepsilon^2 + 1};$$
$$P_{6,8}(\varepsilon) = \frac{-0.354713\varepsilon^6 + 0.280003\varepsilon^4 - 0.721617\varepsilon^2 + 1}{-0.0476736\varepsilon^8 - 0.0872062\varepsilon^6 - 0.168401\varepsilon^4 - 0.255943\varepsilon^2 + 1}.$$

(7.69)

We also constructed the two factor approximants. The first one, $\mathcal{F}_{12}^*(\varepsilon)$, is completely standard, while the second, $\mathcal{F}_{12,s}^*(\varepsilon)$, is "shifted" and conditioned in such a way that the (unknown) shift is supposed to give the sought value,

$$\mathcal{F}_{12}^*(\varepsilon) = \left(1 - 0.867964\varepsilon^2\right)^{0.474676}$$
$$\times \left(1 + (0.0821614 + 0.533783i)\varepsilon^2\right)^{1.35488 + 0.258822i}$$
$$\times \left(1 + (0.0821614 - 0.533783i)\varepsilon^2\right)^{1.35488 - 0.258822i};$$

$$\mathcal{F}_{12,s}^*(\varepsilon) = 0.481814 + 0.518186\left(1 - \varepsilon^2\right)^{0.766642}$$
$$\times \left(1 - (0.074165 + 0.649541i)\varepsilon^2\right)^{1.46148 + 0.0652476i}$$
$$\times \left(1 - (0.074165 - 0.649541i)\varepsilon^2\right)^{1.46148 - 0.0652476i}.$$

(7.70)

Both approximants consume twelve terms from the expansion.

We are interested in the permeability at $\varepsilon = 1$. Thus, we have three estimates,

$$P_{6,6}(1) = 0.51277, \quad \mathcal{F}_{12}^*(1) = 0.50195, \quad \mathcal{F}_{12,s}^*(1) = 0.481814.$$

Their average K_{av} is equal to 0.498845, and corresponding margin of error can be estimated through the variance, which is equal to 0.0128272.

Different formulas for the permeability are shown in Fig. 7.12.

7.4.4.1 Subcritical index

Close to ε_c we simply find that

$$P_{6,6}(\varepsilon) \simeq 0.51277 + 2.30175(1 - \varepsilon),$$

and the correction to constant is trivially linear. Similarly one can calculate

$$\mathcal{F}_{12,s}^*((\varepsilon) \simeq 0.481814 + 1.36825(1 - \varepsilon)^{0.766642},$$

possibly indicating a non-trivial subcritical index of 0.767.

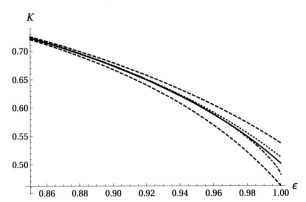

Figure 7.12 Bounds (7.69) are shown with dashed lines. Comparison of the formulas in the vicinity of ε_c: Padé approximant $P_{6,6}$ is shown with dotted line, factor approximant $\mathcal{F}_{12}^*$ from (7.70) is shown with solid line, and shifted factor approximant $\mathcal{F}_{12,s}^*$ from (7.70) is shown with dot–dashed line.

Now we would like to study in more detail the behavior of permeability in the vicinity of ε_c, assuming some deviations from linearity. Let us start with a more general initial approximation for the permeability which holds in the vicinity of $\varepsilon_c = 1$,

$$K_0(\varepsilon) \simeq A_0 + A_1(\varepsilon_c^2 - \varepsilon^2)^{\lambda_0}. \tag{7.71}$$

To find the unknowns we set from the start $A_0 = K_{av}$, and try to satisfy the expansion (7.67) in the second order. Then, $A_1 = 0.501155$, $\lambda_0 = 0.929201$. Expression (7.71) can be explained as a two-fluid model, since there are two components in the flow. One which is getting blocked by the obstacles to flow and another which can not be blocked.

We should recognize here that (7.71) with its parameters is only a crude approximation. In what follows let us attempt to correct the formula $K_0(\varepsilon)$, assuming instead of λ_0 some more general functional dependence $\Lambda^*(\varepsilon)$. As $\varepsilon \to \varepsilon_c$, $\Lambda^*(\varepsilon) \to \lambda_c$, the sought corrected value. The function $\Lambda^*(\varepsilon)$ will be designed in such a way, that it smoothly interpolates between the initial value λ_0 valid at small ε, and the sought value λ_c valid as $\varepsilon \to \varepsilon_c$. The corrected or "dressed" permeability $K^*(\varepsilon)$ is now given as follows:

$$K^*(\varepsilon) = A_0 + A_1(\varepsilon_c^2 - \varepsilon^2)^{\Lambda^*(\varepsilon)}, \tag{7.72}$$

and should be valid for all ε. From (7.72) one can express $\Lambda^*(\varepsilon)$, but only formally since $K^*(\varepsilon)$ is not known. But we can use its asymptotic form (7.67), express $\Lambda^*(\varepsilon)$ as a series for small ε, and apply some resummation procedure (e.g. Padé technique) in order to extend the series to the whole region of ε. Finally we calculate the limit of the approximants as $\varepsilon \to \varepsilon_c$ and find the corrected value $\lambda_c = \Lambda^*(\varepsilon_c)$.

In what follows the $p(\varepsilon) = K_{14}(\varepsilon)$ stands for an asymptotic form of $K^*(\varepsilon)$ as $\varepsilon \to 0$. Corresponding asymptotic expression for Λ^*, just called $\Lambda(\varepsilon)$, can be made explicit from the following relation,

$$\Lambda(\varepsilon) \simeq \frac{\log\left(-\frac{(A_0-p(\varepsilon))}{A_1}\right)}{\log(\varepsilon_c^2 - \varepsilon^2)}, \tag{7.73}$$

which can be explicitly presented as expansion in powers of ε around the value of λ_0,

$$\Lambda(\varepsilon) = \lambda_0 + \Lambda_1(\varepsilon). \tag{7.74}$$

It appears that one can construct now a sequence of diagonal Padé approximants

$$\Lambda_n(\varepsilon) = \lambda_0 + PadeApproximant[\Lambda_1[\varepsilon], n, n]), \tag{7.75}$$

with the sought limit $\Lambda^*(\varepsilon)$. Now we are able to estimate the critical index $\lambda_c = \Lambda^*(\varepsilon_c)$ and reconstruct the whole expression (7.72).

There is a good convergence within the approximations for the λ_c generated by the sequence of Padé approximants, corresponding to their order increasing:

$$\lambda_{c,1} = 0.929201, \quad \lambda_{c,2} = 0.402904, \quad \lambda_{c,4} = 0.631631,$$

$$\lambda_{c,6} = 0.630229, \quad \lambda_{c,8} = 0.702766, \quad \lambda_{c,10} = 0.698385 \quad \lambda_{c,12} = 0.702563.$$

Remarkably, in the highest orders (up to 18-th) the value of index remains practically the same. The final estimate for λ_c can be conjectured to be rational $\frac{2}{3}$.

The function $\Lambda^*(\varepsilon)$ needed to reconstruct the permeability, can be expressed as the Padé approximant. For instance, the approximant corresponding to $\lambda_{c,6}$ has the following form,

$$K_6^*(\varepsilon) = 0.498845 + 0.501155 \left(1 - \varepsilon^2\right)^{\frac{-4.15886\varepsilon^6 + 6.957\varepsilon^4 - 7.18244\varepsilon^2 + 0.929201}{-1.56421\varepsilon^6 + 2.06925\varepsilon^4 - 6.98732\varepsilon^2 + 1}}. \tag{7.76}$$

Formula (7.76), as well as the approximant (7.77), corresponding to $\lambda_{c,8}$,

$$K_8^*(\varepsilon) = 0.498845 + 0.501155 \left(1 - \varepsilon^2\right)^{\frac{0.134578\varepsilon^8 - 0.22113\varepsilon^6 + 0.650924\varepsilon^4 - 0.904689\varepsilon^2 + 0.929201}{-0.0295078\varepsilon^8 - 0.199491\varepsilon^6 + 0.298201\varepsilon^4 - 0.23125\varepsilon^2 + 1}}, \tag{7.77}$$

are confidently located within the Padé-bounds (7.69).

Formulas for the permeability including the subcritical regime, are shown in Fig. 7.13.

We conclude now with some comments. In the case of the transverse Stokes flow through the array of cylinders the lubrication approximation works in a rather broad region of concentrations. Our formulas for permeability of square and hexagonal arrays are derived to respect both, low-concentration and high-concentration limits. In the case of a longitudinal (parallel to cylinders) flow we argue that the long series predict smaller permeability and better filtration ability than it was sought previously.

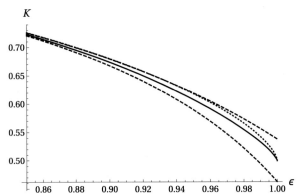

Figure 7.13 Bounds (7.69) are shown with dashed lines. Comparison of the formulas in the vicinity of ε_c: $K_6^*(\varepsilon)$ is shown with dotted line, $K_8^*(\varepsilon)$ is shown with solid line.

In the case of channel with wavy walls, the lubrication approximation works poorly, except when the surfaces are sufficiently close to a plane and for small value of ε. Different methodology, not involving lubrication approximation, is developed above. Closed-form expressions for arbitrary ε work rather well in both situations considered above, when walls can or cannot touch. In the former case, for the first time critical exponents and amplitudes are calculated without invoking the lubrication approximation. In the latter case, we discuss a crossover from the high-permeable to low-permeable state of the channel, characterized with the power low and corresponding exponent for large ε, also not invoking lubrication theory. For the important case of a symmetric sinusoidal three-dimensional channel we also discussed possibility of a nontrivial sub-critical index and calculated its value.

Tiny viscous eddies dominate in the case when walls can touch, their onset over the whole length of the channel explains the quantitative breakdown of the lubrication approximation for the macroscopic quantity, such as permeability. The critical exponent though, is correctly estimated from the lubrication approximation, as the flow can still pass through the narrow channels between the eddies considered as cylinders.

One is also confronted with a challenging task to understand the role of eddies in the nonlinear corrections to Darcy's law (Adler et al., 2013, Chaudhary et al., 2013). For arbitrary Reynolds numbers the conditions on eddies formation are significantly eased compared to the Stokes flow (Malevich et al., 2008). Also Moffat demonstrated the existence of an infinite sequence of eddies near the corner, as the viscosity, usually a damping mechanism, is here responsible for the generation of a geometrical progression of eddies. The damping for very large viscosity has the effect, however, to give a large ratio of the intensities between successive eddies. Is it possible to observe several eddies when the conditions for their formation are eased?

Simple fluids, suspensions and selected random systems

8

8.1 Compressibility factor of hard-sphere and hard-disk fluids. Index function

The structure of many real fluids, especially for high temperatures, is mainly determined by the repulsive forces. Hard-sphere fluid is the simplest and widely used reference model for describing the behavior of real fluids and concentrated colloidal suspensions when particles (colloids) can be viewed as large atoms.

The hard sphere gas of particles of the diameter d_s in the space $\mathbb{R}^d$ is defined by the two-body potential, infinite inside the particle sphere, and zero outside the sphere. Thus it accounts only for the repulsive forces among particles. The effective interaction among sterically stabilized colloidal particles is described almost perfectly by the hard-spheres. The phase behavior of hard spheres is rich, and includes liquid, crystal, and metastable states, with associated phase transitions and crossover.

As the starting approximation it is considered a gas of very small concentration, with particles distributed in space absolutely randomly. Then the concentration is increased and the effect of ordering is studied. Understanding the packing of hard spheres at moderate densities turned out to be crucial to understanding of simple liquids and suspensions. The equation of state (EOS) of a system is the most important thermodynamic relationship. There is no exact theoretical solution. Most of the EOS were obtained from the perturbative expansion discussed below, combined with direct fits to the simulation data.

The state of hard-sphere fluids is described by the compressibility factor

$$\mathcal{Z}(f) = \frac{P}{\rho k_B T}, \tag{8.1}$$

in which P is pressure, ρ is the geometric density (the number of elements per cell), T is temperature, and $f = \frac{\pi \rho}{6} d_s^3$ stands for packing fraction.

The compressibility factor exhibits critical behavior at a finite critical point. This behavior has been found from phenomenological equations (Carnahan and Starling, 1969, Tian et al., 2009, 2010b) as

$$\mathcal{Z}(f) \sim (f_c - f)^{-\beta}, \text{ as } f \to f_c - 0, \tag{8.2}$$

with the parameters corresponding to space filling density $f_c = 1$, and $\beta = 3$ (see celebrated formula (8.4)), although these are not asymptotically exact values. Some proposed equations of state put leading singularity on the real positive density axis, usually at close packing, or "random close packing".

Applied Analysis of Composite Media. https://doi.org/10.1016/B978-0-08-102670-0.00017-2

Since the N-particle problem of a macroscopic gas in a volume V is intractable, it can be reduced to a sum of an increasing number of tractable few particle problems for single particles, pairs of particles, triplets of particles etc., and each group of particles could be considered separately. Density expansions appear, since the number of single particles, pairs of particles, triplets of particles, ..., are proportional to f, f^2, f^3, ... For low packing fraction, the compressibility factor is represented then by the virial expansion

$$Z(f) = 1 + \sum_{n=2}^{\infty} B_n f^{n-1} .$$

The expansion and corresponding coefficients B_n are called virial for historical reasons (Wisniak, 2003). For hard spheres they do not depend on temperature and are defined in terms of integrals whose integrands are products of Mayer functions, a quantity readily related to the interaction potential. Only the first four virial coefficients can be calculated analytically:

$$B_1 = 1, \qquad B_2 = 4, \qquad B_3 = 10,$$

$$B_4 = \frac{2707\pi + [438\sqrt{2} - 4131\arccos(1/3)]}{70\pi} \approx 18.364768 .$$

The higher order virial coefficients, up to the 11th order inclusively, have been calculated (Clisby and McCoy, 2005, 2006, Wheatley, 2013) numerically and considered as exact. The rest of the coefficients are consensual estimates. So there is the following expansion in 3D

$$\begin{aligned}
Z(f) \simeq & 1 + 4f + 10f^2 + 18.3648f^3 + 28.2245f^4 + 39.8151f^5 \\
& + 53.3444f^6 + 68.5375f^7 + 85.8128f^8 + 105.7751f^9 + 127.9263f^{10} \\
& + 150.9949f^{11} + 181.19f^{12} + 214.75f^{13} + 246.96f^{14} + 279.17f^{15}.
\end{aligned}$$
$$(8.3)$$

The values of the virial coefficients increase with the expansion order. One can not even rule out that negative coefficients are going to emerge in higher orders (Clisby, 2017).

In order to get an equation of state, one either employs some version of Padé approximants complemented by some phenomenological terms, or constructs purely heuristic phenomenological equations. To check the validity of the so constructed equations is possible if one forms an equation exactly reproducing the first ten (or less) virial coefficients. Then this equation is expanded in powers of the packing fraction f and one examines how such an expansion reproduces the last virial coefficients from B_{11} to B_{16}. The maximal error in reproducing these last virial coefficients defines the accuracy of the studied equation of state. The analysis of a great number of different equations of states has been accomplished by Tian et al. (2009, 2010b), Gluzman and Yukalov (2016). When all virial coefficients are involved in construction, one can compare the result with detailed fits to data. Also, when one is looking for a compact, yet accurate forms, and (or) is interested in finding unknown critical amplitude and index, it is preferable to employ for verification only exactly known virial coefficients.

One of the most popular equations is the Carnahan–Starling equation (Carnahan and Starling, 1969), which has the structure of a polynomial ratio

$$Z(f) = \frac{1 + f + f^2 - f^3}{(1 - f)^3}. \tag{8.4}$$

This equation reproduces the higher virial coefficients with good accuracy, by design. Its form also appears to be justified as a good interpolation to some approximate solutions from the statistical theory of liquid state (Ziman, 1979). Tian et al., (2009) constructed a phenomenological equation as a Laurent series in powers of f

$$Z(f) = 8.10002 + \frac{5.48979}{(f - 0.926214)^2} + \frac{10.2962}{f - 0.926214} + 2.39485(f - 0.926214)$$
$$- 1.41939(f - 0.926214)^2 - 2.16537(f - 0.926214)^3$$
$$- 1.09717(f - 0.926214)^4 - 0.205088(f - 0.926214)^5. \tag{8.5}$$

It was possible to construct 57 different variants of such equations, the best of which reproduces the higher order virial coefficients with error of 3.2%.

We would like to calculate the critical index β independently on previous estimate $\beta = 3$, minding that it is not asymptotically exact value. Let us start from the initial approximation in the form of factor approximant, but recast it more generally as

$$Z_0(f) = \left(1 - \frac{f}{f_c}\right)^{-\beta_0} R(f), \tag{8.6}$$

where $R(f)$ stands for the regular part and β_0 is initial guess for the index which obviously would need to be corrected. We will attempt to correct $Z_0(f)$ assuming instead of β_0 some functional dependence $\beta(f)$, as suggested in Gluzman et al. (2016a). Similar technique is employed also on pages 37, 193, and in Chapter 3, Subsection 3.4.1.

Thus, we look for the solution in the form extending critical behavior to the whole region of f. But such extension engenders instead of a critical index some functional dependence, or concentration-dependent "index" function $\beta(f)$. Such dependence will be constructed from the known virial coefficients in low-concentration limit. Naturally, as $f \to f_c$, $\beta(f) \to \beta_c$, and gives the sought corrected value.

The function $\beta(f)$ will be designed in such a way, that it smoothly interpolates between the initial value β_0 and the sought value β_c. The corrected functional form for the compressibility (or conductivity as in Gluzman et al. (2016a)) is now given as follows:

$$Z^*(f) = \left(1 - \frac{f}{f_c}\right)^{-\beta(f)} R(f). \tag{8.7}$$

From (8.7) one can express $\beta(f)$, but only formally since true expression for $\mathcal{Z}^*(f)$ is not known. To make the expression (8.7) practical, we can use its asymptotic form (8.3), and express $\beta(f)$ as a series in f. Then we apply to the series some resummation procedure (e.g. Padé technique), and finally calculate the limit of the approximants as $f \to f_c$.

In what follows the ratio $C(f) = \frac{\mathcal{Z}^*(f)}{R(f)}$, stands for an asymptotic form of the singular part of the solution as $f \to 0$, and

$$\beta(f) \simeq -\frac{\log(C(f))}{\log\left(1 - \frac{f}{f_c}\right)}, \tag{8.8}$$

can be easily expanded in powers f, around the value of β_0. Let us choose simply, $R(f) = 1$. From the series for $\beta(f)$ one can construct standard, diagonal Padé approximants $P_{n,n}$, and find their corresponding limits as $f = f_c$. The found values will be our estimates for the critical index,

$$\beta_n = \lim_{f \to f_c} P_{n,n}(f).$$

The corresponding Padé approximants need to appear as holomorphic functions. In such case they do represent not only the critical index, but the whole "index" function $\beta(f)$. The functional dependence $\mathcal{Z}^*(f)$ is reconstructed from the "index function" in the whole region of packing fractions. It appears that one can construct three holomorphic, meaningful diagonal Padé approximants,

$$P_{2,2}(f) = \frac{0.634815 f + 1.7804}{f(f + 0.372048) + 1.04344} + 2.29373,$$

$$P_{4,4}(f) = \frac{f(f(f(2.50129 f - 1.15265) + 6.40288) + 1.71917) + 5.02772}{f(f(f(f - 0.295675) + 2.11456) + 0.429793) + 1.25693},$$

$$P_{7,7}(f) =$$
$$\frac{f(f(f(f(f(f(7.71319 f - 15.0456) - 17.9451) - 25.3543) + 3.73358) + 7.96256) + 7.22236) + 3.30322}{f(f(f(f(f(f(f - 6.33096) - 4.42721) - 5.93058) + 1.55116) + 2.32824) + 1.80559) + 0.825806}, \tag{8.9}$$

and find from here corresponding corrected index,

$$\beta_2 = 3.294, \quad \beta_4 = 3.218, \quad \beta_7 = 3.095.$$

Now we are also in possession of complete expressions for $\mathcal{Z}(f)$. The two higher-order approximants are shown below,

$$\mathcal{Z}_4^*(f) = \left(1 - \frac{f}{f_c}\right)^{-P_{4,4}(f)}, \quad \mathcal{Z}_7^*(f) = \left(1 - \frac{f}{f_c}\right)^{-P_{7,7}(f)}. \tag{8.10}$$

Approximant $\mathcal{Z}_4^*(f)$ appears to predict exceptionally well the two unused virial coefficients, giving the values of 127.746 and 150.965, in 10th and 11th orders, respectively. For comparison, typical Padé approximants generate the values 128.6 and 155 (Guerrero and Bassi, 2008), based on the same input. Even numerical results obtained from

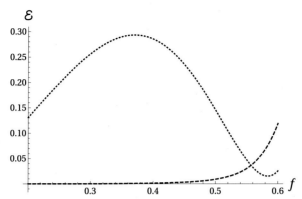

Figure 8.1 The relative error $\mathcal{E}$ of the Carnagan–Starling (8.4), compared with "exact" formula (8.5), is shown with dotted line. The relative error of the $\mathcal{Z}_7^*(f)$, compared with "exact" formula (8.5), is shown with dashed line.

sophisticated differential approximants 127.93 and 152.67 (Clisby and McCoy, 2006), are at the same level of accuracy.

The scheme, due to its simplicity, can always lead to an analytical expression. See Fig. 8.1 for comparison of the different formulas. As the "exact" solution we consider (8.5), and measure deviation of the approximate solutions via respective relative percentage error. We can also use the $DLog$ Padé technique, widely employed in Gluzman et al. (2017). The best result for the critical index evaluates as 2.66. The best for such methodology, complete expression for the compressibility can be reconstructed from an effective critical index, approximated in our case by the approximant $zP_{4,5}(z)$, so that

$$\mathcal{Z}^*(f) = \exp\left(\int_0^{\frac{f}{1-f}} P_{4,5}(z)\,dz\right), \tag{8.11}$$

where

$$P_{4,5}(z) = \frac{z(z(z(2.60319z-0.78798)+1.96547)+3.34326)+1.26698}{z(z(z(z+0.18349)+0.616804)+1.71564)+1.15256)+0.316744}. \tag{8.12}$$

The integral can be calculated in closed form, or numerically. Eq. (8.11) can be compared with other formula for the compressibility, and appears to be less accurate than $\mathcal{Z}_7^*(f)$. See Fig. 8.2 for comparison of the two different methods for calculating critical index. The value of the radial distribution function at the particles contact $G(2, f) \equiv G(d_s)$ (Brady, 1993), is simply expressed through $\mathcal{Z}$,

$$\mathcal{Z}(f) = 1 + 4f G(2, f).$$

All equations presented so far, are valid, at least, for the stable liquid phase of hard spheres, up to $f \approx 0.5$.

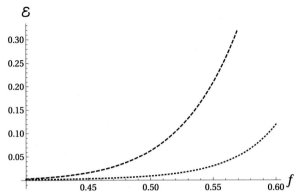

Figure 8.2 The relative error $\mathcal{E}$ of the $\mathcal{Z}_7^*(f)$ given by (8.10), (8.9), compared with "exact" formula (8.5), is shown with dotted line. The relative error (of the $DLog$ Padé approximant (see Eqs. (8.11) and (8.12)), compared with "exact" formula (8.5), is shown with dashed line.

Yelash and Kraska (2001) proposed an improved equation of state, that is more accurate in the high-density metastable fluid region, up to the (rounded) highest threshold $f_c = \frac{\sqrt{2}\pi}{6}$, that can be achieved by any arrangement of spheres,

$$\mathcal{Z}(f) = \frac{3 + 8f + 14f^2 + 14f^3 + \frac{40}{3}f^4}{3 - 4f}. \tag{8.13}$$

Really, they have dwelt on the idea of Korteweg and Boltzmann of introducing singularity at close packing of equal spheres at $f_c \approx 3/4$, as explained in (Lopez de Haro et al., 1998). But (8.13) appears to be as accurate as the (8.4). But the thermodynamics of the system is more complicated because of the freezing phase transition, and the most interesting fluid state becoming metastable.

It is believed that at random close packing (RCP) (see also Subsection 1.2.4) $f = f_c \approx 0.64$, there is a simple pole in $\mathcal{Z}(f)$ (Brady, 1993, Speedy, 1994, Wu and Sadus, 2005, Kamien and Liu, 2007, Torquato, 2018). The pole is not taken into consideration by EOS discussed above and does not follow only from the analysis of virial series. However, this observation does not necessarily imply that it is well defined. The density of random close packing is notoriously difficult to define. The random close packing is currently understood as maximally random jammed packing, in which each particle has a requisite number of contacting particles required for the packing to achieve a level of mechanical stability, corresponding to minimal value of an order metric (Torquato, 2018). For protocols with varying compression rate, the most rapid compression of the liquid ending in mechanically stable packing is considered as the maximally random jammed state (Torquato, 2018).

There is an understanding that the hard spheres EOS may be extended smoothly avoiding the first-order freezing transition, to describe the metastable fluid branch, and end at its first pole at RCP (Rintoul and Torquato, 1996, Kamien and Liu, 2007). Counting of the rate of disappearance of the number of distinct, allowed inherent structures gives another definition of RCP (Kamien and Liu, 2007). There is no longer any need to specify the degree of "disorder" of the states to distinguish ordered and disordered configurations. The fact that a sharp transition at RCP appears to exist, implies that the density of states at RCP is a property of three-dimensional space. According to Rintoul and Torquato (1996), Wu and Sadus (2005), Kamien and Liu (2007), it is possible (but not rigorous) in Monte-Carlo simulations and theoretically, to extend the virial expansion valid for the stable liquid phase, to the metastable region of concentrations up to the very pole. Many different algorithms that yield the RCP pole around 0.64, all designed to avoid crystallization and follow a metastable branch (Kamien and Liu, 2007). Thus, continuous equations based on virial expansions, with the first pole at random close packing consider the low density gas and the high density metastable branch as the same phase.

Speedy (1994) suggested to describe the high density metastable fluid branch empirically. As the random close packing is approached,

$$\mathcal{Z}(f) \simeq \frac{2.67}{1 - 1.543 f}, \tag{8.14}$$

and $G(2, f) \simeq \frac{0.668}{0.648 - f}$.

Liu (2006) suggested the following empirical equation for the entire stable and metastable regions up to the random close packing,

$$\mathcal{Z}(f) = 1 + \frac{0.31416 f}{1 - 1.57336 f} + 4.16371 0^{10} f^{40} - 2.34521 0^{11} f^{42} + 3.66841 0^{11} f^{44}$$
$$+ \frac{3.68584 f}{-0.16012 f^4 - 0.172284 f^3 + 1.9499 f^2 - 2.5848 f + 1}. \tag{8.15}$$

It does posses the pole, so that $\mathcal{Z}(f) \simeq \frac{0.12691}{0.635584 - f}$, but the amplitude seems to be too small, when compared with other estimates in the spirit of (8.14).

Bonneville (2016) deduced a simple semi-empirical formula for all f,

$$\mathcal{Z}(f) = 1 + 4f + \frac{10 \left(0.4063 f^2 - 1.7343 f + 1 \right) f^2}{(1 - 1.56986 f)(1 - f)^2}, \tag{8.16}$$

with pole in the physical region $\mathcal{Z}(f) \simeq \frac{1.1792}{0.637 - f}$, as well as another pole of order 2 at space filling $f = 1$.

In order to take the pole in the form of (8.14) into account, we suggest the simple Padé approximant, smoothly interpolating between (8.3) and (8.14)

$$\mathcal{Z}^*(f) = \frac{0.120412(6.50777 - f)\left(f^2 - 1.33224 f + 0.496627 \right)\left(f^2 + 0.541459 f + 0.56102 \right)}{(0.648088 - f)\left(f^2 - 1.32508 f + 0.767885 \right)\left(f^2 - 1.21195 f + 0.417981 \right)\left(f^2 + 0.311786 f + 1.0496 \right)}. \tag{8.17}$$

The Padé approximant (8.17) agrees with formula (8.13) up to $f \approx 0.55$, and agrees with (8.14) starting from $f \approx 0.6$.

Free volume theories of liquid state tend to overstate the degree of ordering ascribed to liquids, but become correct at close packing (Fisher, 1964). The key quantity, free volume is supposed to reflect on volume difference between solid and liquid states. Close to the close packing, it is expressed only through the distance between neighboring cells and sphere diameter, dimension of space and some geometrical factor (Ziman, 1979). As a consequence, there is a simple pole in compressibility for all dimensions. Local density of the free volume models demonstrates qualitative features typical for liquids (Frenkel, 1946). As the close packing is preceded by RCP in three dimensions, one may as well apply similar idea near the RCP (Ziman, 1979).

Good agreement with the free-volume considerations at high densities, was also established through the Monte-Carlo (MC) simulations in a hard-sphere mixture at high densities, with a heuristic mapping of the MC results to the monodisperse hard-sphere system (Santos et al., 2011, 2014). Such modifications help to avoid the freezing transition, and extrapolate the low-density fluid to all densities. Simulations for seventeen mixtures in the high-density region were performed to infer the equation of state of the pure hard-sphere system in the metastable region. There is a good collapse of the inferred curves up to the f_c (Santos et al., 2014). Similar problems and solutions extend to the two-dimensional case of hard disks (Santos et al., 2017). The collapse of the curves is still reasonable both in the stable and metastable regions up to the values of $f \approx 0.85$. Free volume compressibility was taken in the form (Santos et al., 2011, Torquato, 2018)

$$Z(f) = \frac{3 f_c}{f_c - f}, \quad f_c = 0.637, \tag{8.18}$$

close to (8.14). Jamming threshold experimentally is obtained as an infinite-pressure limit of the free volume model (Torquato, 2018). It is rather straightforward to modify our own interpolant (8.17), according to the typical parameters of the pole (8.18):

$$Z^*(f) = \frac{0.0892792(9.93287 - f)(f^2 - 1.14267 f + 0.359265)(f^2 + 0.541974 f + 0.562227)}{(0.637 - f)(f^2 - 1.38457 f + 0.772197)(f^2 - 1.10503 f + 0.34106)(f^2 + 0.304939 f + 1.06771)}.$$
$$\tag{8.19}$$

The two approximants (8.17), (8.19) are still close numerically.

Compressibility can be probed directly in sedimentation experiments by scanning of the concentration profile. The sedimentation rate can be also measured directly in transient sedimentation regime, from the settling profile and from equilibrium measurements of compressibility (Buzzaccaro et al., 2008). The phase transition (Paulin and Ackerson, 1990), and metastable phase (Davis et al., 1989), can be observed in sedimentation, leading to different sedimentation velocity, defined in Section 8.6.

Introduction of metastable liquid branches for composites made of the hard-spheres, is somewhat arbitrary. To obtain the metastable branch, one must introduce additional constraints that effectively exclude crystalline states. Different constraints give different metastable branches. Nevertheless, the ambiguity is expected to be exponentially small in the distance from the freezing point and as the distance between freezing point and the close packing is not so large, one might expect to obtain a meaningful result in this way. Indeed, different possible continuation of the liquid equation of state (e.g. Carnahan–Starling) differ by less than 10% in the dense region around

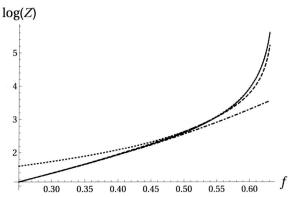

Figure 8.3 The Carnahan–Starling formula (8.4) is shown with dot–dashed line. Our suggestion (8.19) is shown with solid line. Free-volume pole given by (8.18) is shown with dotted line. Formula (8.16) is shown with dashed line.

0.64 in 3D. The ambiguity becomes smaller with increasing dimensionality of space (Parisi and Zamponi, 2010). Eq. (8.4) can be compared with other formula for the compressibility for the metastable phase, see Fig. 8.3 for comparison of the different EOS.

The metastable branch is used to study the dynamics in granular matter made from millimetric particles sheared in a Couette geometry, as noted in Kamien and Liu (2007). The flow properties are quantitatively predicted from a locally Newtonian, continuum model. Along the metastable branch, the effective viscosity $\mu_e(f)$ apparently diverges as $(f_c - f)^{-1.75}$ (Losert et al., 2000). The so-called positional order parameter for hard spheres diverges at f_c with the same critical exponent (Cohen et al., 2016). This observation possibly suggests a connection between the physics of jamming in granular matter and the thermodynamics of simple fluids (Cohen et al., 2016). Granular systems made up of macroscopic particles are different from nanometric colloids, since thermal fluctuations are negligible for such materials, and granular systems are athermal (Messina et al., 2015). Therefore, we expect that the effective viscosity can be modeled by the well known expansion in volume fraction, as in Section 8.3.

The idea of RCP does not seem to apply easily in two dimensions for monodisperse disks (O' Hern et al., 2006). The very same, modern state of art protocols are able to find the RCP in 3D, but not in 2D, where the metastable branch is conventionally believed to extend up to the close packing (Torquato and Stillinger, 2010). But recent attempt to find maximally random jammed state at $f \approx 0.826$ in 2D, raised the question of finding protocol strongly biased towards formation of disordered packings, thus frustrating crystallization (Atkinson et al., 2014). The pair correlation function calculated by Atkinson et al. (2014) includes spikes typical to the local geometries on triangular lattice, but they appear strongly suppressed. Remarkably, it is found that the most disordered jammed configuration is not the one that shows up the most frequently in any known protocol.

In two dimensions, the hard disks packings are used to model molecular monolayer films, molecules adsorbed on substrates and even organization of epithelial cells

(Atkinson et al., 2014). In two-dimensional case of hard disks the following expansion is available (Clisby and McCoy, 2006), as conveniently presented in Maestre et al. (2011):

$$\mathcal{Z}(f) \simeq 1 + 2f + 3.12802 f^2 + 4.25785 f^3 + 5.3369 f^4 + 6.36296 f^5$$
$$+ 7.35186 f^6 + 8.3191 f^7 + 9.27215 f^8 + 10.2163 f^9 + 11.15 f^{10} \quad (8.20)$$
$$+ 12.08 f^{11} + 13.03 f^{12} + 13.93 f^{13} + 14.91 f^{14} + 15.86 f^{15},$$

where $f \equiv \frac{\pi \rho}{4} d_s^2$. The last six coefficients in (8.20) are estimated, while the rest are numerically exact (Clisby and McCoy, 2006). McCoy (2001) comments "that since the kissing number (maximum number of spheres which can touch a given sphere) is equal to 12 in three dimensions that the virial expansion cannot possibly include the effects of the geometry of hard spheres until at least the 12th coefficient has been computed." In two-dimensional case the kissing number equals 6 and one can expect that virial series of the 9th order in f, with exact and numerically exact coefficients will work somewhat better for the extrapolation from low-density region to all f. The compressibility factor exhibits critical behavior at a finite critical point. This behavior has been found from phenomenological equations as

$$\mathcal{Z}(f) \simeq (f_c - f)^{-\beta}, \text{ as } f \to f_c - 0, \quad (8.21)$$

with the parameters $f_c = 1$ and $\beta = 2$ (Santos et al., 1995, Mulero et al., 2009), although these are not a unique candidate values to be included into consideration. Nevertheless, according to Clisby 2017, the singularity which determines radius of convergence for the 2D virial series is indeed located at $f = 1$.

It was noted in Santos et al. (1995) that even the function $\mathcal{Z}(f) = (1 - f)^{-2}$, is a decent predictor of all virial coefficients. There is also a good approximation for all f given in Mulero et al. (2009),

$$\mathcal{Z}(f) = \frac{1 + \frac{f^2}{8} - \frac{f^4}{10}}{(1 - f)^2}, \quad (8.22)$$

which can be used for comparison. Its accuracy is almost as good as of a cumbersome fit to various numerical data (Tian et al., 2010a), shown below

$$\mathcal{Z}(f) =$$
$$1 + 2f + 3.12802 f^2 + 4.25785 f^3 + 5.3369 f^4 + 6.36296 f^5$$
$$+ 7.35186 f^6 + 8.3191 f^7 + 9.27215 f^8 + 10.2163 f^9 - \frac{3.6046 \times 10^6 f^{46}}{1 - 1.10266 f} \quad (8.23)$$
$$+ \frac{44.0358 f^{47} - 44.9756 f^{46} - 10.2026 f^{11} + 11.1424 f^{10}}{(1-f)^2}.$$

Let us apply the method of index function in 2D, following the same technique as for three-dimensional case. It appears that we can construct the following meaningful diagonal Padé approximants for the index function

$$P_{1,1}(f) = \frac{2.26352(f + 1.81884)}{f + 2.0585}, \quad P_{2,2}(f) = \frac{1.4175(f^2 - 3.25637 f + 5.75405)}{f^2 - 2.56899 f + 4.07819}, \quad (8.24)$$

and find corresponding critical index,

$$\beta_1 = 2.086, \quad \beta_2 = 1.98,$$

in line with expectations. The two approximants for compressibility are shown below,

$$Z_1^*(f) = (1-f)^{\frac{0.542463}{f+2.0585}-2.26352}, \quad Z_2^*(f) = (1-f)^{\frac{0.974354f-2.37554}{f(f-2.56899)+4.07819}-1.4175}. \quad (8.25)$$

In construction of $Z_2^*(f)$ we employed only the terms up to the 5th order inclusively. It turns out that $Z_2^*(f)$ is rather good in predicting all the rest of virial coefficients, with maximal error of 0.77%. Corresponding average error equals just 0.35%.

We can also use the $DLog$ Padé technique. The best result for the critical index is equal to 1.87, and the best complete expression for the compressibility can be reconstructed from the effective critical index, approximated in our case by the approximant $zP_{3,4}(z)$, where

$$P_{3,4}(z) = \frac{z(z(1.8663z+5.13902)+5.55649)+1.95291}{z(z(z(z+3.51348)+5.13043)+3.6297)+0.976457}, \quad (8.26)$$

leading to the closed-form expression,

$$Z_8^*(f) = 0.0282855e^{-0.817734\tan^{-1}\left(0.200318+\frac{1.32177}{f-1}\right)}$$
$$\times \left(\frac{1}{1-f}-0.355088\right)^{0.0193324} \left(\frac{f(f-1.49088)+2.17056}{(1-f)^2}\right)^{0.216027} \quad (8.27)$$
$$\times \left(1+\frac{5.82491}{1-f}\right)^{1.41492}.$$

The approximant $Z_8^*(f)$ employs the virial terms up to 8th order inclusively, and predicts the remaining virial coefficients with maximal error of 1.14%. Corresponding average error equals 0.58%. Overall, the approximant $Z_2^*(f)$ remains closer to the "benchmark" formula (8.22), than the $DLog$ Padé approximant.

It is believed that more accurate threshold is the highest threshold for the two-dimensional arrays of hard disks, $\frac{\pi}{\sqrt{12}} \approx 0.9069$, attained for the regular hexagonal array of disks. It is also believed that as $f \to 0.9069$, there is a simple pole in compressibility (Santos et al., 1995, Maestre et al., 2011),

$$Z(f) \simeq \frac{A}{0.9069-f}. \quad (8.28)$$

The value of amplitude in 2D can be found from the free volume theory. It is a correct limit-case as close packing fraction is approached (Modes and Kamien, 2008, Lopez de Haro et al., 2008). The general expression for arbitrary D is presented in Loeser et al. (1991).

From the cell (free volume/area) theory of liquid state (Modes and Kamien, 2008, Lopez de Haro et al., 2008), one deduces that $A = \frac{\pi}{\sqrt{3}} \approx 1.8138$, and

$$Z(f) \simeq \frac{1.8138}{0.9069-f}. \quad (8.29)$$

Also, close estimate for the amplitude, $A = 1.99518$, could be obtained for the hard–disc fluid in negative curvature, the concept leading to the special protocol designed to resolve a disordered state close to the threshold (Modes and Kamien, 2007, 2008). The model confirms validity of free-volume considerations for the disordered packings of hard disks. Looking at curvatures near, but not at, zero guarantees avoiding crystallization, and generate disordered, monodisperse configurations that may be carried into flat space. Tentative conclusion is drawn that introducing curvature does not change the location of the leading singularity in the virial expansion.

The critical amplitude A could be estimated from various EOS discussed in Santos et al. (1995). Typically, it is much larger, within the range from 3.4 to 4.9. For instance, there is a particularly simple EOS

$$Z(f) = \frac{1.01065}{(1.1144 - f)(0.9069 - f)}, \tag{8.30}$$

suggested by Santos et al. (1995), with amplitude $A \approx 4.87$, much larger than expected from the free-volume theory.

The amplitude can be estimated directly from the virial expansion (8.20), by means of the technique extensively discussed by Gluzman et al. (2017). Let us first extract the singularity from the original series, dividing it by $(0.9069 - f)^{-1}$, and to the ratio apply the diagonal Padé approximants, so that the sought approximation

$$Z_n^*(f) = \frac{1}{0.9069 - f} \; P_{n,n}.$$

We obtain from here the following sequence of approximations for the amplitude:

$$A_1 = 3.863, \quad A_2 = 3.9264, \quad A_5 = 1.986, \quad A_6 = 1.837.$$

The last two approximations give rather close description of the simulation data (Erpenbeck and Luban, 1985, Tian et al., 2010a). The approximant

$$Z_5^*(f) = \frac{0.29328(f-3.0363)(f-0.944853)(f+0.502832)(f^2-1.71568f+3.29942)}{(0.9069-f)(f+0.502833)(f^2-2.50877f+2.92551)(f^2-2.03771f+1.04632)}, \tag{8.31}$$

nicely reproduces the remaining virial coefficients with average error of just 0.17%, and the maximal error of 0.29%. The amplitude A_5 is also close to the values anticipated above from free-volume considerations. The higher order approximant corresponding to A_6 is shown below as well,

$$Z_6^*(f) = \frac{0.292515(f-3.08378)(f-0.93901)(f+0.106254)(f+0.490796)(f^2-1.71059f+3.30394)}{(0.9069-f)(f+0.106254)(f+0.490797)(f^2-2.52211f+2.9555)(f^2-2.03699f+1.0441)}. \tag{8.32}$$

More ambitiously, we can attempt to calculate the critical index, expected to be equal unity. In a nutshell, we define "random close packing" as the point where pressure has a simple pole, then calculate sequence of approximate critical indices and

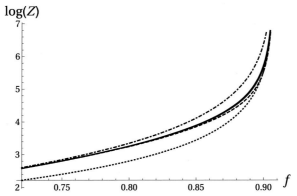

Figure 8.4 The best Padé approximants (8.31) and (8.32) are very close and appear as "fat" solid line, are compared to several approximations: the cell theory, represented by the formulas (8.29) (dotted); $DLog$ Padé approximant (8.34) (dot–dashed); and approximant (8.33) (dashed) obtained through the index function.

look how it converges (to unity in perfect scenario). Such approach allows to deal with finite numbers, in lieu of taking an infinite-pressure limits to find metastable states. The method of index function produces several values, apparently convergent to unity,

$$\beta_1 = 1.817, \quad \beta_2 = 1.493, \quad \beta_3 = 1.15, \quad \beta_4 = 1.084.$$

One can hope that exact higher order virial coefficients will help further to confirm (or not) the convergence. The best approximant corresponding to the last value is shown below,

$$\mathcal{Z}_4^*(f) = (1 - 1.10266f)^{\frac{f(f(f(5.16388-1.37325f)-9.55501)+3.23178)+2.29711}{f(f(f(f-3.14785)+5.19152)-1.76571)-1.26646}}, \tag{8.33}$$

with the critical amplitude 0.9. Formula (8.33) is also reasonably good in predicting higher-order virial coefficients with average error of 1.44%, and describes the simulation data very well.

We can also find the index by applying the $DLog$ Padé technique. The best result for the critical index is equal to 1.039. The best complete expression for the compressibility can be reconstructed as well,

$$\mathcal{Z}_6^*(f) = 0.00432472e^{4.09998\,\tan^{-1}\left(\frac{1.48074}{0.9069-f}+0.546488\right)}$$
$$\times \left(-\frac{0.9069}{f-0.9069} - 0.510121\right)^{0.00425348} \left(\frac{f(f-3.06003)+3.64103}{(0.9069-f)^2}\right)^{0.517616}, \tag{8.34}$$

where the critical amplitude equals 3.55.

In Fig. 8.4 the best Padé approximants (8.31) and (8.32) are compared to the cell theory, represented by formulas (8.29), $DLog$ Padé approximant (8.34), and approximant (8.33) obtained through the index function. More elegant formula (8.33) produces values close to the (8.31) up to $f \approx 0.85$. Formula (8.29) is valid only asymptotically, and (8.34) gives results close to the simple EOS (8.30).

Lets us construct the final approximation with the free volume condition imposed. Asymptotically it should behave at high concentrations as

$$\mathcal{Z}(f) \simeq \frac{1.8138}{0.9069 - f} + B, \tag{8.35}$$

where the second amplitude B will be found in the course of calculations.

The way how we proceeded typically is to look for multiplicative corrections to some plausible "zero-order" approximate solution. We can look for additive corrections in a similar fashion, as exemplified in Gluzman et al. (2017). To this end, subtract the already known part, $\mathcal{Z}_0(f) = \frac{1.8138}{0.9069 - f}$, from the virial series (8.20), only to get some new series $\mathcal{Z}_b(f)$. The Padé approximants applied to the series $\mathcal{Z}_b(f)$, are supposed to give the value of B. To calculate the correction one has to find the value of the corresponding approximant as $f \to f_c = 0.9069$. The following sequence of approximations for the amplitude B shows the best convergence,

$$B_n = PadeApproximant[\mathcal{Z}_b(f \to f_c, n, n+1], \tag{8.36}$$

and can be calculated readily,

$$B_3 = 42.2592, \quad B_4 = 36.5568, \quad B_5 = 36.5813, \quad B_6 = 36.5806, \quad B_7 = 37.2005,$$

implying $B \approx 36.5$. This predicted value is much larger than 1.9 quoted in Loeser et al. (1991).

Let us bring below the two approximants, constructed as sum of $\mathcal{Z}_0(f)$ and Padé approximants corresponding to the two starting terms from the sequence B_n,

$$\mathcal{Z}_3^*(f) = \frac{0.27598(f-3.95994)(f-0.933238)(f^2-1.6049f+3.36229)}{(0.9069-f)(f^2-2.70939f+3.53141)(f^2-2.06815f+1.07075)},$$
$$\mathcal{Z}_4^*(f) = \frac{0.292495(f-3.08666)(f-0.938207)(f+0.493098)(f^2-1.71061f+3.30485)}{(0.9069-f)(f+0.493099)(f^2-2.52355f+2.9576)(f^2-2.03674f+1.04366)}. \tag{8.37}$$

Both approximants, shorter and longer, are good in predicting all remaining virial coefficients with average accuracy of 0.78% and 0.14% respectively, and in describing the numerical data of Erpenbeck and Luban (1985), Tian et al. (2010a).

8.2 "Sticky" rods and disks. Mapping to Janus swimmers

Consider the 1D and 2D sticky-hard-rods (disks) model introduced by Baxter (1968), Post and Glandt (1986), Santos (2014). The fluid is made of impenetrable particles of diameter d_s, which interact through a square-well potential provided that a certain limit is taken. In this limit the width of the well Δ becomes zero and the depth υ is considered infinite. The stickiness parameter $\mathcal{A} = \Delta \exp(\upsilon\beta)$, (where $\beta = \frac{1}{T}$, or inverse temperature) should remain finite while the limit is taken. The model has proved to provide a starting point for the study of active systems with short-range attraction quantified by the parameter $\mathcal{A}$ (Ginot et al., 2015).

We will exploit the known exact solution for the compressibility factor of the one-dimensional system of "sticky" rods (Santos, 2014, Rohrmann and Santos, 2014). It can be expressed by the following exact formula

$$Z = \frac{\sqrt{\frac{4\mathcal{A}f}{1-f} + 1} - 1}{2\mathcal{A}f}, \tag{8.38}$$

where $\mathcal{A}$ stands for the dimensionless parameter of "stickiness", which measures the strength of surface adhesion due to attraction. From (8.38) the following asymptotic expressions as $f \to 1$ can be deduced,

$$Z = \frac{1}{1-f}, \text{ for } \mathcal{A} = 0, \tag{8.39}$$

and

$$Z = \frac{1}{\sqrt{\mathcal{A}}\sqrt{1-f}}, \text{ for } \mathcal{A} \neq 0. \tag{8.40}$$

Thus, the exact solution suggests *two* critical exponents in the two different cases, in sharp difference to the conventional case of hard rods. The conventional case is fully described by the formula (8.39), for zero "stickiness", while the typical to hard rods critical exponent is equal to unity.

The very same value of unity holds for higher dimensions, as explained in the preceding section. While for higher dimension, for arbitrary finite "stickiness" there could be conjectured a different power-law of the type of (8.40), with the activity critical exponent β:

$$Z \sim (f_c - f)^{-\beta}, \tag{8.41}$$

with $f_c = 1$, $\beta = 1/2$ for rods. As $\mathcal{A} \to 0$ there is the Taylor expansion to (8.38), truncated in 4th order,

$$Z \simeq 1 + (1 - \mathcal{A})f + \left(2\mathcal{A}^2 - 2\mathcal{A} + 1\right)f^2 + \left(-5\mathcal{A}^3 + 6\mathcal{A}^2 - 3\mathcal{A} + 1\right)f^3$$
$$+ \left(14\mathcal{A}^4 - 20\mathcal{A}^3 + 12\mathcal{A}^2 - 4\mathcal{A} + 1\right)f^4. \tag{8.42}$$

Let us try to reconstruct the function (8.38) from the knowledge of asymptotic regimes. All three asymptotic regimes should be taken into account for proper reconstruction of the original formula. Similar problem was discussed in Subsection 1.1.3.

Let us consider a general ansatz for the sought function, motivated by various scaling laws (Stanley, 1971, Olsson and Teitel, 2007). It suggests to look for the following form in the vicinity of the critical point, as $f \to 1$, with unknown amplitude $A(\mathcal{A})$

$$Z \simeq \frac{1 + f\frac{A(\mathcal{A})}{\sqrt{1-f}}}{1-f}. \tag{8.43}$$

Let us apply to the expansion in (8.43), the resummation technique of low-order root approximant.

$$Z^* = \frac{\left(\frac{\mathcal{P}(\mathcal{A})f}{\sqrt{1-f}} + 1\right)^m}{1-f}, \tag{8.44}$$

with unknown amplitude $\mathcal{P}(\mathcal{A})$ and unknown index m. After ensuring all asymptotic relations in leading orders, the following approximant arises,

$$Z^* = \frac{1}{\sqrt{\mathcal{A}}\sqrt{1-f}f + (1-f)}, \tag{8.45}$$

which reproduces qualitatively all regimes of the original solution and gives reasonable accuracy in the critical region.

One can proceed alternatively, and consider different root approximant, based on the expansion near f_c for $\mathcal{A} \neq 0$,

$$Z^* = \mathcal{P}_0 (1-f)^{-1/2} \left(\mathcal{P}_1(1-f)^{1/2} + 1\right)^{m_1}. \tag{8.46}$$

In order to satisfy the asymptotic conditions in the leading orders, we can simply find

$$\mathcal{P}_0 = \frac{1}{\sqrt{\mathcal{A}}}, \quad \mathcal{P}_1 = \frac{1}{\sqrt{\mathcal{A}}} - 1, \quad m_1 = -1.$$

From the knowledge of the two non-trivial starting terms from (8.42), and of the general form of leading term from (8.40), one can attempt to calculate critical index β for rods. In particular, one can construct and calculate the simplest root approximant,

$$Z_1^*(f, \mathcal{A}) = \frac{\left(\frac{f\mathcal{P}_1}{f_c - f} + 1\right)^{m_1}}{f_c - f} = \left((1-f)\sqrt[3]{\frac{3\mathcal{A}f}{1-f} + 1}\right)^{-1}, \tag{8.47}$$

and find reasonable value, $\beta = 2/3$, for the critical index. Although the virial coefficients depend on "stickiness", the critical index does not.

The estimate can be improved if all terms from (8.42) and correct form of the higher-order corrective terms in (8.40), are taken into account. After some power-counting the following root approximant

$$Z_3^*(f, \mathcal{A}) = \frac{\left(\left(\left(\frac{f\mathcal{P}_1}{f_c - f} + 1\right)^{3/2} + \mathcal{P}_2\left(\frac{f}{f_c - f}\right)^2\right)^{5/4} + \mathcal{P}_3\left(\frac{f}{f_c - f}\right)^3\right)^{m_3}}{f_c - f}$$

$$= \left((1-f)\left(\left(\frac{5\mathcal{A}^2 f^2}{6(f-1)^2} + \left(\frac{10\mathcal{A}f}{3-3f} + 1\right)^{3/2}\right)^{5/4} - \frac{625\mathcal{A}^3 f^3}{216(f-1)^3}\right)^{4/25}\right)^{-1}, \tag{8.48}$$

brings rather good approximate value of $\beta = 13/25 \approx 0.52$, for the critical index. The approximant $\mathcal{Z}_3^*(f)$ can be compared with the exact solution for arbitrary $\mathcal{A}$ and f. The maximal error remains around 3.5%, and the point where it is reached, moves to smaller f with $\mathcal{A}$ increasing.

Most interesting, the sticky particles can be connected to the active colloids. Active colloids are composed of colloidal-scale particles locally converting chemical energy into motility, mimicking micro-organisms. Exploring the potential of this new class of systems requires the development of concepts relevant for intrinsically non-equilibrium systems. The activity modifies the equation of state in a way which can be described by the introduction of both an effective temperature for the dilute system, and of an effective adhesive interaction at finite f. The sedimentation experiments are performed to probe the non-equilibrium equation of state of a two-dimensional assembly of active Janus micro-spheres, together with computer simulations of a model of self-propelled hard disks (Ginot et al., 2015).

Janus active colloids represent an interesting, verifiable experimentally model to study the statistical physics of active colloids. The Janus particles, which are colloidal objects with two sides differing in their physical or chemical properties, permit the simplest realization of autonomous swimmers. They are self-propelled by generating local gradients in concentration and electrical potential via surface reactions, and can be controlled using external stimuli. Despite the complex mechanisms responsible for self-propulsion at the micro-scale, the experimental equation of state reduces to a very simple two-dimensional model of self-propelled hard disks.

The self-propulsion profoundly affects the equation of state, but these changes can be mapped to the equilibrium model. Indeed, active systems can exist in homogeneous states, and since homogeneity is required in standard thermodynamics, active matter systems are amenable, in principle, to an effective thermodynamic description (Ginot et al., 2015, Bertin, 2015, Takatori and Brady, 2015). Active colloids behave, in the dilute limit, as an ideal gas with an activity-dependent effective temperature T_{eff}, directly proportional to the persistence time of the self-propulsion. At finite density, increasing the activity is similar to increasing adhesion (attraction) between equilibrium particles. The effective adhesion is quantified in the form relating activity and effective adhesion, both in experiments and simulations. The effective adhesion $\mathcal{A}$ induced by the self-propulsion mechanism, is simply $\mathcal{A} \approx \sqrt{T_{eff}/T_0}$, where T_0 is the thermal bath temperature for passive colloids, which also enters into general compressibilty definition (Ginot et al., 2015).

For "sticky disks" (2D case) there are only two virial coefficients available in the small f expansion,

$$\mathcal{Z}(f) \simeq 1 + c_1(\mathcal{A})f + c_2(\mathcal{A})f^2,$$
$$c_1 = 2 - \mathcal{A}, \qquad c_2 = \tfrac{25}{8} - \tfrac{25\mathcal{A}}{8} + \tfrac{4\mathcal{A}^2}{3} - 0.122\mathcal{A}^3. \tag{8.49}$$

Although in 2D case there is no exact solution, and therefore not much guidance in the high-concentration limit, we are still going to consider the model with two critical exponents, based on intuition gained from 1D system. For passive disks, with $\mathcal{A} = 0$,

$$\mathcal{Z} = (f_c - f)^{-1} \tag{8.50}$$

and for adhesive disks, with $\mathcal{A} \neq 0$, we assume as $f \to f_c$,

$$\mathcal{Z} \sim (f_c - f)^{-\beta}, \tag{8.51}$$

with $f_c \approx 0.9069$.

The unknown activity critical index β can be evaluated by means of the ansatz (8.44), extended naively to the two-dimensional case, with the critical index $\beta = (1 + \frac{m}{2})$. Parameters $\mathcal{P}$, m are to be calculated from the asymptotic expression (8.49).

$$\mathcal{P}(\mathcal{A}) = \mathcal{A}(1.37868 - 0.232364\mathcal{A}) + \frac{0.172235}{\mathcal{A} - 0.897342} + 0.144517,$$
$$\tag{8.52}$$
$$m(\mathcal{A}) = \frac{\mathcal{A}(4.09836\mathcal{A} - 7.35526) + 3.30009}{\mathcal{A}(\mathcal{A}(\mathcal{A} - 6.8306) + 4.70222) - 0.183133},$$

and

$$\mathcal{Z}^* = \frac{\left(\frac{\mathcal{P}(\mathcal{A})f}{\sqrt{1-f}} + 1 \right)^{m(\mathcal{A})}}{1 - f}. \tag{8.53}$$

The activity critical index β also can be calculated from the asymptotic expressions (8.49), (8.50) by means of the approximant (8.47), in its general form. In the 2D case it can be easily obtained for arbitrary f and $\mathcal{A}$,

$$\mathcal{Z}_1^*(f, \mathcal{A}) = \frac{\left(\frac{f\left(\mathcal{A}(1.31293 - 0.221284\mathcal{A}) + \frac{0.164021}{\mathcal{A} - 0.897342} + 1.13763 \right)}{0.9069 - f} + 1 \right)^{\frac{\mathcal{A}(4.09836\mathcal{A} - 7.35526) + 3.30009}{\mathcal{A}(\mathcal{A}(\mathcal{A} - 6.8306) + 0.183133) + 3.87204}}}{1 - 1.10266f}. \tag{8.54}$$

Compressibility dependence (8.54) on concentration for various values of activity parameter is shown in Fig. 8.5. Notable is a smooth transition to the non-monotonous behavior for high activity, compared with monotonous behavior for passive disks, due to the negative contribution to the total pressure from particles activity.

Of course, we can proceed differently by generalizing (8.46) by replacing power $-1/2$ with unknown activity index β. One should extract first the singularity pertaining to the activity exponent, and then arrange the powers so that the passive power-law could prevail in another limit,

$$\mathcal{Z}_1^*(f) = \mathcal{P}_0(f_c - f)^{-\beta} \left(\mathcal{P}_1(f_c - f)^{\beta} + 1 \right)^{\frac{\beta-1}{\beta}}. \tag{8.55}$$

The unknown parameters can be found numerically from asymptotic equivalence to the (8.49). E.g., in the four cases corresponding to strong activity we found the fol-

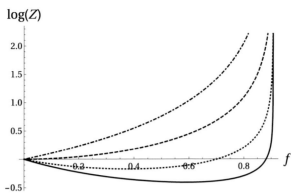

Figure 8.5 Log of compressibility dependence (8.54) on concentration with varying activity parameter $\mathcal{A} = 1$ (dot–dashed); $\mathcal{A} = 2$ (dashed); $\mathcal{A} = 3$ (dashed); $\mathcal{A} = 3.5$ (solid). The compressibility remains positive. It becomes non-monotonic for larger activity parameters.

lowing expressions:

$$\mathcal{Z}_1^*(f) = \frac{0.879254}{\left(1 - 0.81774(0.9069 - f)^{0.976105}\right)^{0.0244799}(0.9069 - f)^{0.976105}}, \text{ for } \mathcal{A} = 1,$$

$$\mathcal{Z}_1^*(f) = \frac{0.576081}{\left(1 - 0.788733(0.9069 - f)^{0.73413}\right)^{0.362157}(0.9069 - f)^{0.73413}}, \text{ for } \mathcal{A} = 2,$$

$$\mathcal{Z}_1^*(f) = \frac{0.325117}{\left(1 - 0.832122(0.9069 - f)^{0.590849}\right)^{0.69248}(0.9069 - f)^{0.590849}}, \text{ for } \mathcal{A} = 3,$$

$$\mathcal{Z}_1^*(f) = \frac{0.222838}{\left(1 - 0.840885(0.9069 - f)^{0.525146}\right)^{0.904233}(0.9069 - f)^{0.525146}}, \text{ for } \mathcal{A} = 3.5,$$

(8.56)

with the activity critical exponent β not too far from $1/2$ in the last two cases. The critical index appears to be dependent on $\mathcal{A}$, unlike the 1D case where it is constant.

For qualitative comparison we bring the expression for the compressibility factor of active swimmers suggested by Takatori and Brady (2015),

$$\mathcal{Z}_{TB}(f) = 1 - f - 0.2f^2 + \frac{4Pe}{\pi\left(1 - \frac{f}{0.9}\right)}f, \tag{8.57}$$

with the reorientation Peclet number (the ratio of the swimmer size to its run length) set as Pe $= 0.1$, corresponding to the homogeneous state. The formula (8.57) is valid for small Pe and arbitrary f. Small Pe means that the reorientation time is large, causing the particles to obstruct each others paths. The self-motion of active matter generates an athermal negative swim pressure at intermediate f, which prevents self-propelled body from swimming away in space. For $f \to 0.9$ there is a divergence with critical exponent which is equal to unity, as for passive case. At $f = 0$ there is an ideal gas of swimmers. Our formulas also anticipate a region of negative pressure for non-small activity parameter for intermediate values of f.

For illustration purposes only, the results for compressibility for all concentrations and $\mathcal{A} = 3.5$ obtained with formulas (8.53), (8.54), (8.56), are compared with refer-

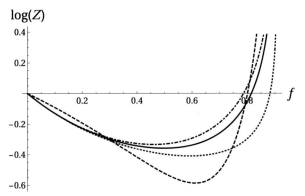

Figure 8.6 Results for log of the compressibility factor in 2D case for all concentrations, for the activity parameter $\mathcal{A} = 3.5$, obtained with formula (8.54) (dotted), formula (8.53) (dot–dashed), and formula (8.56) shown with solid line. The two formulas are compared with reference expression (8.57) (dashed) from Takatori and Brady (2015).

ence expression (8.57), as shown in Fig. 8.6. There is a qualitative agreement between all formulas. Possible validity of the independent, activity-induced critical exponent in the compressibility, is the most intriguing message. The activity turns out to be relevant, or at least marginal, variable. In the latter case the activity index depends continuously on the activity parameter.

8.3 3D elasticity, or high-frequency viscosity. Critical index

Consider an elasticity problem for perfectly rigid spherical inclusions embedded at random into an incompressible matrix. In the case of isotropic components we have two shear moduli, G_1, G, for the particles and matrix respectively. The limiting, jamming regime when $\frac{G_1}{G}$ is close to infinity and f tends to the random close packing fraction $f_c \approx 0.637$, can be hardly investigated numerically. The elasticity problem is analogous to the problem of high-frequency effective viscosity of a suspension (Batchelor and Green, 1972). The suspension consists of hard spheres immersed into a Newtonian fluid of viscosity μ, when only hydrodynamic interactions between pairs of suspended particles are considered. Brownian motion of a suspension is not taken into account. Such suspension (composite) has a Newtonian elastic behavior and the ratio of the effective shear modulus $G_e \equiv \mu_e$ to that of the matrix is (Wajnryb and Dahler, 1997)

$$G_e(f) = 1 + c_1 f + c_2 f^2 + O(f^3),$$
$$\text{with } c_1 = \frac{5}{2}, \quad c_2 = 5.0022. \tag{8.58}$$

Cluster expansion is involved into derivation, and the second order term takes into account interactions within the two-particle clusters. The central point of cluster approach is to account precisely for the interactions among all the particles inside the progressively larger clusters that are supposed to represent the composite material with greater accuracy. The value of the third-order coefficient estimated in Cichocki et al. (2003) appears to be questionable. For the coefficient c_2 Batchelor and Green give slightly higher estimate, $c_2 = 5.2$ (Batchelor and Green, 1972). The estimates for critical index presented below, are only weakly sensitive to such variation.

It is expected that in the vicinity of the 3D threshold that

$$G_e(f) \simeq A(f_c - f)^{-S},$$

where (Losert et al., 2000, Gluzman and Yukalov, 2017b) $f_c \approx 0.637$, $S \approx 1.7$. Here S is the critical index for 3D elasticity (viscosity). Let us employ various special resummation techniques well suited for very short series and estimate the critical index S. The strategy consists in employing various conditions on the fixed point in order to compensate for a shortage of the coefficients. Of course, there is a tacit assumption involved, that the metastable branch of the suspension is considered. Thus, smooth interpolation between the low-and-high-concentration regimes can be performed.

From the Eq. (1.101) on page 43, we find that inverse 3D conductivity is

$$[\sigma(f)]^{-1} \simeq 1 + 2.52f + 4.83f^2, \text{ as } f \to 0. \tag{8.59}$$

Comparing (8.58) and (8.59), we anticipate some close results for the viscosity critical index S and conductivity critical index t. Closeness of the two expansions is the best known to us explanation of widely accepted analogy between the two quantities (Bicerano et al., 1999).

Let us get back to the original suggestion for calculation of the critical index (Yukalov and Gluzman, 1997a, Gluzman and Karpeev, 2017). In this case one has to apply resummation technique to the inverse series

$$[G_e(f)]^{-1} \simeq 1 - c_1 f + (c_1^2 - c_2)f^2 \approx 1 + \tilde{c}_1 f + \tilde{c}_2 f^2, \tag{8.60}$$

where $\tilde{c}_1 = -5/2$, $\tilde{c}_2 = 1.2478$. Then, one should construct two different approximants, for instance the simplest pair given below by the lowest-order expressions, with m being a control parameter,

$$R_1^*(f) = \left(1 - \frac{\tilde{c}_1}{m}f\right)^{-m}, \quad R_2^*(f) = 1 + \tilde{c}_1 f \left[1 - \frac{\tilde{c}_2 f}{\tilde{c}_1(1+m)}\right]^{-(1+m)}. \tag{8.61}$$

In general, the control parameters m_i for the two approximants may differ, but we set $m_2 = m_1 \equiv m$, to have the same behavior at infinity.

From the first order approximant we estimate threshold as a function of m,

$$f_1^c(m) = \frac{m}{\tilde{c}_1}. \tag{8.62}$$

One would like to have the two solutions to differ from each other minimally. The minimal difference condition between the two approximations is reduced to the condition on stabilizer m. It will be determined as a minimizer of the following expression

$$\left| 1 + \tilde{c}_1 f_1^c(m) \left[1 - \frac{\tilde{c}_2 f_1^c(m)}{\tilde{c}_1(1+m)} \right]^{-(1+m)} \right|, \qquad (8.63)$$

and is located at $m = -1.57646$, leading to the critical index $\mathcal{S} = -m$. Even more reasonable is the corresponding value of threshold $f_1^c(m) = 0.631$, which is very close to the RCP value of 0.637. The original coefficients c_1, c_2, entail practically correct threshold! It gives more credibility to the estimate of critical index $\mathcal{S}$, which does not require any knowledge of the threshold value, but only general assumption of its existence.

8.3.1 Modification with iterated roots. Condition imposed on thresholds

Following the discussion in Section 1.3, one can also select the iterated root as the second approximant, so that

$$\mathcal{R}_1^*(f) = \left(1 + \frac{c_1}{m} f\right)^m, \quad \mathcal{R}_2^*(f) = \left((\mathcal{P}_1(m)f + 1)^2 + \mathcal{P}_2(m)f^2\right)^m, \qquad (8.64)$$

where

$$\mathcal{P}_1(m) = \frac{c_1}{2m}, \quad \mathcal{P}_2(m) = \frac{-2c_1^2 m + c_1^2 + 4c_2 m}{4m^2}.$$

The second-order approximant is still simple enough to allow for an explicit expression for the threshold,

$$f_2^c(m) = \frac{\sqrt{2c_1^2 m^3 - c_1^2 m^2 - 4c_2 m^3} + c_1 m}{c_1^2 m - c_1^2 - 2c_2 n}.$$

Let us demand the threshold to be independent on approximation, i.e.

$$f_1^c(m) = f_2^c(m), \qquad (8.65)$$

and find the stabilizer m [1].

It appears that equivalently we can require the approximants to be the same $\mathcal{R}_1^*(f) = \mathcal{R}_2^*(f)$, meaning that the sequence of iterated approximants converges in one single step, in the fastest possible manner. The control parameter m can be found uniquely in this case to guarantee the equality. Indeed, one can just express the second order approximant in another form,

$$\mathcal{R}_2^*(f) = \left(1 + \frac{c_1}{m} f + b(m) f^2\right)^m,$$

---

[1] See Subsection 1.3.3 for more formal explanation of the minimal difference condition.

where

$$b(m) = \frac{-c_1{}^2(m-1) + 2c_2 m}{2m^2}.$$

Let us set $b(m) = 0$ and find the sought control parameter

$$m = \frac{c_1{}^2}{c_1{}^2 - 2c_2}.$$

The critical index is the same with control parameter, $S = -m = 1.66471$. The found values of critical index agrees well for the different schemes. The found value of threshold $f_2^c = 0.66589$ is also quite reasonable.

8.3.2 Condition imposed on the critical index

What if we simply require that the critical index by itself do not depend on the approximation, since such property should hold for the two sequential approximations as the true fixed point is reached? Of course, in this case we would need to introduce an additional parameter to be determined from this requirement. E.g., one can use the value of threshold as a parameter f_0, and introduce it to the original series (8.60) through the transformation

$$z = \frac{f}{f_0 - f} \Leftrightarrow f = \frac{z f_0}{z + 1}.$$

As usual we have to construct the two low-order approximations,

$$\mathcal{R}_1^*(z) = (1 + p_1(f_0)z)^{m_1(f_0)}, \quad \mathcal{R}_2^*(z) = ((\mathcal{P}_1(f_0)z + 1)^2 + \mathcal{P}_2(f_0)z^2)^{m_2(f_0)/2}.$$

$$(8.66)$$

The parameters of the first-order approximant are defined uniquely, e.g.

$$m_1(f_0) = \frac{c_1{}^2 f_0}{c_1{}^2 f_0 + 2c_1 - 2c_2 f_0}.$$

There are two solutions for the parameters of the second order approximant, and we bring here only the branch which turns to be relevant,

$$m_2(f_0) =$$
$$\frac{3c_1{}^3 f_0{}^2 + 6c_1{}^2 f_0 + c_1 f_0 \sqrt{c_1{}^4 f_0{}^2 + 12c_1{}^3 f_0 - 12c_1{}^2 c_2 f_0{}^2 + 12c_1{}^2 - 24c_1 c_2 f_0 + 36c_2{}^2 f_0{}^2 - 6c_1 c_2 f_0{}^2}}{2(c_1{}^3 f_0{}^2 + 3c_1{}^2 f_0 - 3c_1 c_2 f_0{}^2 + 3c_1 - 6c_2 f_0)}.$$

$$(8.67)$$

The critical index thus can be expressed in two different ways. After finding the explicit expressions the problem consists in minimization of the difference between the two expressions for the critical index

$$|m_2(f_0) - m_1(f_0)|,$$

with respect to the parameter f_0. It turns out that the difference possesses minimum at a very reasonable value of threshold $f_0 = 0.656746$. From the first order approximant

$$\mathcal{R}_1^*(f) = \left(\frac{1 + 1.01373f}{0.656746 - f}\right)^{1.61963},$$

one can find the critical index $\mathcal{S} = 1.61913$.

8.3.3 Conditions imposed on the amplitude

Assume that we do know in advance the correct value of threshold $f_c = 0.637$, and try to use the knowledge in the course of estimating the index. Let us apply the transformation, $z = \frac{f}{f_c - f} \Leftrightarrow f = \frac{zf_c}{z+1}$ to the original series (8.58), with the resulting series

$$\mathcal{R}(z) = 1 + b_1 z + b_2 z^2 + O(z^3) \quad b_1 = 1.5925, \quad b_2 = 0.437238. \tag{8.68}$$

The set of approximations to $\mathcal{R}(z)$ including the two starting terms from (8.68), can be written down and the expression for the renormalized quantity $\mathcal{R}_1^*$ can be readily obtained:

$$\mathcal{R}_1^* = \left(1 - \frac{b_1 z}{s_1}\right)^{-s_1} \longrightarrow \left(\frac{s_1}{-b_1}\right) z^{-s_1}, \text{ as } z \to \infty. \tag{8.69}$$

The control parameter/stabilizer s_1 should be negative, if we want to reproduce in the limit $z \to \infty$, the correct power-low behavior of the effective shear modulus. For comparison, one would also need to construct a different set of approximations. It can be accomplished simply by leaving the constant term from (8.68) outside of the renormalization procedure leading to the approximant

$$\mathcal{R}_2^* = 1 + b_1 z \left[1 - \frac{b_2 z}{b_1(1 + s_2)}\right]^{-(1+s_2)} \longrightarrow \left(-\frac{b_2}{1 + s_2}\right)^{-(1+s_2)} b_1^{2+s_2} z^{-s_2}, \text{ as } z \to \infty. \tag{8.70}$$

Demand now that both (8.69) and (8.70) have the same power-law behavior as $z \to \infty$, so that

$$s_2 = s_1 = s.$$

Require now the fulfillment of the convergence for the two available approximations, imposed in the form of the minimal-difference condition imposed on critical amplitude. One obtains the condition on the negative s to be determined from the minimum of the expression:

$$\left|\left[\left(\frac{-b_2}{1+s}\right)^{-(1+s)} b_1^{(2+s)} - \left(\frac{s}{-b_1}\right)^s\right]\right|. \tag{8.71}$$

Generally speaking, it is sufficient to ask for an extremum of this difference. The nontrivial zero of (8.71) is located at the point $\mathbf{s} = -1.60483$. The final formula which respects (8.58), has the following form:

$$\mathcal{R}_2^*(z) = 1 + b_1 z \left[1 - \frac{b_2 z}{b_1 (1 + \mathbf{s})} \right]^{-(1+\mathbf{s})}. \tag{8.72}$$

Expressed in the original variable it gives the effective shear modulus

$$G_e^*(f) = 1 + \frac{1.5925 \left(\frac{0.637 - 0.546056 f}{0.637 - f} \right)^{0.604834} f}{0.637 - f} \tag{8.73}$$

The critical index $\mathcal{S} = -\mathbf{s} = 1.60483$, and the critical amplitude can be estimated as well, $A = 0.4789$.

Formula (8.73) can be compared with analytical expression from Cichocki et al. (2003),

$$G_e(f) = 1 + \frac{5 \left(0.63 f^2 + 1.0009 f + 1 \right) f}{2 \left(-0.63 f^3 - 1.0009 f^2 - f + 1 \right)}, \tag{8.74}$$

as well as to the theoretical results from Beenakker (1984), given in the numerical form. Experimental data are quoted in Beenakker (1984) for moderately large f. There is a good agreement with (8.74) up to $f = 0.45$ and with Ref. (Beenakker, 1984) till $f = 0.3$. In the interval $0.3 < f < 0.45$ both formulas (8.73) and (8.74) are in a much better agreement with various experimental data than theory of Beenakker (1984). On the other hand, (8.74) definitely fails in the region of high-concentrations, since it predicts an incorrect threshold and index, while (8.73) captures this region close to f_c and predicts index $\mathcal{S}$. As it is noted in Chapter 1, applications of formulas derived by SCM, like the formula (8.74), are restricted because of the "default" approximations within $O(f)$ in its derivation.

8.3.4 Minimal derivative condition

In order to receive the critical index in a different way, we may start with selecting the iterated root for the second approximant,

$$\mathcal{R}_2^*(z) = ((\mathcal{P}_1(\mathbf{s})z + 1)^2 + \mathcal{P}_2(\mathbf{s})z^2)^{\mathbf{s}}, \tag{8.75}$$

where

$$\mathcal{P}_1(\mathbf{s}) = \frac{b_1}{2\mathbf{s}}, \quad \mathcal{P}_2(\mathbf{s}) = \frac{-2b_1^2 \mathbf{s} + b_1^2 + 4b_2 \mathbf{s}}{4\mathbf{s}^2}.$$

The minimal difference condition on the fixed point leads to the simplest $DLog$ Padé approximant, so that one has to find some other way to proceed. Based on the shape of the (8.75) the control parameter $\mathbf{s}$ can be also interpreted as the "critical" index in the vicinity of the (quasi)threshold $f_e^c(\mathbf{s})$. Beneficially for us, the second-order approximant (8.75) is still simple enough to allow for an expression for the quasi-threshold to be found explicitly

$$f_e^c(\mathbf{s}) = \frac{\sqrt{-\mathcal{P}_2(\mathbf{s})} - \mathcal{P}_1(\mathbf{s})}{\mathcal{P}_1(\mathbf{s})^2 + \mathcal{P}_2(\mathbf{s})}. \tag{8.76}$$

It seems natural when the correct critical index at infinity does not depend on the position of a quasi-threshold. Without actually inverting (8.76), one can find the value of $\mathbf{s}$ from the minimal derivative (sensitivity) condition imposed on the quasi-threshold[2]

$$\frac{\partial f_e^c(\mathbf{s})}{\partial \mathbf{s}} = 0,$$

and the solution leads to the critical index $\mathcal{S} = 2\mathbf{s} = 1.788$. This estimate is in line with estimates of Losert et al. (2000), Gluzman and Yukalov (2017b), and is smaller than 2, the value expected in the general case of 3D viscous suspensions (Brady, 1993, Wang and Brady, 2015).

Altogether, through applying different definitions of the critical point, we are left with five fairly close estimates for the critical index $\mathcal{S}$

$$1.57646, \quad 1.60483, \quad 1.61963, \quad 1.66471, \quad 1.788.$$

The solutions are centered around the value of 1.65. Such "framing" of the unknown solution is what one should realistically expect from the methodology based on a short series. Margins of error could be found as well, so that $\mathcal{S} = 1.65 \pm 0.13$. Slightly different estimate, 1.726 ± 0.06, was obtained by Gluzman and Yukalov (2017b).

We stress that techniques based on the short series will work not always. It is necessary to have the dispersion of results to remain tolerable, so that a meaningful conclusion could be inferred. Our solution is reasonable also with respect to a sufficient criterion, such as the predicted threshold.

Scatter is considerable for various experiments on the viscosity of non-Brownian particles in the vicinity of the jamming transition where thermal fluctuations are weak, with the critical index data positioned in the segment [1.67, 2.55] (Kawasaki et al., 2015). With strong thermal fluctuations the colloidal glass transition is observed at for the values of threshold f_c in the range $0.58 - 0.60$, with $\mathcal{S} \approx 2.2 - 2.6$ (see Wang and Brady (2015) and references therein).

Thermal (or Brownian) contribution arises from the so-called "Brownian" (or thermal) forces generated by particles statistical distribution function (Batchelor, 1977). For the dilute Brownian suspensions of small particles in the case of relatively strong Brownian motion, Wajnryb and Dahler (1997) found that $c_2 = 5.9147$, as it includes both the hydrodynamic contribution from (8.58), and additional Brownian contribution. Let us estimate the critical index with given threshold $f_c = 0.6$. The same methodology as above can be applied again. Just like in Subsection 8.3.3, we found

---

[2] See Subsection 1.3.3 for more formal explanation of the minimal derivative condition.

that $S \approx 2.18$. And acting just like in Subsection 8.3.4, we calculate $S \approx 2.66$. Both estimates for the index are in agreement with Wang and Brady (2015).

For practical purposes, notwithstanding the threshold value, it is sufficient to consider a power-law divergence with the accepted value of $S = 2$, as conjectured in Wang and Brady (2015). But why does an approach based on a short series, has a chance to work for the considered phenomena?

1. The first and second order coefficients c_1 and c_2, contain the information about the interaction of the inclusion with the background flow alone.
2. The asymptotic behavior near the random dense packing limit encodes the many-particle interactions.
3. The geometry of the problem enters through the value of random closest packing of hard spheres, which characterizes the typical particle arrangements in the dense limit.

It is likely that any higher order terms in the concentration expansion carry little information not already contained in the expansions. For passive suspensions low concentration expansions are relatively well known, as are the asymptotics near the random dense packing critical point. As long as no other critical phenomena arise in the intermediate range of concentrations, the two asymptotic expansions can be used to effectively reconstruct viscosity (shear modulus) in the full range of concentrations using methodology of Gluzman et al. (2013).

Another valid case of holes/cavities/air bubbles, randomly and without overlaps embedded into elastic matrix, was considered in Chen and Acrivos (1978). Holes are defined just as in Section 3.4, dedicated to the effective elastic constants for the hexagonal array of soft fibers. The effective elastic moduli of an incompressible elastic matrix containing holes at small concentrations f, was found in the linear approximation,

$$G_e(f) \simeq 1 + c_1 f = 1 - \frac{5}{3} f. \tag{8.77}$$

It is also expected that a power-law

$$G_e(f) \propto (f_c - f)^{\mathcal{T}}, \tag{8.78}$$

holds in the vicinity of the 3D-threshold. Here $\mathcal{T}$ is the critical index for elasticity, and $f_c \approx 0.64$, $\mathcal{T} \approx 3.75$ (Torquato, 2002). The value of index is obtained numerically and considered to be confirmed experimentally (Benguigui, 1984, Deptuck et al., 1985, Feng et al., 1987). It is valid for a lattice and continuum systems, unless some special geometries are involved (Feng et al., 1987).

The problem of approximate reconstruction based on the two asymptotics can be solved by analogy to (Gluzman et al., 2013). Method of derivation is the same as used in Eqs. (3.7)–(3.8) from (Gluzman et al., 2013), leading to the closed-form expression in terms of the so-called exponential integral [3]

$$G_e^*(f) = 1.72097 e^{3.75 Ei(1.9644 f - 1.25722)}. \tag{8.79}$$

[3] $Ei(z) = -\int_{-z}^{\infty} \frac{e^{-t}}{t} \, dt.$

Although one can construct some other expressions with same asymptotics (Gluzman et al., 2013), the formula (8.79) appears to be close to their mean. From the formula (8.79) we find the expansion for small f

$$G_e^*(f) = 1 - \frac{5}{3}f - 1.55f^2 + \dots.$$

Thus, our results disagree with the theoretical prediction from Chen and Acrivos (1978), that $c_2 \approx 0.5$. For the inverse moduli we find

$$\left[G_e^*(f)\right]^{-1} = 1 + \frac{5}{3}f + 4.33f^2 + 9.8f^3 + \dots.$$

Surprisingly, we find that estimated coefficients are of the same order of magnitude as in the problem of rigid inclusions considered above in Section 8.3.

8.4 Diffusion coefficients

It is expected that the problem of a translation short-time self-diffusion coefficient $\mathcal{D}_s^t$ in concentrated suspensions can be studied based on the expression for the effective viscosity (Mazur and Geigenmiller, 1987). The coefficient $\mathcal{D}_s^t$ characterizes self-diffusion on a time scale such that the root-mean-square displacement (MSD) $\delta r(t)^2$ along of a single-tagged particle trajectories remains much smaller than the average distance between particles. It appropriate to say the diffusion of a particle happens in a cage formed by the surrounding particles. But most often we are interested in the diffusion unfolding on long times. Both short- and long-time diffusion coefficients for any space dimension d are determined from the initial, small-time and long-time behavior of the MSD for the tagged particle as

$$\mathcal{D} = \frac{\delta r(t)^2}{2dt},$$

as $t \to 0$, or $t \to \infty$, respectively. In both cases we consider quantities normalized by the corresponding diffusion coefficient at infinite dilution $\mathcal{D}_0$. The region of moderate concentrations is typically studied, so that the hydrodynamic interactions with other particles play a role, but not yet the direct hard-sphere interactions. Case of $\delta r(t)^2 \sim t$ corresponds to normal diffusion.

One can also be interested in situations corresponding to anomalous diffusion, when $\delta r(t)^2 \sim t^\beta$ corresponds to anomalous diffusion. Case of $1 < \beta < 2$ corresponds to a super-diffusion, when the mean-square displacement grows faster than linearly with time. For instance, measurements of single-particle trajectories in cytoplasm and their random-walk interpretations discover $\beta \approx 1.5$ (Regner et al., 2013). There is also a crossover in porous media, from anomalous super-diffusion to classic diffusion resulting from processes with a finite correlation length (Koch and Brady, 1988).

Case of $\beta = 2$ is called ballistic. Case of $0 < \beta < 1$ is called sub-diffusion, when the mean-square displacement grows slower than linearly with time. In the case of cytoplasmic experiments with nocodazole-treated extract, it is observed that $\beta \approx 0.6$ in the short time analysis, but $\beta \approx 0.98$ is observed in the long time case (Regner et al., 2013). Similar crossover has been noted in a number of studies of the intra-cellular anomalous diffusion (see Regner et al. (2013) and references therein).

There is a famous example of a temporal crossover in diffusion. Wu and Libchaber (2000) reported a groundbreaking experiment with bacteria moving freely within a two-dimensional fluid film seeded with passive polystyrene beads. They studied the estimates for the mean-square displacement of the beads from their recorded trajectories, as the passive beads are being pushed around by the bacteria. To fit the experimental data for all times, they use the following form

$$\delta r(t)^2 \sim t \left(1 - e^{-\frac{t}{t_c}}\right),\tag{8.80}$$

which follows from Langevin equation for the bead motion with a force term correlated in time over the crossover scale t_c. For $t < t_c$, the motion turns out to be ballistic with $\delta r(t)^2 \sim t^2$, and for $t > t_c$ it is diffusive with $\delta r(t)^2 \sim t$. It could not account for nontrivial exponents, possibly recorded in the experiment at short times, and corresponding *ad hoc* expression was suggested by Gregoire et al. (2001)

$$\delta r(t)^2 \sim \frac{t}{\left(\frac{t}{t_c}\right)^{1-\alpha} + 1},\tag{8.81}$$

with $\alpha > 1$. It does interpolate between a super-diffusive behavior t^α at short times and a standard diffusive behavior at long times.

But the crossover described in Wu and Libchaber (2000), holds for the model system of self-propelling Janus particles with swimming velocity V (Howse et al., 2007, Palacci et al., 2010). As $t < t_c$, the motion is ballistic with $\delta r(t)^2 \simeq V^2 t^2$. At longer times $t > t_c$, the motion becomes diffusive with $\delta r(t)^2 \simeq D_{eff} t$, where $D_{eff} = D_0 + \frac{V^2 t_c}{6}$. At times long the particle undergoes a random walk whose step length is the product of the propelled velocity estimated in the interval between $V \approx 0.3\,\mu\mathrm{m\,s}^{-1}$ and $3.3\,\mu\mathrm{m\,s}^{-1}$, and of the typical crossover diffusion time $t_c \approx 0.9$ s, leading to an enhancement of the effective diffusion coefficient over the classical passive value $D_0 \approx 0.34\,\mu\mathrm{m}^2\,\mathrm{s}^{-1}$.

From the knowledge of two asymptotic regimes one can restore the behavior for all times using the two approximations (8.80), (8.81). Typically, they are employed for fitting data, but can be used as approximants as well. Functional dependencies in the form of root approximants are suited even for more general situations, with t^α for small times, and t^β for long times

$$\delta r(t)^2 \sim t^\alpha \left(\frac{t}{t_c} + 1\right)^{\beta - \alpha} = \left(\frac{t_c}{t} + 1\right)^{\beta - \alpha} t^\beta t_c^{\alpha - \beta},$$

as well as to the case with two (or more) distinct crossover scales and indices,

$$\delta r(t)^2 \sim t^\alpha \left(\left(\frac{t}{t_{c_1}} + 1\right)^{m_1} + \left(\frac{t}{t_{c_2}}\right)^2\right)^{\frac{\beta - \alpha}{2}}$$

$$= t^\beta t_{c_2}^{\frac{\alpha-\beta}{2}} \left(\left(\frac{t}{t_*} \right)^{m_1-2} \left(\frac{t_{c_1}}{t} + 1 \right)^{m_1} + 1 \right)^{\frac{\beta-\alpha}{2}}, \tag{8.82}$$

with $t_{c_1} < t_{c_2}$, $m_1 < 2$, and $t_*^{m_1-2} = \frac{t_{c_1}^{m_1}}{t_{c_2}^2}$. For $t < t_{c_1}$ and $t > t_*$ it reproduces the sought behaviors, and at $t_{c_1} < t < t_*$ there is a regime with confluent singularities, $\delta r(t)^2 \sim t^\beta (1 + \mathcal{P} t^{m_1-2} + \ldots)$. Similar expressions were employed for polymers, where analogous crossovers are known (Yukalov and Gluzman, 1999a), (Valle et al., 2005).

Crossover has been observed also in the motion of tracer beads in the flow near activated solid-fluid interface, with surfaces covered by "bacterial carpets", as well as for "auto-mobile beads" propelled by adsorbed bacteria (Darnton et al., 2004).

It was suggested in Mazur and Geigenmiller (1987) to take into account the influence of the hydrodynamic interactions by using an effective viscosity μ_{eff} in a Stokes–Einstein like expression for the translations self-diffusion coefficient $\mathcal{D}_s^t$ (with sphere radius r_s)

$$\mathcal{D}_s^t = \frac{k_b T}{6\pi \mu_{eff} r_s}, \tag{8.83}$$

and the simple formula $\mathcal{D}_s^t = \mathcal{D}_0^t \frac{1-f}{\frac{3f}{2}+1}$, was deduced, with the first-order coefficient $c_1 = -5/2$, corresponding to the number found by Einstein for the effective viscosity, see (6.2). The formula turns out to be reasonably accurate only for small concentrations. It is not surprising since such a replacement is equivalent to application of a self-consistent method valid within the first order approximation in f.

There is also some asymptotic information which should be incorporated into the final formula. Experiment and dynamics simulation show that the short-time diffusion, or self-diffusion coefficient goes to zero at random close packing (Brady, 1993). The question concerning the value of critical index for the coefficient $\mathcal{D}_s^t$ in the vicinity of RCP, still remains open. Currently, the most plausible is empirical Brady's linear law (Brady, 1993),

$$\mathcal{D}_s^t \propto 0.85 \left(1 - \frac{f}{f_c} \right) \approx 1.328 \, (0.64 - f), \tag{8.84}$$

with $f_c \approx 0.64$. It could be accepted based on available experimental and numerical data. The particles are assumed to interact as perfect Brownian hard spheres, or through the short-range inter-particle forces. In addition, for small f, the following expansion for the self-diffusion coefficient can be obtained directly (Batchelor, 1976),

$$\mathcal{D}_s^t / \mathcal{D}_0^t \simeq 1 - 1.83 f, \tag{8.85}$$

and could be incorporated into such approach as well. For comparison, the empirical Pearson–Shikata formula (Pearson and Shikata, 1994)

$$\mathcal{D}_s^t = \mathcal{D}_0^t (1 - 1.83 f + 0.4217 f^2),$$

gives also the estimate for the second-order coefficient $c_2 \approx 0.42$. In contrast, the theory of Clercx and Schram (1992), Cichocki et al. (1999) gives negative $c_2 \approx -0.2$.

The problem of approximate reconstruction based on the two asymptotics (8.84) and (8.85), can be solved by analogy to Eqs. (3.7)–(3.8) from Gluzman et al. (2013), leading to the closed-form expression in terms of the exponential integral,

$$\mathcal{D}_s^t = \mathcal{D}_0^t(3.01399e^{Ei(0.158029-0.24692f)}). \tag{8.86}$$

Although one can construct some other expressions with same asymptotics following Gluzman et al. (2013), the formula (8.86) appears to be close to their mean. From the formula (8.86) we find the expansion for small f

$$\mathcal{D}_s^t = \mathcal{D}_0^t(1 - 1.83f + 0.470694f^2 - 0.0910495f^3 + \ldots),$$

and the critical amplitude $A = 1.3255$. Thus, our results are in a good agreement with Brady's and Pearson–Shikata fits, while disagree with the theoretical prediction for c_2 (Clercx and Schram, 1992, Cichocki et al., 1999).

Consider also the case of the long-time diffusion of tracers in the random media. Its minimal model variant corresponds to the problem of site percolation diffusion studied within the Lorenz 2D gas model (Nieuwenhuizen et al., 1986). Through the diffusion coefficient for the tracer $\mathcal{D}_l$, one can express the macroscopic conductivity. The diffusion ceases to exist at the critical density of the excluded sites. If f stands for the concentration of conducting or not excluded sites in the Lorenz model, then $x = 1 - f$ is the concentration of excluded sites. For the Lorenz model explained above, the long-time diffusion coefficient for small x is expressed as an expansion

$$\mathcal{D}_l/\mathcal{D}_0 \simeq 1 - 2.14159x - 0.85571x^2.$$

In the vicinity of the site percolation threshold x_c the diffusion coefficient goes to zero,

$$\mathcal{D}_l/\mathcal{D}_0 \sim (x_c - x)^t \qquad (x \to x_c - 0),$$

with the best known value of $x_c = 0.4073$, $t = 1.310$. The critical index t is the very same critical exponent for the conductivity.

Let us use the method explained in Subsection 8.3.2, with minimization condition imposed on the two different explicitly written expressions for the critical index as function of the variation parameter x_0, which has the meaning of an approximate position of the percolation threshold. It appears that such difference possesses zero at $x_c = x_0 \approx 0.437$, and the critical index is calculated very accurately as $t \approx 1.31$. The complete formula for all x can be deduced in the course of calculations,

$$\mathcal{D}_l/\mathcal{D}_0 = \left(1 + \frac{0.714413x}{0.437167 - x}\right)^{-1.31049}. \tag{8.87}$$

It turns out that such threshold and index hold for the realistic model of tracer propagation in the disordered heterogeneous media (Skinner et al., 2013).

The scatterers should not necessarily be restricted to the regular lattice positions. They could be distributed randomly and independently in whole space (Hofling et al., 2006). In this case there is an analogy between Lorenz gas and continuum percolation when particles can overlap. In 2D the value of critical index remains the same as in the lattice model (Bauer et al., 2010) despite overlaps. But in 3D the value of critical index for continuum percolation $t = 2.88$ is considerably larger than its lattice counterpart $t \approx 2$ (Hofling et al., 2006). It is also understood that the dynamics of tracers ensembles in the soft heterogeneous media is different from the dynamics of their hard-sphere counterparts. Mapping to Lorenz model is possible only when tracers all have exactly the same energy, in a sharp contrast with the hard-sphere systems. For an ideal gas of tracers confined in a soft matrix the localization transition in the Lorentz model becomes rounded, and the typical power-laws and a sharp transition point where all particles movements are blocked, cannot be identified (Schnyder, 2014).

Similar ideas are applied to 3D systems with overlapping scatterers. For instance, the diffusion in cytoplasm is heavily influenced by effects of excluded volume due to internal membranes and cytoskeletal structures (Novak et al., 2009). Numerical modeling and homogenization are used to find the long-time diffusion coefficient for all volume fractions of obstacles. The problem is conveniently reduced to the standard case of the effective conductivity of non-conducting inclusions of various shapes, such as disks and cylinders.

8.5 Non-local diffusion

Mind that Onsager put forward a definitive regression hypothesis, which asserts that "The relaxation of macroscopic non-equilibrium disturbances is governed by the same laws as the regression of spontaneous microscopic fluctuations in an equilibrium system". In particular, the average regression of density fluctuations described by corresponding auto-correlation function will obey the same law as the corresponding macroscopic irreversible diffusion process. It is accepted practice nowadays to deduce the diffusion coefficients and alike, not from macroscopic observation of the diffusion process, but rather by statistical analysis of the observed regression of density fluctuations. In this practical sense, Onsager's regression hypothesis is universally accepted, and leads to the powerful technique, discussed below.

To study various crossovers in the diffusion one has to recognize that we are dealing with a non-stationary, spatially non-uniform process. In such case the diffusion coefficient is a function of space and time, or of the wave vector $\mathbf{q}$ (the wave number $q = |\mathbf{q}|$) and frequency ω, when expressed in terms of the Fourier transform. To study the complete dependence on q, ω, or q, t, one can use even more direct method based on the so-called intermediate scattering function (Segre and Pusey, 1997). It is directly measurable in dynamic light scattering experiments through interaction of coherent light with density fluctuations in a small volume. The fluctuations can be of polymers in solution, or of nanoparticles in suspension. The intermediate scattering function $F(q,t)$ is the auto-correlation function of the spatial Fourier components of spontaneous fluctuations in the number density of particles. For a N-particle system with coordinantes $\vec{r}_i(t)$, the space- and- time dependent local number density

$$\rho(\mathbf{r},t) = \sum_{i=1}^{N} \delta\left(\mathbf{r} - \vec{r}_i(t)\right),$$

and the intermediate scattering function is the auto-correlation function of the Fourier components of the local density $\rho_{q,t}$ is given as follows:

$$F(q,t) = \frac{1}{N} \langle \rho_q(t)\rho_{-q}(0)\rangle.$$

The averaging is performed over statistical ensemble. Also $F(q,t)$ is considered as normalized to unity as $t = 0$.

Number-vector-and-time dependent diffusion coefficient is defined as follows:

$$\mathcal{D}(q,t) = -\frac{1}{q^2}\frac{\partial \log(F(q,t))}{\partial t}, \tag{8.88}$$

describing a diffusive response in the suspending liquid, that propagates instantaneously, to the long-time processes associated with structural rearrangement or diffusion over larger distances. The corresponding short-time and long-time diffusion coefficients are defined respectively as follows:

$$\mathcal{D}_s = \mathcal{D}(q,0), \quad \mathcal{D}_l = \mathcal{D}(q,\infty).$$

They are typically obtained from fits to the functional forms (Segre and Pusey, 1997). Such as initial rapid Brownian decay at short times

$$F(q,t) \simeq 1 - \mathcal{D}_s(q)q^2t, \quad t \to 0,$$

and final, very slow Brownian decay at the longest times,

$$F(q,t) \sim \exp\left(-\mathcal{D}_l(q)q^2t\right), \quad t \to \infty.$$

To capture also an intermediate region of non-exponential behavior one can develop an approximate form

$$F(q,t) = \exp\left(-\mathcal{D}_l(q)q^2t + (\mathcal{D}_l - \mathcal{D}_s)q^2t\exp(-D_cq^2t)\right),$$

with fitting parameter D_c. Or, alternatively, we suggest the following approximant

$$F(q,t) = \exp\left(-\mathcal{D}_s(q)q^2t\frac{1+\mathcal{D}_l(q)q^2t}{1+\mathcal{D}_s(q)q^2t}\right),$$

which satisfy both asymptotic requirements without additional parameters. One can think about $F(q, t)$ as a time-dependent structure factor, since the static structure factor $S(q) = F(q, 0)$. It can be computed as discussed by Zhang (2016).

Interesting that in passive suspensions, if there is a peak of structure factor at the position q_m, then there is a minimum of the effective diffusion $D_s(q)$, or the de-Gennes narrowing–the initial diffusive decay of density fluctuations is slowest at q_m. It would be challenging to answer the question, how does the effect just mentioned, transforms due to the particles activity in biological suspensions, at short and moderate times.

One can apply Fourier transformation to $F(\vec{q}, t)$ also in the temporal domain, and obtain the dynamic structure factor

$$S(q, \omega) = \int_{-\infty}^{\infty} F(q, t) \exp(ti\omega) \, dt,$$

which is directly measurable in light scattering experiments, giving the diffusion coefficients from studying the central peak (Froba and Leipertz, 2005). The behavior of density fluctuations spectrum is found from the simple pole $\omega = -i\omega_c$, in the dynamic structure factor.

We shall discuss below the behavior of density–density correlations in liquid systems near their critical point T_c, where ω_c can be reconstructed from some general asymptotic information. The distance of the temperature from the critical point is measured by $|\frac{T-T_c}{T_c}|$. Only long-time relaxation properties are considered. The characteristic frequency is proportional to inverse characteristic relaxation time $\tau = \frac{1}{\omega_c}$ of the density fluctuations. The characteristic freguency divided by q^2, gives q-dependent, non-local diffusion coefficient (see, e.g., Kostko et al. (2004)).

The dynamic scaling hypothesis states that the characteristic frequency obeys the following ansatz

$$\omega_c(\zeta, q) = q^z \Omega \left(\frac{q}{\zeta}\right) \equiv D_T q^2 \omega_c \left(\frac{q}{\zeta}\right),$$

where z is the dynamical critical index, ζ stands for the inverse correlation length where D_T is a thermal diffusivity (Stanley, 1971). Let us find an approximation for the non-dimensional function ω_c (or Ω) and estimate index z. A non-trivial feature of simple liquids (or binary solutions in the vicinity of their critical point), is the presence of a mesoscopic structural length scale ζ^{-1} intermediate between the atomic and macroscopic scales. Asymptotic expansions in the hydrodynamic regime ($\frac{q}{\zeta} \ll 1$) and in the fluctuation regime ($\frac{q}{\zeta} \gg 1$) are available in general form for isotropic fluid (Stanley, 1971):

$$\omega_c(\zeta, q) \simeq D_T q^2 \left(1 + B \left(\frac{q}{\zeta}\right)^2 + \ldots\right), \quad B > 0 \quad (\frac{q}{\zeta} \ll 1),$$

$$\omega_c(\zeta, q) \simeq A q^z \left(1 + A' \left(\frac{\zeta}{q}\right)^2 + \ldots\right), \quad (\frac{q}{\zeta} \gg 1). \tag{8.89}$$

The value of amplitude B could be accurately estimated as $B = 3/5$ (Kawasaki, 1970). Assume that the value of A is known. Applying the technique of self-similar approximants in its simplest form, one can reconstruct the analytical expression for the characteristic frequency for arbitrary $\frac{q}{\zeta}$ obtaining the following crossover approximation (Gluzman and Yukalov, 1998),

$$\omega_c(\zeta, q) = D_T q^2 \left(1 + C \left(\frac{q}{\zeta}\right)^2\right)^{\frac{z}{2} - 1}, \tag{8.90}$$

where C is constant,

$$C = \zeta^2 \left(\frac{A}{D_T}\right)^{\frac{2}{z-2}}. \tag{8.91}$$

If now we plug into expression (8.91) the dependencies of $D_T \sim \epsilon^{\gamma - a}$, and $\zeta \sim \epsilon^\nu$, then we recover the well known relation

$$z = 2 + \frac{(\gamma - a)}{\nu}, \tag{8.92}$$

between the dynamic critical index z, and three other critical indices $\gamma \approx 1.239$, $\nu \approx 0.63$ and a, which represents one of the central results of the dynamical scaling hypothesis. Here γ stands for the critical index for heat capacity, a is the critical index for thermal conductivity, and ν is defined as the correlation length critical index. The combination $a' = \gamma - a$ is called the critical index for thermal diffusivity. Indeed, from (8.91) it follows then that

$$A = A(\epsilon) = D_T \zeta^{2-z} C^{\frac{z}{2} - 1},$$

but in such a way that physical requirement $A(0)$=constant, is satisfied only when (8.92) is valid. It was convincingly argued (see e.g. Hohenberg and Galperin (1977)) that the index a' is determined from the analogy with the concentration (long time) diffusion coefficient, which in turn can be estimated from the analog of Stokes formula (8.83) with r_s replaced with the correlation length $R = \zeta^{-1}$ (see Hohenberg and Galperin (1977), Kostko et al. (2004) for the complete expression). Then the value of a' is dominated by the correlation length index, if the effective viscosity is assumed to remain finite at the critical point as $\epsilon = 0$. Thus, $a' \approx \nu$. From the scaling relation (8.92) the dynamic critical index could then be estimated as $z \approx 3$. But, in fact, there is also a small correction to z originating supposedly from weakly divergent viscosity. The divergence is characterized with the index $x_\mu \approx 0.068$ (measured in units of the critical index ν), so that $z \equiv z(x_\mu) = 2 + (1 + x_\mu)$. The discussion of different estimates for x_μ can be found in Bhattacharjee et al. (2012).

Thence it takes longer time for density fluctuations with short wavelength to relax, compared with hydrodynamic, long wavelength regime. The unknown C can be estimated from the asymptotic equivalence with the expansion in the hydrodynamic regime $C = \frac{2B}{z-2}$, for instance $C \approx 1.2$. To calculate the thermal diffusivity D_T one can simply use the diffusion coefficient $\mathcal{D}_l = \mathcal{D}(0, \infty)$ approximated by Stokes formula (Hohenberg and Galperin, 1977, Kostko et al., 2004).[4]

The non-dimensional part ω_c from (8.90) is approximated by simple expression

$$\omega_c = \omega_{nd} = \sqrt{1 + 1.2x^2}, \tag{8.93}$$

with $x = \frac{q}{\zeta}$. And its asymptotic behavior as $x \to 0$

$$\omega_{nd}(x) \simeq 1 + 0.6x^2 - 0.18x^4 + \ldots, \tag{8.94}$$

while as $x \to \infty$

$$\omega_{nd}(x) \simeq 1.09545x + 0.456435x^{-1} - 0.0950907x^{-3}, \tag{8.95}$$

are in agreement with the expressions (8.89), as they obey the dynamic scaling in all orders. It is known that in practice, the simple dependence $\sqrt{1 + x^2}$ (Ackerson et al., 1975, Ackerson and Straty, 1978, Doyeux et al., 2016), serves as a reasonable description in the light scattering experiments.

The best known theoretical result for the non-dimensional part

$$\omega_c = \omega_{kaw}(x) = \frac{3\left(\left(x^3 - \frac{1}{x}\right)\tan^{-1}(x) + x^2 + 1\right)}{4x^2}, \tag{8.96}$$

is due to Kawasaki (1970) (see also (Hohenberg and Galperin, 1977)), with the following asymptotic behavior as $x \to 0$

$$\omega_{kaw}(x) \simeq 1 + \frac{3x^2}{5} - \frac{x^4}{7} + \frac{x^6}{15}, \tag{8.97}$$

and as $x \to \infty$

$$\omega_{kaw}(x) \simeq \frac{3\pi}{8}x + \frac{1}{x^2} - \frac{3\pi}{8x^3}. \tag{8.98}$$

While (8.97) is in agreement with the dynamic scaling in the limit given by the first equation in (8.89), the second limit (8.98) does agree with the leading term in (8.89), but violates the dynamic scaling by means of x^{-2} term in (8.98). Besides, the term $O(x^{-1})$ is altogether absent. Similar features persist in higher orders.

Thus, Kawasaki approximation (Kawasaki, 1970) is "only" correct as far as the leading term is considered for $x \to \infty$. But the dynamic scaling can be restored in all orders, if we employ some form of approximant extending (8.93). Such approximant will keep asymptotically a few terms from the correct Kawasaki limit $x \to 0$,

---

[4] Formula (8.90) can be applied also when $B < 0$, and may describe a monotonous, or non-monotonous crossover from normal diffusion to super-diffusion, with possible relevance to porous media (Koch and Brady, 1988). E.g., with $\beta = 1.5$, $z = 1.33$, and there would be anomalous, faster than normal super-diffusive behavior of the density fluctuations. As $\beta = 2$, in the ballistic regime, we find $z = 1$, and there is a propagating sound-like mode, instead of relaxation.

and the leading term as $x \to \infty$, while canceling out the terms violating scaling. It turns out that already simplest factor approximants give rather good approximation to the Kawasaki function, while restoring the dynamic scaling in all orders. In fact, the low-order factor approximants

$$F_2^*(x) = \left(1.07619x^2 + 1\right)^{0.557522}, \quad F_3^*(x) = \frac{\left(x^2+1\right)^{0.714286}}{\left(0.6x^2+1\right)^{0.190476}},$$

even manage to predict the expected value of $\frac{z-1}{2} \approx 0.534$ with good accuracy. Based on this observation, one can suggest a simple factor approximant to the non-dimensional part of ω_c, just generalizing the simplest forms presented above to an arbitrary $z(x_\mu)$. For instance, the second-order factor approximant satisfies just one nontrivial coefficient from (8.97), and is designed to give the correct critical index, as shown below

$$F_2^*(x, x_\mu) = \left(1 + \frac{6x^2}{5(x_\mu+1)}\right)^{\frac{x_\mu+1}{2}}. \tag{8.99}$$

The next-order factor approximant satisfies just two nontrivial coefficients from (8.97), and also is designed to give the correct critical index $z = 2 + (1 + x_\mu)$. The Kawasaki-case corresponds to $x_\mu = 0$, in the formula

$$F_3^*(x, x_\mu) = \left(x^2 + 1\right)^{\frac{13-113x_\mu}{19-175x_\mu}} \left(\frac{16x^2}{35-175x_\mu} + 1\right)^{\frac{7(1-5x_\mu)^2}{350x_\mu-38}}. \tag{8.100}$$

The higher-order approximant obeys asymptotically the three nontrivial coefficients from (8.97), and also gives the correct critical index z. It also has a closed form, but with lengthier expressions for the parameters,

$$F_4^*(x, x_\mu) = \left(x^2 P_1\left(x_\mu\right) + 1\right)^{m_1(x_\mu)} \left(x^2 P_2\left(x_\mu\right) + 1\right)^{\frac{1}{2}(x_\mu+1)-m_1(x_\mu)}, \tag{8.101}$$

where

$$P_1(x_\mu) = \frac{4123x_\mu + \sqrt{7}\sqrt{(961x_\mu-61)(2527x_\mu-227)} - 623}{7910x_\mu-910},$$

$$P_2(x_\mu) = \frac{4123x_\mu - \sqrt{7}\sqrt{(961x_\mu-61)(2527x_\mu-227)} - 623}{7910x_\mu-910},$$

$$m_1(x_\mu) =$$
$$\frac{x_\mu\left(-589\sqrt{7}x_\mu + \sqrt{(961x_\mu-61)(2527x_\mu-227)} + 856\sqrt{7}\right) + \sqrt{(961x_\mu-61)(2527x_\mu-227)} - 67\sqrt{7}}{4\sqrt{(961x_\mu-61)(2527x_\mu-227)}}.$$

$$\tag{8.102}$$

In the higher-order we find numerically, using five coefficients and critical index,

$$F_6^*(x, 0) = \frac{\left(1.00883x^2+1\right)^{0.687631}}{\left(0.177485x^2+1\right)^{0.0513135}\left(0.620581x^2+1\right)^{0.136318}}. \tag{8.103}$$

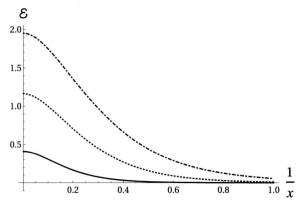

Figure 8.7 Factor approximants to the Kawasaki function are compared with reference expression (8.96), while $x_\mu = 0$. The relative error is shown along the y-axis, dependent on the parameter $x^{-1} = \frac{\zeta}{q}$. The error obtained with formula (8.103), is shown with solid line. The error obtained with formula (8.101), is shown with dotted line; and the error for the formula (8.100), is shown with dot–dashed line.

The three approximants are compared with reference expression (8.96), as shown in Fig. 8.7.

If one is interested in even more accurate approximation of the Kawasaki function, it makes sense to impose the condition on the correct amplitude as $x \to \infty$, and construct the following approximant

$$\mathcal{F}_6^*(x, 0) = \frac{\left(1.01306x^2+1\right)^{0.667659}}{\left(0.106805x^2+1\right)^{0.031412}\left(0.535956x^2+1\right)^{0.136247}}. \tag{8.104}$$

Now, the Kawasaki function is reconstructed with maximal error of just 0.07%. At large x, we find

$$\mathcal{F}_6^*(x, 0) = 1.1781x + 0.130454x^{-1} + \dots.$$

Approximants (8.100), (8.101) invariably respect the dynamic scaling, and can be used also with a non-zero x_μ, when Kawasaki function is not rigorously applicable. In such case we rely on the fact that Kawasaki approximation works well at small x, at least in leading order (Stanley, 1971). Kawasaki function is often used as a part of the corrected formulas with corrected value of the critical index z, such as (Paladin and Peliti, 1982),

$$\omega_{pp}(x) = \left(x^2 + 1\right)^{x_\mu} \omega_{kaw}(x)^{1-x_\mu}, \quad x_\mu = 0.07. \tag{8.105}$$

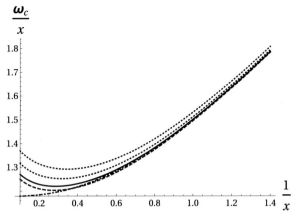

Figure 8.8 Factor approximant (8.99) divided by x is shown with dashed line, as $x_\mu = 0.068$. Factor approximant (8.101) divided by x is shown with solid line, as $x_\mu = 0.068$. They are compared with expressions (8.106) (dotted), and (8.105) (also shown with dotted line corresponding to the upper curve). Kawasaki formula (8.96), is shown with dot–dashed line. The non-dimensional ratio $\frac{\omega_c}{x}$ is shown along the y-axis, dependent on the parameter $x^{-1} = \frac{\zeta}{q}$.

Another expression is given in Burstyn et al. (1983), Kostko et al. (2004), Sengers et al. (2009)[5]

$$\omega_{bsbg}(x) = \omega_{kaw}\left(1 + \frac{1}{4}x^2\right)^{x_\mu/2} \qquad x_\mu = 0.068. \qquad (8.106)$$

Factor approximants (8.99), (8.101) are compared with the two approximations based on Kawasaki function and with reference Kawasaki function itself, as shown in Fig. 8.8. Factor approximants deviate from (8.96) much less than two other formulas. Possibly, such difference is above the typical experimental error of 10%.

8.6 Sedimentation. Particle mobility

Sedimentation is fundamental problem of studying how a suspension moves under gravity. The suspension is build of a small rigid spheres ith random positions, which are falling through Newtonian fluid under gravity. A homogeneous mixture of solid particles and a fluid is allowed to stand in a container. The particles settle out under gravity at a rate which depends in particular on concentration, the dependence originating from the hydrodynamic interaction between particles.

The basic quantity of interest is the sedimentation velocity U, which is the averaged velocity of suspended particles, measured with respect to the velocity U_0 with which a single particle would move in the suspending fluid under the given force field

[5] Formulas (27), (28) from Sengers et al. (2009) contain some misprints, corrected in (8.106).

in absence of any other particles. The ratio U/U_0 is the collective mobility, or sedimentation coefficient. The dependence of the collective mobility at very low packing fractions is determined by the single-particle mobility, but the ratio decreases at high packing fractions. The problem of the sedimentation recalls that of a Darcy flow in a porous medium. However, the relation is not simple because the physics of the particle interactions are quite different in the two problems.

Is there a simple formula for the sedimentation coefficient valid for arbitrary concentration of particles in the suspension? At present time, we do not have a rigorous formula originating from the theory developed in previous chapters. Below, we follow empirical and semi-empirical observations and develop formulas based on some physical intuition. Mind that according to Stokes, $U_0 \sim \frac{1}{\mu_0}$, where μ is an ambient fluid viscosity. Then one might expect the solution in the form of

$$U \sim \frac{1}{\mu_e(f)}, \tag{8.107}$$

where $\mu_e(f)$ is the effective viscosity for the movement of the particle in a suspension. The dependence on f arises from the interaction between particles, exerted by means of the velocity distribution generated in the fluid surrounding each moving particle (Batchelor, 1972)[6]. The Maude-Whitmore formula from Batchelor (1972), for the observed mean settling velocity of identical spheres valid over a wide range of volume fractions f is given as follows:

$$\frac{U}{U_0} = (1 - f)^5. \tag{8.108}$$

Even the most recent attempt to correlate various data (Gilleland et al., 2011),

$$\frac{U}{U_0} = (1 - f)^{5.4}, \tag{8.109}$$

is confined to the same simple class. Slightly different value of 5.5 was measured (Buzzaccaro et al., 2008).

Another formula of the same type (Brady and Durlofsky, 1988, Ladd, 1990) for the sedimentation velocity is based on Rotne–Prager hydrodynamics,

$$\frac{U}{U_0} = \frac{(1 - f)^3}{2f + 1}. \tag{8.110}$$

It was improved by Beenakker and Mazur as quoted in Hayakawa and Ichiki (1995),

$$\frac{U}{U_0} = \frac{(1 - f)^3}{(2f + 1)\left(\frac{5f}{2} + 1\right)}, \tag{8.111}$$

[6] Stokes demonstrated that terminal settling velocity of a sphere in a fluid was inversely proportional to the fluid's viscosity and directly proportional to the density difference of fluid and solid, the radius of the sphere involved, and the force of gravity. Stokes equation is valid, however, only for very small spheres.

and by Hayakawa and Ichiki (1995)

$$\frac{U}{U_0} = \frac{(1-f)^3}{1.492 f (1-f)^3 + 2f + 1}. \tag{8.112}$$

Motivated by these empirical and semi-empirical formulas, we are going to look for a simple expression satisfying for large f the power law

$$\frac{U}{U_0} \sim (1-f)^\beta,$$

with positive $\beta \approx 3$. On the other hand, for small f the following single-term expansion formula was found in Batchelor (1972)

$$\frac{U}{U_0} = 1 - 6.55 f + O(f^2), \tag{8.113}$$

and already c_2 is unknown. Its value will be estimated below.

Let us employ various resummation techniques tailored for very short series, and estimate the dependence for all f as in Gluzman et al. (2013). We would like to suggest an accurate expression satisfying (8.113), under additional asymptotic condition in the form of power-law. Let us find for U/U_0 the simplest approximant with fixed position of singularity and given critical index, so that

$$\frac{U}{U_0} = \left(1 + \frac{2.18333 f}{1-f}\right)^{-3}. \tag{8.114}$$

The value of the second-order coefficient $c_2 \approx 22.05$, can be found. One can also estimate $c_3 \approx -53.425$.

Another simple formula can be obtained similarly to (Gluzman et al., 2013, eq. 3.6),

$$\frac{U}{U_0} = e^{-3.55 f} (1-f)^3, \tag{8.115}$$

and it gives close value for $c_2 \approx 19.95$. Both estimates for c_2 do not disagree much with the value of 21.918 given in Cichocki et al. (2002), although the method of Cichocki et al. (2002) is questionable in principle, see Appendix A.4.

Three other formulas are shown in Fig. 8.9, and compared with (8.114) and simulation results of Ladd (1990). The new formula (8.114) is better in comparison with simulation results, and we will also employ the estimated value of c_2 for further estimates. Barnea–Mizrahi formula (Barnea and Mizrahi, 1973),

$$\frac{U}{U_0} = \frac{(1-f)^2}{(\sqrt[3]{f} + 1) \exp(\frac{5f}{3(1-f)})}, \tag{8.116}$$

also works rather well in the region of high concentrations, but gives different and apparently worse results as $f < 0.2$, when compared to the data of Ladd (1990).

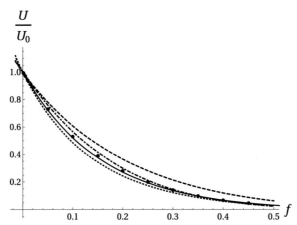

Figure 8.9 Results for the sedimentation coefficient up to the value of $f = 0.5$ obtained with formula (8.114), are shown with solid line. They are compared with (8.111) (dotted). Formula (8.109) is shown with dot–dashed line, and formula (8.110) is shown with dashed line. Simulation results from Ladd (1990) are presented as well with dots.

All formulas presented above are expected to be applicable up to $f \approx 0.5$, where the suspension made of hard spheres freezes. If one to consider a metastable branch of the suspension, one may look for a formula up to the jamming threshold of 0.637. Assume also that in the vicinity of jamming threshold,

$$U \sim (0.637 - f)^{1+\kappa},$$

with unity to be expected as from the usual contribution from the (inverse) radial distribution function at the particles contact $G(2, f)$ (Brady, 1993, Losert et al., 2000), and the value of index κ coming from the hydrodynamic interactions in the suspension. Let us calculate κ.

One can suggest a simple root approximant

$$r(z) = \left[1 + b_1 z (1 + b_2 z)^\kappa\right]^{-1}, \quad z = \frac{f}{0.637 - f},$$

which, by design, takes into account the unity contribution to the index. After imposing the asymptotic equivalence with (8.114), we find

$$\frac{U}{U_0}(f) = \left(1 + \frac{4.17235 f \left(1 + \frac{1.4814 f}{0.637 - f}\right)^{0.6937}}{0.637 - f}\right)^{-1}. \tag{8.117}$$

Thus, we estimate $\kappa \approx 0.7$. We find, just as suggested in (8.107), that the total value of the critical index equals 1.7, and is close to the direct estimates for the critical index for effective viscosity /shear modulus obtained in Subsection 8.3 and in (Gluzman and Yukalov, 2017b).

Table 8.1 Sequences of critical indices for
the index $\beta = 1 + \kappa$, describing sedimentation.
β_k obtained from the optimization conditions
$\Delta_{kn}(\beta_k) = 0$.

β_k	$\Delta_{k\,k+1}(\beta_k) = 0$	$\Delta_{k\,3}(\beta_k) = 0$
β_1	1.97	1.87
β_2	1.77	1.77

Let us also employ the methodology used to construct the table of indices, with the method of construction in great detail explained on page 45. It is based on considering iterated root approximants as functions of the critical index by itself. The index is to be found by imposing optimization condition in the form of minimal difference on critical amplitudes. From one known coefficient c_1 and two higher-order coefficients estimated from (8.114), one can get three estimates for the critical index β, shown in Table 8.1. The result is still fairly close to 1.7.

Simple average of the three values from Table 8.1 gives a bit higher estimate $\beta \approx 1.87 \pm 0.1$. The value of critical index obtained in Section 8.3 from the direct expansion for the viscosity, well agrees with the value deduced in this section from the expansion for sedimentation rate, as anticipated by the relation (8.107).

8.7 Polymer coil in 2D and 3D

The calculation of the expansion function Υ of a flexible polymer coil is of long standing interest in polymer science (see also Subsubsection 1.3.3.4). This quantity defines the ratio of the mean square end-to-end distance $\langle R^2 \rangle$ of the chain to its unperturbed value $\langle R^2 \rangle_0 = Nl^2$, where N is the number of segments with the length l each, so that Nl is the contour length of the chain, $\Upsilon(g) = \frac{\langle R^2 \rangle}{\langle R^2 \rangle_0}$, as a function of a dimensionless interaction parameter g. The latter is $g = \left(\frac{3}{2\pi} \right)^{3/2} \frac{b\sqrt{N}}{l^3}$, for the three-dimensional coil, where b is the effective binary cluster integral of segments (Muthukumar and Nickel, 1984, 1987), which measures the excluded volume strength due to the repulsion between segments.

An isolated polymer chain is represented as a random-flight chain with δ-function interaction between the two segments. A path of this flight of N-steps represents a conformation of a polymer with excluded volume. Excluded volume means that polymer segments can not occupy space that is already occupied by another part of the same polymer. Because of such condition there is a swelling effect in $\langle R^2 \rangle$, quantified by $\Upsilon(g)$. Although the subject of interest is geometrical quantity, it can be calculated via statistical mechanics formalism.

When the excluded volume interaction is very weak, a perturbation theory around the random walk with intersections allowed, leads to an asymptotic series for the swelling of an isolated chain. In three-dimensional case (Muthukumar and Nickel,

1984),

$$\Upsilon(g) \simeq 1 + \frac{4}{3}g - 2.075385g^2 + 6.296880g^3 - 25.057251g^4$$
$$+ 116.134785g^5 - 594.71663g^6, \text{ as } g \to 0. \tag{8.118}$$

The asymptotic formula for the strong coupling limit is of the form

$$\Upsilon(g) \simeq 1.5310g^{0.3544} + 0.1843g^{-0.5756}, \text{ as } g \to \infty. \tag{8.119}$$

Consider the interpolation problem, with $c_1 = 4/3$, and the leading amplitude $b_1 = 1.53092$, both values borrowed from the expansions. Then we can apply the recursion from Chapter 1, given by (1.133) on page 54, with $\beta = 0.3544$, $\gamma = -0.93$. Here only the low-order approximant is constructed

$$RA_1^* = \left(\frac{g}{g_0} + 1\right)^{0.3544} e^{1.07527\alpha_1\left(\frac{1}{g_0^{0.93}} - \frac{1}{(g+g_0)^{0.93}}\right)}. \tag{8.120}$$

There is a complex-conjugate pair of solutions, and the real-valued solution is given as half of their sum. Explicitly, in terms of the coupling constant g we obtain the expression

$$\Upsilon_1^*(g) =$$
$$(0.46607 + 0.081406i)e^{\frac{0.035498 - 0.037121i}{(g+(0.22726+0.12061i))^{0.93}}}(1 + (3.43322 - 1.82211i)g)^{0.3544}$$
$$+ (0.46607 - 0.081406i)e^{\frac{0.035498 + 0.037121i}{(g+(0.22726-0.12061i))^{0.93}}}(1 + (3.43322 + 1.82211i)g)^{0.3544}. \tag{8.121}$$

For comparison one can consider the following root approximant (Yukalov and Gluzman, 1999a),

$$R_2^*(g) = \left(11.0631g^2 + (6.5866g + 1)^{1.07}\right)^{0.1772}. \tag{8.122}$$

Eq. (8.122) uses as an input the strong-coupling expansion. The $DLog$ additive approximant (8.121) deviates from Eq. (1.109) on page 46 by less than 0.08%, while root approximant (8.122) brings maximal error of 0.35%. For sake of comparison we calculated also the additive approximant

$$A_2^*(g) = 0.844308(5.47981g + 1)^{0.3544} + \frac{0.155692}{(3.41847g + 1)^{0.5756}},$$

and found that it brings the maximal error of 0.8%.

The low-order approximant (8.121) can be corrected with the higher-order terms from the weak-coupling expansion, encapsulated within the Padé approximant. The corrected expression is given as follows,

$$Cor^*(g) = \Upsilon_1^*(g)P_{2,2}(g), \tag{8.123}$$

where the Padé approximant is given below,

$$P_{2,2}(g) = \frac{303.695g^2 - 9.82611g + 1}{303.637g^2 - 9.82611g + 1}.$$

(8.124)

The corrected approximant (8.123) is the most accurate. Its maximal error is only 0.06%. More information on closed-form expressions can be found in Caracciolo et al. (2008), D'Adamo and Pelissetto (2017).

For two-dimensional polymer coil the following expansion as $g \to 0$ is known, (Muthukumar and Nickel, 1984),

$$\Upsilon(g) \simeq 1 + \tfrac{g}{2} - 0.121545g^2 + 0.0266314g^3 - 0.132236g^4,$$

(8.125)

where for the two-dimensional coil $g = \frac{bN}{\pi l^2}$. Define also that $\Upsilon(g) \equiv \alpha^2(g)$. For arbitrary g in two-dimensional case one can use generalized Flory expression (Grosberg and Khokhlov, 1994),

$$\alpha^4(g) - \alpha^2(g) = const \times g,$$

(8.126)

where the $const = \tfrac{1}{2}$, is taken in order to make the formula explicit. I.e.,

$$\alpha^2(g) = \frac{1}{2}\left(1 + \sqrt{2g+1}\right),$$

(8.127)

and is compatible with the leading term of the expansion for small g. As $g \to \infty$

$$\alpha^2(g) \simeq 0.747659\sqrt{g},$$

with exact critical index $1/2$ (Grosberg and Khokhlov, 1994), and critical amplitude $A \approx 0.747659$. Interesting enough, (8.127) is nothing else but the limit value of the convergent continued fraction

$$1 + (g/2)/(1 + (g/2)/(1 + (g/2)/(1 + ...))),$$

as shown in Wall (1948).

In Subsection 1.3.3 on page 47, we found that iterated root approximants can reasonably well predict the critical index. It seems natural then to construct interpolations based on roots, namely

$$\mathcal{R}_1^*(g) = \sqrt{g+1},$$
$$\mathcal{R}_2^*(g) = \sqrt[4]{0.013819g^2 + (g+1)^2},$$

(8.128)

$$\mathcal{R}_3^*(g) = \sqrt[6]{\left(0.013819g^2 + (g+1)^2\right)^{3/2} - 0.184119g^3}.$$

The approximants (8.128) are conditioned on the exact critical index at large g, as well as on the small g expansion (8.125). They give the following, rather close estimates for the critical amplitude A:

$$1, \ 1.00344, \ 0.970718,$$

all of them pointing to $A \approx 1$, and are significantly larger than suggested by the modified Flory formula.

Root approximants written in general form on page 29, could provide an estimate for the correction to scaling index γ and corresponding amplitude (Yukalov and Gluzman, 1999a). Given the expansion (8.125) and exact critical index for the two-dimensional coil, we can construct the following approximant

$$\mathcal{R}^*(g) = \left((1 + \mathcal{P}_1 g)^{m_1} + \mathcal{P}_2 g^2 \right)^{1/4} , \tag{8.129}$$

and calculate m_1, $\mathcal{P}_1$, $\mathcal{P}_2$. From the two solutions we select the one least deviating asymptotically from $\mathcal{R}_2^*(g)$, namely

$$\mathcal{R}^*(g) = \sqrt[4]{0.770492 g^2 + (1.75667 g + 1)^{1.13852}} . \tag{8.130}$$

From (8.130) one can find the expansion in strong-coupling limit

$$\mathcal{R}^*(g) \simeq 0.936897 \sqrt{g} \left(1 + \frac{0.616249}{g^{0.861484}} \right) = 0.936897 \sqrt{g} + \frac{0.577362}{g^{0.361484}} , \tag{8.131}$$

with non-trivial $\gamma = 0.861$, not very different from the three-dimensional value. It is rather difficult to distinguish its contribution from the trivial term with $\gamma = 1$. Corresponding value for critical amplitude A equals 0.937.

Correction to scaling index can be as well calculated from additive approximants introduced on page 51, based on the same input as for roots,

$$\mathcal{A}_2^*(g) = \mathcal{P}_1 (1 + \lambda g)^{1/2} + \mathcal{P}_2 (1 + \lambda g)^{m_2} , \tag{8.132}$$

with the result

$$\mathcal{A}_2^*(g) = \frac{0.29239}{(1.84001 g + 1)^{0.280678}} + 0.70761 (1.84001 g + 1)^{0.5} , \tag{8.133}$$

leading to the estimate $\gamma = 0.786$, and $A = 0.9599$, compatible with the result from roots. But the two estimates from self-similar approximants are larger than the value of $1/2$ obtained by Shannon et al. (1996a,b). Many more different results are quoted in Caracciolo et al. (2005).

Using the correct critical index and $\gamma = \frac{1}{2}$, as well as the two starting coefficients from (8.125), we construct the $DLog$ additive approximant

$$\mathcal{R}\mathcal{A}_2^*(g) = 1.02804 e^{-\frac{0.0280379}{\sqrt{g + 1.02766}}} \sqrt{0.973086 g + 1}, \tag{8.134}$$

and find the critical amplitude $A = 1.014$, in line with the other estimates, except from the generalized Flory.

Using the correct critical index, same γ and the three coefficients from (8.125), we can construct the third-order $DLog$ additive approximant

$$\mathcal{R}\mathcal{A}_3^*(g) = 0.747445 e^{\frac{0.0380131}{g + 0.650466} + \frac{0.187639}{\sqrt{g + 0.650466}}} \sqrt{1.53736 g + 1}, \tag{8.135}$$

with the critical amplitude $A = 0.927$, agreeing with the other two estimates from
(8.133), (8.131), suggesting that A is located in the interval $0.93 \div 0.96$. All three
approximants involving correction to scaling are based on the same information from
the weak-coupling limit.

Mind that Padé approximants (and alike), even applied with supposedly correct
variable $\sqrt{g}$, are unable to work without the value of amplitude known in advance, being purely interpolation instrument. Thus, in two dimensions the problem of verifying
an accurate expression for the swelling for all g remains open. It depends on knowledge of the correction to scaling index, since the critical index by itself is known. Also
the quality of the series in two dimensional case have been questioned (Des Cloizeaux
et al., 1985). Could it be improved and developed to higher orders?

Exactly analogous approach, based on dynamical scaling discussed above in Section 8.5 applies to polymer solution (De Gennes, 1977, Grosberg and Khokhlov,
1994). At small qR there is a normal diffusion of the coils considered as "blobs"
of size R and their internal structure is irrelevant. The individual blob diffuses with
$D_l \sim R^{-1}q^2$. At large qR there is anomalous, slower-than-normal sub-diffusive be-
havior [7] $D_l \sim q^3$, with $\beta = \frac{2}{z} \approx \frac{2}{3}$, which takes into account the intra-coil dynamics.

8.8 Factor approximants and critical index for the viscosity of a 3D suspension

In an isotropic Newtonian fluid viscosity μ is a scalar quantifying the rheology of the
fluid – the relation between the macroscopic strain-rate ϵ_{ij} and stress σ_{ij} in the fluid.
Einstein (1906) had started a long line of research on the analytical expressions for the
effective viscosity of suspensions. His seminal work remains one of his most quoted
and not less important than Maxwell's, in the field of effective properties. Effective
viscosity for isotropic suspensions μ_e is the multiplier, by which the average rate of
strain is multiplied to result in the average stress.

The fundamental proposition going back to Einstein (1906), states that the viscous
energy dissipation rate of the suspension must be equal to the dissipation rate of the
effective homogeneous fluid. Such definition gives the effective viscosity, and is applicable for any volume fractions, up to the threshold were lubrication forces dominate.
The effective viscosity then corresponds to a quantity that can be measured experimentally. Such defined effective viscosity may become negative as well, as is the case
for bacterial suspensions.

According to the theory of Wajnryb and Dahler (1997), one can produce an approximation similar to (8.58)

$$\frac{\mu_e}{\mu} \simeq 1 + c_1 f + c_2 f^2 \tag{8.136}$$

---

[7] As defined on page 292.

for the low deformation rate, effective viscosity of a suspension. An elegant expression exists for the first order coefficient c_1 in the expansion of the effective viscosity in the volume fraction of hyperspheres in $\mathbb{R}^d$ as a function of dimension d (Belzons et al., 1981, Brady, 1984),

$$c_1 = \frac{d+2}{2}, \qquad (8.137)$$

so that for $d = 2$ the first order coefficient in the expansion $c_1 = 2$. The higher order terms are calculated in Chapter 4 from Eqs. (4.41)–(4.46). In particular, we find $c_1 = c_2 = 2$. The higher order coefficients depend on the location of particles. This implies that it is impossible to write a universal formula for c_3, and a universal formula for 2D suspensions valid within the precision $O(f^3)$. The value $c_1 = \frac{5}{2}$ derived by Einstein (1906), does follow from (8.137) in 3D case. The effective viscosity increases with f because of energy dissipation on individually considered spheres and does not depend on their size.

The behavior of the viscosity in the dense regime is not well understood. Experiments show that the shear viscosity grows rapidly with increasing volume fraction, because it is getting harder to shear a dense suspension, since particles have no room to move past each other. Such simple reasoning suggests that many-body correlations should appear and govern the dynamics in dense particle suspensions. In the case of 3D passive suspensions of spherical particles, the effective viscosity of random suspensions of the hard spheres with the stick boundary conditions, is expressed as the expansion in the volume fraction of hard spheres f. The expansion (8.136) is available only up to the second order term inclusively, as $f \to 0$.

For the second-order coefficient c_2, several close estimates are available. In particular, Batchelor (1977) gives $c_2 = 6.2$, and Wajnryb and Dahler (1997) give $c_2 = 5.9147$[8]. Both results for c_2 include the hydrodynamic contribution considered above in the chapter, as well as additional contributions from the strong Brownian motion of the particles. Obviously, we come back to formula (8.58), when the Brownian contribution is omitted.

Furthermore, there is a critical point characterized by a power-law divergence of the viscosity in the vicinity of the maximal volume fraction value

$$\frac{\mu_e}{\mu} \simeq A(f_c - f)^{-S},$$

as $f \to f_c$. The value of the critical exponent $S \simeq 2$ is currently accepted for $f_c \simeq 0.637$, based on some theoretical arguments and experiments (Brady, 1993, Bicerano et al., 1999). Even with varying threshold values, it is suggested to consider such divergence as an universal power law with the value of $S = 2$ (Wang and Brady, 2015). But "universality" of the value may be very well related to a certain protocol, while

---

[8] As demonstrated in Chapter 6, the development of the Batchelor methodology to the effective conductivity (Jeffrey, 1973), yields incorrect value for the second order coefficient. We can't guarantee that application of the method developed in the book, will not give also the different coefficient c_2 for the effective viscosity.

the whole spectrum of values, between the value in regular case $S = 1$ (Franke and Acrivos, 1967), and fully random systems can be observed (see Preface, Fig. 0.1).

As discussed by Gluzman et al. (2013), to achieve a good estimate of c_2, based solely on the given data c_1, S and f_c, one would clearly need some additional information about the system, but where to find the source? One might expect such improvement from a meaningful change of variables transformation. It is demonstrated in Haines and Mazzucato (2012) that the next order correction to the Einstein formula for a suspension in a finite-size container is at least of order $f^{3/2}$. In order to incorporate this information into our study we use a direct resummation approach by applying the self-similar approximants directly to the expansions/expressions for the effective viscosity in the dilute limit and close to f_c. This will satisfy the limits and also allow the terms of order $f^{3/2}$ to appear in the expansions at least implicitly. Since we do not know the amplitude for such a term, we can only require it to be very small. The factor approximants are uniquely able to address this issue together with the following variable transformation

$$z = \frac{(f/f_c)^{1/2}}{1 - (f/f_c)^{1/2}},$$

which admits an inverse transformation $f = f_c \frac{z^2}{(1+z)^2}$. The factor approximant (8.138) was developed by Gluzman et al. (2013), so that both factors within it are going to balance each other. This will completely suppress all the terms formally appearing after the re-expansion of the order of $f^{1/2}$ and $f^{3/2}$,

$$\mathcal{F}^*(f) = \left(1 + \mathcal{P}_1 \frac{(f/f_c)^{1/2}}{1 - (f/f_c)^{1/2}}\right)^{S-m} \left(1 + \mathcal{P}_2 \frac{(f/f_c)^{1/2}}{1 - (f/f_c)^{1/2}}\right)^m, \qquad (8.138)$$

where

$$\mathcal{P}_2 = \frac{3S + (S(9S - 8c_1 f_c))^{1/2}}{2S}, \quad m = -\frac{2c_1 S}{\mathcal{P}_2^2 S - 2c_1}, \quad \mathcal{P}_1 = \frac{\mathcal{P}_2 m f_c}{m - S}. \qquad (8.139)$$

From (8.138) one can evaluate the second order coefficient in the expansion at small concentrations. Evaluation with the random close packing $f_c = 0.636$ and with given critical index, yields $c_2 \approx 5.9$ (Gluzman et al., 2013), remarkably close to the result of Wajnryb and Dahler (1997), obtained after exceptionally long and tedious calculations.

Most intriguing, one can also use (8.138) to calculate the critical index S directly from the knowledge of c_1, c_2 and f_c. Let us expand the right-hand side of (8.138) in powers of f and consider $\mathcal{P}_1$, $\mathcal{P}_2$, m, S as unknowns. The solution to the system of four equations is unique and we find after some calculations

$$S = \frac{3c_1 - \sqrt{-3c_1^2 + 8c_1 c_2 f_c - 4c_1^3 f_c}}{2c_1}. \qquad (8.140)$$

With $c_2 = 5.9147$ (Wajnryb and Dahler, 1997), we estimate $S \approx 2.036$, in close agreement with accepted value.

We can construct yet another simple factor approximant, based on the threshold f_c and two known coefficients in the expansion,

$$\frac{\mu_e(f)}{\mu} = (1 - 1.56986 f)^{-2.00267} (1 + f)^{-0.643908},$$

and, again, perfectly estimate the critical index, $S = 2.00267$. In reverse situation, using the value of index $S = 2$, one obtains an excellent value of $f_c = 0.63647$ for the threshold. Such estimates do look encouraging [9].

The method of index function, developed in preceding chapters, can be applied to calculate S as well. Let us recall the popular Krieger–Dougherty (KD) formula (Krieger and Dougherty, 1959),

$$\frac{\mu_e(f)}{\mu} = \left(1 - \frac{f}{f_c}\right)^{-c_1 f_c}.$$

The critical index can be expressed as (see (Stanley, 1971))

$$S = - \lim_{f \to f_c} \frac{\log \frac{\mu_e(f)}{\mu}}{\log(1 - f/f_c)}.$$

Let us look for the solution in more general form,

$$\frac{\mu_e(f)}{\mu} = \left(1 - \frac{f}{f_c}\right)^{-\mathcal{S}(f)}. \tag{8.141}$$

The index function $\mathcal{S}(f)$ will be constructed in such a way that as $f \to 0$, $\mathcal{S}(f) \to c_1 f_c$, or the KD estimate for the critical index. And as $f \to f_c$, $\mathcal{S}(f) \to S$, the sought critical index to be found using the series.

To perform actual calculations one has to express $\mathcal{S}(f)$ from (8.141) and expand the function $\frac{\log(\mu_e(f))}{\log(1-f/f_c)}$ for small f, using the known series for the effective viscosity. When only linear term in the expansion is considered available, we simply have a constant index function, and return to the KD formula. With the second-order term, we have a linear form of the index function. To the found new series we apply some resummation technique, e.g., the simplest Padé approximant $P_{0,1}$, to find

$$\mathcal{S}(f) = \frac{c_1 f_c}{1 + \frac{f(c_1^2 f_c + c_1 - 2c_2 f_c)}{2c_1 f_c}},$$

and evaluate the critical index

$$S = \mathcal{S}(f_c) = \frac{2c_1^2 f_c}{c_1^2 f_c + 3c_1 - 2c_2 f_c}.$$

---

[9] Following Subsection 1.3.4, one can use the factor approximant to estimate the upper bound $S \approx 2.08$, while root approximant gives the lower bound $S \approx 1.71$.

Then we estimate $S \approx 2.02$. Overall, the index function monotonically increases from the KD value away from f_c, to the calculated critical index in the vicinity of f_c.

We stress again that techniques based on the short series work not always. In the case of viscosity, the dispersion of results for critical index obtained from factor-based methods, do remain tolerable. The finding seems to justify the whole body of work on the short series for viscosity, and give a physical meaning to the series as a valuable source of estimating S.

Appendix

A.1 Equations of viscous flow and elasticity

There are different physical mechanisms behind analogous governing equations of viscous fluids and solids. But their basic mathematical equations are similar. The main joint feature is the biharmonic equation to which the elastic displacement and the fluid velocity satisfy.

Consider 2D viscous slow flow perpendicular to the unidirectional fibers in the plane (x_1, x_2), see, e.g., Fig. 1.3. Let $\mathbf{u}(x_1, x_2) = (u_1(x_1, x_2), u_2(x_1, x_2))$ denote the transverse velocity vector, and $p = p(x_1, x_2)$ the pressure. It is assumed that p can include the gravitational force $-\rho g x_2$. Let μ denote the viscosity and ∇^2 the 2D Laplace operator. The slow flow of a viscous fluid is governed by the Stokes equations in a domain D

$$\mu \nabla^2 u_1 = \frac{\partial p}{\partial x_1}, \quad \mu \nabla^2 u_2 = \frac{\partial p}{\partial x_2}, \tag{A.1}$$

$$\frac{\partial u_1}{\partial x_1} + \frac{\partial u_2}{\partial x_2} = 0. \tag{A.2}$$

It follows from Eq. (A.2) that there exists such a function $\Psi(x_1, x_2)$, called the streamline function, that

$$u_1 = \frac{\partial \Psi}{\partial x_2}, \quad u_2 = -\frac{\partial \Psi}{\partial x_1}. \tag{A.3}$$

Substitute (A.3) into (A.1), so that

$$\mu \nabla^2 \frac{\partial \Psi}{\partial x_2} = \frac{\partial p}{\partial x_1}, \quad -\mu \nabla^2 \frac{\partial \Psi}{\partial x_1} = \frac{\partial p}{\partial x_2}. \tag{A.4}$$

Calculate the partial derivatives $\frac{\partial}{\partial x_1}$ of the first equation (A.4), $\frac{\partial}{\partial x_2}$ of the second equation (A.4) and add the results. Then, we arrive at the Laplace equation for the pressure

$$\nabla^2 p = 0. \tag{A.5}$$

Calculation of the partial derivatives $\frac{\partial}{\partial x_2}$ of the first equation (A.4), $\frac{\partial}{\partial x_1}$ of the second equation (A.4) and subtraction of the results yields the biharmonic equation for the streamline function

$$\nabla^4 \Psi = 0. \tag{A.6}$$

Eqs. (A.5)–(A.6) yields the same Kolosov–Muskhelishvili formulas (3.7) used in the 2D elasticity[1]. Instead of the elastic displacement (3.17) the velocity is expressed through the complex potentials by analogous formula

$$\mathbf{u}(x_1, x_2) \equiv u_1(x_1, x_2) + iu_2(x_1, x_2) = \frac{1}{2\mu}[\varphi(z) - z\overline{\varphi'(z)} - \overline{\psi(z)}], \qquad (A.7)$$

where $z = x_1 + ix_2$. The pressure has the form

$$p = \frac{\sigma_{11} + \sigma_{22}}{2} = 2\operatorname{Re}\varphi'(z). \qquad (A.8)$$

Here, σ_{jk} denote the components of stress tensor in viscous fluid. Instead of the deformations the velocity deviations are considered in the fluid mechanics

$$\epsilon_{jk} = \frac{1}{2}\left(\frac{\partial u_j}{\partial x_k} + \frac{\partial u_k}{\partial x_j}\right). \qquad (A.9)$$

It follows from (A.2) that $\epsilon_{11} + \epsilon_{22} = 0$. The stresses and the velocity deviations are related by equation

$$\sigma_{jk} = -p\delta_{jk} + 2\mu\epsilon_{jk}, \qquad (A.10)$$

where δ_{jk} denotes the Kronecker symbol.

This implies that all the results obtained for the complex potentials of elastic problems can be extended to the complex potentials of viscous fluid problems by substitution $\kappa = 1 \Leftrightarrow v' = \frac{1}{2} \Leftrightarrow v = 1$ (see relations on page 103). In particular, the effective shear modulus G_e coincides with the effective viscosity μ_e for $\kappa = 1$ for macroscopically isotropic composites and suspensions. The difference in the conditions (3.14)–(3.15) of the elasticity and fluid theories can be in the additional, given terms taking into account the different balance of the external forces (Mikhlin, 1964, Karakin, 1985).

The elasticity equations including the value $1 - 2v'$ or $1 - v$ in denominators have to be replaced by other equations. The 2D Hook's law linearly expresses the deformations ϵ_{jk} by the stresses σ_{jk}

$$E\epsilon_{11} = \sigma_{11} - v\sigma_{22}, \quad E\epsilon_{22} = \sigma_{22} - v\sigma_{11}, \qquad (A.11)$$

$$E\epsilon_{12} = (1 + v)\sigma_{12} \Leftrightarrow 2G\epsilon_{12} = \sigma_{12}.$$

It can be written as a reversed dependence in the theory of elasticity, for instance, the first two equations yield

$$\sigma_{11} = \frac{E}{(1 - v)^2}(\epsilon_{11} + v\epsilon_{22}), \quad \sigma_{22} = \frac{E}{(1 - v)^2}(\epsilon_{22} + v\epsilon_{11}). \qquad (A.12)$$

---

[1] In Richard Feynman own words "The same equations have the same solutions."

In the fluid mechanics, such an inversion is impossible, since the dependence (A.10) as a relation between the stresses and the velocity deviations is degenerate. The elasticity relations (A.12) contain an indeterminate limit as $\nu \to 1 \Leftrightarrow \nu' \to \frac{1}{2}$. They have to be replaced by Eqs. (A.10). The shear modulus G in the last equation (A.11) becomes the viscosity μ in (A.10).

A.2 Eisenstein's series

Following (Czapla et al., 2012a), we present constructive formulas for the Eisenstein–Rayleigh sums S_m and the Eisenstein functions $E_m(z)$, corresponding to the lattice $\mathcal{Q}$ defined by the fundamental complex-numbered translation vectors ω_1 and ω_2.

The Eisenstein–Rayleigh lattice sums S_m are defined as

$$S_m = {\sum_{m_1,m_2}}' (m_1\omega_1 + m_2\omega_2)^{-m} = {\sum_P}' P^{-m}, \quad m = 2, 3, \ldots, \tag{A.13}$$

where the expression $P = m_1\omega_1 + m_2\omega_2$ is introduced for brevity. m_1 and m_2 run over all integers except $m_1 = m_2 = 0$, which is denoted by prime in the sums. The series (A.13) is conditionally convergent for $m = 2$, hence, its value depends on the order of summation. Here, the order is fixed by using the Eisenstein summation (Weil, 1999)

$$\sum_{p,q}^{e} := \lim_{M_2\to+\infty} \lim_{M_1\to+\infty} \sum_{q=-M_2}^{M_2} \sum_{p=-M_1}^{M_1}. \tag{A.14}$$

The sums (A.13) are slowly convergent if computed directly. But they can be easily calculated if expressed through the rapidly convergent series

$$S_2 = \left(\frac{\pi}{\omega_1}\right)^2 \left(\frac{1}{3} - 8\sum_{m=1}^{\infty} \frac{mq^{2m}}{1 - q^{2m}}\right), \quad q = \exp\left(\pi i \frac{\omega_2}{\omega_1}\right), \tag{A.15}$$

$$S_4 = \left(\frac{\pi}{\omega_1}\right)^4 \left(\frac{1}{45} + \frac{16}{3}\sum_{m=1}^{\infty} \frac{m^3 q^{2m}}{1 - q^{2m}}\right), \tag{A.16}$$

$$S_6 = \left(\frac{\pi}{\omega_1}\right)^6 \left(\frac{2}{945} - \frac{16}{15}\sum_{m=1}^{\infty} \frac{m^5 q^{2m}}{1 - q^{2m}}\right). \tag{A.17}$$

S_{2n} $(n \geq 4)$ can be calculated by the recurrence formula

$$S_{2n} = \frac{3}{(2n+1)(2n-1)(n-3)} \sum_{m=2}^{n-2} (2m-1)(2n-2m-1) S_{2m} S_{2(n-m)}. \tag{A.18}$$

The rest of sums vanish. The following formula was proved in Mityushev (1997b)

$$S_2 = \frac{2}{\omega_1}\zeta\left(\frac{\omega_1}{2}\right). \tag{A.19}$$

Application of formula (5) from Hurwitz and Courant (1964, page 210) for $\zeta\left(\frac{\omega_1}{2}\right)$ yields (A.15). It is worth noting that Akhiezer (see Akhiezer (1990, page 204)) deduced a similar formula, which, after substitution into (A.19) yields

$$S_2 = \left(\frac{\pi}{\omega_1}\right)^2 \left(\frac{1}{3} - 8 \sum_{m=1}^{\infty} \frac{q^{2m}}{(1-q^{2m})^2}\right). \tag{A.20}$$

One can see that expansion of the expressions (A.15) and (A.20) in q leads to the same double series. Formulas (A.16) and (A.17) follow from formulas (5) from Hurwitz and Courant (1964, page 210), for invariants of the Weierstrass functions $g_2 = 60\,S_4$ and $g_3 = 140\,S_6$. The review of other formulas for S_2 can be found in Yakubovich et al. (2016).

For the hexagonal array,

$$S_2 = \pi, \quad S_4 = 0, \quad S_6 \approx 3.80815, \quad S_8 = 0, \tag{A.21}$$

while for the square array,

$$S_2 = \pi, \quad S_4 \approx 3.151211, \quad S_6 = 0, \quad S_8 \approx 4.2557732. \tag{A.22}$$

Let $\zeta(z)$ denote the Weierstrass ζ-function for which (Akhiezer, 1990)

$$\zeta(z + \omega_j) - \zeta(z) = \delta_j \quad (j = 1, 2), \tag{A.23}$$

where $\delta_j = 2\zeta\left(\frac{\omega_j}{2}\right)$. Using the Eisenstein summation we introduce the Eisenstein function

$$E_1(z) = \sum_{m_1, m_2 \in \mathbb{Z}} \frac{1}{z + m_1\omega_1 + m_2\omega_2}. \tag{A.24}$$

The Weierstrass function $\zeta(z)$ and Eisenstein function are related by formula $E_1(z) = \zeta(z) - S_2 z$. It follows from Legendre's identity $\delta_1\omega_2 - \delta_2\omega_1 = 2\pi i$ (see Akhiezer (1990), Weil (1999)) and (A.23) that the jumps of $E_1(z)$ have the form

$$E_1(z + \omega_1) - E_1(z) = 0, \quad E_1(z + \omega_2) - E_1(z) = -\frac{2\pi i}{\omega_1}. \tag{A.25}$$

The high-order Eisenstein functions are related to the Weierstrass function $\wp(z)$ by the identities (Weil, 1999)

$$E_2(z) = \wp(z) + S_2, \quad E_m(z) = \frac{(-1)^m}{(m-1)!} \frac{d^{m-2}\wp(z)}{dz^{m-2}}, \quad m = 3, 4, \ldots. \tag{A.26}$$

Every function (A.26) is doubly periodic and has a pole of order m at $z = 0$. The Eisenstein functions of the even-order $E_{2m}(z)$ can be presented in the form of the series (Weil, 1999)

$$E_{2m}(z) = \frac{1}{z^{2m}} + \sum_{k=1}^{\infty} s_k^{(m)} z^{2(k-1)}, \tag{A.27}$$

where

$$s_k^{(m)} = \frac{(2m+2k-3)!}{(2m-1)!(2k-2)!} S_{2(m+k-1)}.$$ (A.28)

The function $E_1(z)$ can be expanded into Laurent's series

$$E_1(z) = \frac{1}{z} - \sum_{k=1}^{\infty} S_{2k} z^{2k-1}.$$ (A.29)

The Eisenstein functions are related by the formula

$$E'_m(z) = -m E_{m+1}(z).$$ (A.30)

Relation (A.30) implies the following form of the nth derivative of E_k

$$\frac{d^n E_k}{dz^n}(z) = (-1)^n \frac{(k+n-1)!}{(k-1)!} E_{k+n}(z).$$ (A.31)

To make the formulas for the effective conductivity more transparent it is convenient to adopt the following definition

$$E_n(0) := S_n.$$ (A.32)

Such an expression appears often in the sums involving the terms $E_n(a_k - a_m)$, which has to be calculated by (A.32) for $a_k = a_m$.

The Eisenstein functions determine the 2D structural sums $e_{n_1,...,n_q}^{(j_1,...,j_q)(l_1,...,l_q)}$ and $e_{p_1,p_2,p_3,...,p_n}^{f_0,f_1,f_2,...,f_n}$ given by (1.12) and (1.15), respectively. These sums are used for mono- and poly-disperse conductivity problems governed by the Laplace equation and for elasticity problems.

A.3   Eisenstein–Natanzon–Filshtinsky series

Works by Eisenstein (1847)[2], by Rayleigh (1892), by Natanzon (1935) and by Filshtinsky (1964)[3] strongly influenced the constructive 2D theory of composites.

Their contributions concern the lattice sums, analytic (meromorphic) and bi-analytic doubly periodic functions, and their relations to the Weierstrass elliptic functions. They were developed further in the recent works, e.g. Yakubovich et al. (2016), Chen et al. (2018), Drygaś and Mityushev (2018). We tried to match the great names with their contributions. It was difficult since all of them did not refer to each other and

[2] All details and lengthy reference to Eisenstein's work of 1847 are given on pages 3–4, of the book by A. Weil (1999).
[3] Leonid Anshelovich Filshtinsky (1930–2019) begun his outstanding research work in 1961. He published 12 books and about 380 papers, mainly in Russian.

worked independently, with the exception of Filshtinsky's paper based on Natanzon's paper.

The Eisenstein–Natanzon–Filshtinsky lattice sums are defined as

$$S_m^{(j)} = \sum_{m_1,m_2} \frac{(\overline{m_1\omega_1 + m_2\omega_2})^j}{(m_1\omega_1 + m_2\omega_2)^m}, \quad m = 2, 3, \ldots, \quad j = 0, 1 \ldots (m \geq j+2), \quad \text{(A.33)}$$

where m_1, m_2 run over all integers except $m_1 = m_2 = 0$. The series (A.33) is conditionally convergent for $m = j+2$, hence, its value depends on the order of summation. The series (A.33) converges absolutely for $m > j+2$.

It follows from Grigolyuk and Filishtinskii (1992) that

$$2\alpha(k+1)S_{2k+3}^{(1)} = ((2k+3)(k+2) - 1)\, S_{2k+4} - 2k\beta S_{2k+2}$$
$$- \sum_{l=1}^{k-1}(2l+1)S_{2l+2}S_{2(k-l)+2}, \quad \text{(A.34)}$$

where

$$\alpha = \frac{\pi}{\omega_1 \operatorname{Im}\omega_2}, \quad \beta = S_2 - \alpha. \quad \text{(A.35)}$$

For the normalized ($|Q_0| = \omega_1 \operatorname{Im}\omega_2 = 1$) hexagonal lattice we arrive at the relation

$$2\pi(k+1)S_{2k+3}^{(1)}$$
$$= ((2k+3)(k+2) - 1)\, S_{2k+4} - \sum_{l=1}^{k-1}(2l+1)S_{2l+2}S_{2(k-l)+2}. \quad \text{(A.36)}$$

For the hexagonal array the exact values from (Yakubovich et al., 2016, Drygaś and Mityushev, 2018, Chen et al., 2018)

$$S_3^{(1)} = \frac{\pi}{2}, \quad S_4^{(1)} = 0, \quad S_5^{(1)} = \frac{3\sqrt{3}\Gamma^{18}\left(\frac{1}{3}\right)}{20480\pi^7}, \quad \text{(A.37)}$$

are known, while for the square array,

$$S_3^{(1)} = \frac{\pi}{2} + \frac{\Gamma^8(\frac{1}{4})}{384\pi^3}, \quad S_4^{(1)} = 0, \quad S_5^{(1)} = 0. \quad \text{(A.38)}$$

Using the Eisenstein summation we introduce the Eisenstein–Natanson–Filshtinsky function

$$E_3^{(1)}(z) = \sum_{m_1,m_2\in\mathbb{Z}} \frac{\overline{z + m_1\omega_1 + m_2\omega_2}}{(z + m_1\omega_1 + m_2\omega_2)^3}. \quad \text{(A.39)}$$

The Eisenstein–Natanson–Filshtinsky functions are related to the Weierstrass functions by formula (Grigolyuk and Filishtinskii, 1992, Drygaś and Mityushev, 2016)

$$E_3^{(1)}(z) = -\frac{1}{2}\bar{z}\wp'(z) + \frac{1}{6\pi}\wp''(z) + \frac{1}{2\pi}\zeta(z)\wp'(z) + S_3^{(1)}. \tag{A.40}$$

To make the formulas for the effective elasticity more transparent it is convenient to adopt the following definition

$$E_m^{(1)}(0) := S_m^{(1)}. \tag{A.41}$$

Such an expression appears often in the sums involving the terms $E_n^{(1)}(a_k - a_m)$, which has to be calculated by (A.32) for $a_k = a_m$.

The general Eisenstein–Natanson–Filshtinsky functions can be presented in the form of the series

$$E_m^{(j)}(z) = \frac{\bar{z}}{z^m} + \sum_{p=0}^{\infty}(-1)^p\frac{(m+p-1)!}{p!(m-1)!}\left[\sum_{k=0}^{j}\frac{j!}{k!(j-k)!}\bar{z}^{j-k}S_{m+p}^{(k)}\right]z^p, \tag{A.42}$$

where $m + p \geq j + 2$. One can find computationally effective formulas for $S_{m+p}^{(k)}$ in Drygaś and Mityushev (2018).

The Eisenstein–Natanson–Filshtinsky functions determine the 2D structural sums $e_{(n)}^{(j)(l)}$ given by (1.18). These sums are used in elasticity problems governed by the bi-Laplace equation.

The values $S_2 = \pi$ and $S_3^{(1)} = \frac{\pi}{2}$ and the corresponding functions (A.26), (A.39) are used in computation of the local elastic fields following the Eisenstein–Rayleigh approach in Chapters 3 and 4. The values $S_2 = 0$ and $S_3^{(1)} = 0$ with the corresponding modified Eisenstein–Natanson–Filshtinsky functions and the structural sums are taken in computation of the effective constants following the Maxwell approach.

A.4 Cluster approach and its limitations

Theoretical difficulties in derivation of approximate analytical formulas for disperse random composites were discussed by Batchelor and Green (1972), Jeffrey (1973), O'Brein (1979), where the authors confronted a conditionally convergent integral. The integral was estimated by a renormalization based on physical intuition, but not on a rigorous mathematical investigation. Brein writes "the renormalization quantity arises naturally from the macroscopic boundary integral referred to earlier, so there is no uncertainty about its choice" (O'Brein, 1979). Beginning with these papers many authors have been trying to extend the "renormalization method". However, attempts to replace an absolutely divergent integral by an integral equal to a constant can lead to misleading assertions outlined in Gluzman et al. (2017, Chapter 3)[4] and to wrong formulas, e.g. Chapter 6.

Extensions of Maxwell's approach from single- to n-inclusions problems are called cluster methods as summarized in Mityushev (2018)[5]:

i) Based on the field around a finite cluster without clusters interactions one can deduce a formula for the effective constants only for dilute clusters.

ii) The uncertainty arising in various self-consistent cluster methods when the number of elements n in a cluster C_n tends to infinity is analyzed by means of the conditionally convergent series discussed in (Rayleigh, 1892, Yakubovich et al., 2016).

In order to illustrate the assertion i) we consider the clusters C_3 and C_4 displayed on the left parts of Figs. A.1 and A.2. The cluster C_4 is obtained from the clusters C_3 by addition of the fourth disk. The clusters C_3 and C_4 form two different composites displayed on the right parts of Figs. A.1 and A.2. The normalized effective conductivity of the dilute clusters C_3 and C_4 are given by two different formulas

$$\sigma_{C_3} = \frac{1 + \alpha_3 f}{1 - \alpha_3 f} + O(f^2), \quad \sigma_{C_4} = \frac{1 + \alpha_4 f}{1 - \alpha_4 f} + O(f^2), \tag{A.43}$$

where the coefficients α_3 and α_4 sometimes called shape factors are computed for single clusters C_3 and C_4, respectively, by Maxwell's approach. It is worth noting that the local interior concentration f_{loc} in the cluster should not be confused with the concentration of the clusters f in composite. In Maxwell's approach, the concentration f is small and prescribed a priori. The authors use self-consistent methods to replace f_{loc} by f, and to consider clusters with high f_{loc}. They wrongly assert that they calculate the effective constants for high concentration f. Actually, they compute a set of the effective conductivities σ_{C_n} up to $O(f^2)$ for different composites. Various modifications, like Mori–Tanaka, difference and other methods follow this methodologically wrong line.

The next assertion ii) complicates the study when n tends ∞, since the following limit depends on the geometrical way of the cluster sequence $\{C_1, C_2, \ldots\}$

$$\alpha_\infty = \lim_{n \to \infty} \alpha_n. \tag{A.44}$$

Hence, actually nobody knows where we arrive by application of such a method. A change of the order of elements in a cluster yields another result.

The proper investigation of the limit (A.44) requires an advanced asymptotic analysis by the Eisenstein summation applied in the bulk of this book. The main theoretical result presented below consists in proper justification of the different Eisenstein–Rayleigh and Maxwell approaches concerning conditionally convergent series arisen in the constructive homogenization. It is demonstrated that determination of the local

---

[4] The term "renormalization" is used in a different context in the present book (see page 20) and in O'Brein (1979).

[5] This Appendix is based on the paper (Mityushev, 2018).

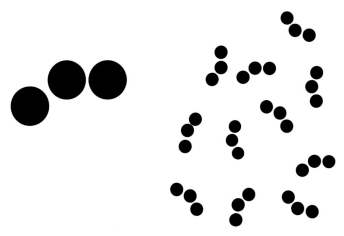

Figure A.1 A single cluster $\mathcal{C}_3$ generated by 3 disks (left) and a dilute composite of $\mathcal{C}_3$ clusters (right).

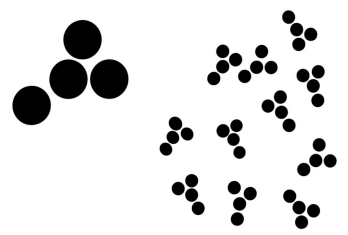

Figure A.2 A single cluster $\mathcal{C}_4$ generated by 4 disks (left) and a dilute composite of $\mathcal{C}_4$ clusters (right).

fields by an expansion method in concentration or in contrast parameters has to be based on the Eisenstein summation. In the same time, computation of the effective elastic constants has to be based on the symmetric summation. These methods give different results for the same series. We concentrate our attention to the 2D conductivity problem. Elasticity and 3D problems can be investigated by the same method Mityushev and Drygaś (2019).

We begin with discussion of a finite cluster. Let $z = x_1 + ix_2$ denote a complex variable in the extended complex plane $\widehat{\mathbb{C}} = \mathbb{C} \cup \{\infty\}$. Consider non-overlapping disks $|z - a_k| < r$ $(k = 1, 2, \ldots, n)$, denoted below by D_k, of conductivity σ embedded in the host material of the normalized unit conductivity occupying the domain D, the complement of all the disks $|z - a_k| \leq r$ to $\widehat{\mathbb{C}}$. The potentials $u(z)$ and $u_k(z)$ are harmonic in D and D_k, respectively, except at infinity where $u(z) \sim x = \mathrm{Re}\, z$ and continuously differentiable in the closures of the considered domains. The singularity of $u(z)$ determine the external flux applied at infinity.

The perfect contact condition between the components is expressed by two real relations

$$u_k(z) = u(z), \quad \sigma \frac{\partial u_k}{\partial \mathbf{n}}(z) = \frac{\partial u}{\partial \mathbf{n}}(z), \quad |z - a_k| = r \ (k = 1, 2, \dots, n), \quad \text{(A.45)}$$

where $\frac{\partial}{\partial \mathbf{n}}$ denotes the outward unit normal derivative to $|z - a_k| = r$. Introduce the contrast parameter $\varrho = \frac{\sigma - 1}{\sigma + 1}$. Two real Eqs. (A.45) are reduced to the $\mathbb{R}$-linear complex condition (Gluzman et al., 2017)

$$\varphi(z) = \varphi_k(z) - \varrho \overline{\varphi_k(z)}, \quad |z - a_k| = r \ (k = 1, 2, \dots, n), \quad \text{(A.46)}$$

where $\varphi(z)$ and $\varphi_k(z)$ are analytic in D and D_k, respectively, except at infinity where $\varphi(z) \sim z$, and continuously differentiable in the closures of the considered domains. The harmonic and analytic functions are related by the equalities $u(z) = \text{Re } \varphi(z)$ in D, $u_k(z) = \frac{2}{\sigma + 1} \text{Re } \varphi_k(z)$ in D_k.

Consider Schottky group of inversions and their compositions with respect to the circles $|z - a_k| = r$, $k = 1, 2, \dots, n$ (plus the identity element)

$$z_{(k)}^* = \frac{r^2}{z - a_k} + a_k, \ z_{(k_1 k_2 \dots, k_m)}^* := (z_{(k_2 \dots, k_{m-1})}^*)_{k_1}^*, \quad (k_{j+1} \neq k_j). \quad \text{(A.47)}$$

Exact solution of the considered problem for any $|\varrho| < 1$ was found in the form of the absolutely and uniformly convergent Poincaré type series (Mityushev, 1993)

$$\varphi(z) = z + \varrho \sum_{k=1}^{n} \overline{z_{(k)}^*} + \varrho^2 \sum_{k=1}^{n} \sum_{k_1 \neq k} z_{(k_1 k)}^* + \varrho^3 \sum_{k=1}^{n} \sum_{k_1 \neq k} \sum_{k_2 \neq k_1} \overline{z_{(k_2 k_1 k)}^*} + \cdots \quad \text{(A.48)}$$

The set of inclusions $C_n = \cup_{k=1}^{n} D_k$ forms a finite cluster on the plane. Its important characteristic is the dipole moment $M^{(n)}$ (Bigon et al., 1998) equal to the coefficient of $\varphi(z)$ on z^{-1} which can be extracted from (A.48). For our purposes it is sufficient to use asymptotic formula

$$M^{(n)} = n\varrho r^2 \mathcal{M}^{(n)}, \quad \text{(A.49)}$$

$$\mathcal{M}^{(n)} = 1 - n\varrho r^2 e_2^{(n)} + n^2 \varrho^2 r^4 e_{22}^{(n)} - n^3 \varrho^2 r^6 [2e_{33}^{(n)} + \varrho e_{222}^{(n)}] + O(r^8).$$

Here, the multiple sums arisen from (A.48) are shortly written as

$$e_p^{(n)} = \frac{1}{n^p} \sum_{k, k_1} \frac{1}{(a_k - a_{k_1})^p}, \quad e_{pp}^{(n)} = \frac{1}{n^{p+1}} \sum_{k, k_1, k_2} \frac{1}{(a_k - a_{k_1})^p (a_{k_1} - a_{k_2})^p}, \quad \text{(A.50)}$$

$$e_{ppp}^{(n)} = \frac{1}{n^{p+2}} \sum_{k, k_1, k_2, k_3} \frac{1}{(a_k - a_{k_1})^p (a_{k_1} - a_{k_2})^p (a_{k_2} - a_{k_3})^p} \quad (p = 2, 3).$$

For simplicity, we consider macroscopically isotropic composites when $M^{(n)} \in \mathbb{R}$. Following Maxwell's formalism consider a large disk $\mathbb{D}$ of radius R_0 containing $N(R_0)$ equal clusters $C^{(n)}$. Mind that n (number of disks per cluster) and $N(R_0)$ (number of clusters in $\mathbb{D}$) are independent. Let the disk $\mathbb{D}$ be occupied by a homogenized medium with an unknown effective conductivity $\sigma_e^{(n)}$. Its dipole moment is equated to the sum of cluster moments (Maxwell, 1873)

$$\frac{\sigma_e^{(n)} - 1}{\sigma_e^{(n)} + 1} R_0^2 = N(R_0)M^{(n)}. \tag{A.51}$$

The total number of small disks in the disk $\mathbb{D}$ is equal to $nN(R_0)$. Introduce the concentration of small disks in the plane $f = nr^2 \lim_{R_0 \to \infty} N(R_0)R_0^{-2}$ assuming that it exists. Of course, f does not depend on the local fraction of disks in the cluster. Substituting (A.49) into (A.51) we find

$$\sigma_e^{(n)} = \frac{1 + f\varrho\mathcal{M}^{(n)}}{1 - f\varrho\mathcal{M}^{(n)}} + O(f^2). \tag{A.52}$$

Example. Let m be a natural number. Consider a square cluster consisting of $n = (2m + 1)^2$ disks $D_{m_1 m_2}$ with the centers $a_{m_1 m_2} = m_1 + im_2$ when $m_{1,2} = -m, -m + 1, \ldots, -1, 0, 1, \ldots, m - 1, m$. It follows from the symmetry and confirmed by computations that $e_2^{(n)} = e_{22}^{(n)} = e_{222}^{(n)} = 0$. The expression $\beta(n) = -2n^3 e_{33}^{(n)}$ takes the values $\beta(1) = 13.0$, $\beta(2) = 8.89$, $\beta(3) = 7.04$, $\beta(4) = 5.95$, $\beta(5) = 5.22$, $\beta(6) = 4.69$, $\beta(7) = 4.29$, $\beta(8) = 3.97$ and tends to zero as $n \to \infty$ (slowly but still does). Therefore, the effective conductivity of these different composites, represented by clusters C_n, can be calculated as (A.52) with $\mathcal{M}^{(n)} = 1 + \beta(n)\varrho^2 r^6 + O(r^8)$ within the first order approximation in f.

The following natural question can be stated. Is the limit $\lim_{n \to \infty} \sigma_e^{(n)}$ correctly defined? The answer is negative since $e_2^{(n)}$ is conditionally convergent. It was shown in McPhedran and McKenzie (1978), Mityushev (1997a) that the limit $\lim_{n \to \infty} e_2^{(n)}$ depends on the shape of exterior curve enclosing the inclusions. In order to answer the question the regular square array is considered for definiteness.

Without loss of generality the linear geometrical scale can be normalized in such a way that for any n the value $n\pi r^2$ is equal to the fraction of inclusions. It does not change the result since the value $\mathcal{M}^{(n)}$ from (A.49) is dimensionless. We have $e_{22} = e_2^2$ and $e_{222} = e_2^3$ for the regular square array. The sum $e_3^{(n)}$ converges absolutely to zero for the regular square array (Gluzman et al., 2017). Hence, it does not depend on the method of summation.

Introduce for brevity the undetermined values $X = \frac{e_2}{\pi}$ and A_m ($m = 1, 2, 3$) in $X_3(f) = \sum_{m=1,2,3} A_m f^m$. The limit of (A.49) becomes

$$\mathcal{M}_3^{(\infty)} = 1 - f\varrho X + (f\varrho X)^2 - (f\varrho X)^3 - X_3(f) =: 1 + \sum_{m=1,2,3} b_m f^m, \tag{A.53}$$

where $\mathcal{M}_3^{(\infty)}$ denotes the third order approximation for $\mathcal{M}^{(\infty)} = \mathcal{M}_3^{(\infty)} + O(f^4)$. The correction $\mathcal{X}_3(f)$ will be explained below. The limit formula

$$\sigma_e = \frac{1 + f\varrho\mathcal{M}_3^{(\infty)}}{1 - f\varrho\mathcal{M}_3^{(\infty)}} + O(f^5), \tag{A.54}$$

following from (A.52) can be asymptotically transformed into equation

$$\sigma_e = 1 + 2\varrho f + 2f^2(b_1\varrho + \varrho^2) + 2f^3(b_2\varrho + 2b_1\varrho^2 + \varrho^3) \tag{A.55}$$
$$+ 2f^4(b_3\varrho + b_1^2\varrho^2 + 2b_2\varrho^2 + 3b_1\varrho^3 + \varrho^4) + O(f^5).$$

As justified by Rayleigh (1892), Movchan et al. (1997), the effective conductivity of the square array has the form

$$\sigma_e = 1 + 2\varrho f + 2\varrho^2 f^2 + 2\varrho^3 f^3 + 2\varrho^4 f^4 + O(f^5). \tag{A.56}$$

Comparison of (A.55) and (A.56) yields $b_m = 0$, hence $\mathcal{M}_3^{(\infty)} = 1$.

The simplest formal way to satisfy (A.56), is to set $\mathcal{X}_3(f) = 0$ and $X = e_2 = 0$ by definition. However, the limit $n \to \infty$ in the local field (A.48) has to be calculated by means of the Eisenstein summation (Weil, 1999) which leads to another result $e_2 = \lim_{n\to\infty} e_2^{(n)} = \pi$ for macroscopically isotropic composites. In this case, the value $\mathcal{X}_3(f) = \sum_{m=1,2,3}(-f\varrho)^m$ must be introduced.

We demonstrate that from the local field in a finite cluster, one can only deduce a formula for the effective conductivity valid only for a dilute case. This result has cast doubts on declarations concerning extension of Maxwell's formalism of finite clusters to high concentrations. While study of infinite clusters leads to conditionally convergent series. Its interpretation can be found in McPhedran and McKenzie (1978, Sec. 2.4) and Mityushev (1997a). The limit $e_2 = \lim_{n\to\infty} e_2^{(n)}$ depends on the shape of exterior curve γ enclosing the inclusions. The term $\mathcal{X}_3(f)$ is related to the total charge density over γ. The Eisenstein summation (A.14) transforms the terms of (A.50) into the Eisenstein functions for a periodic structure (Drygaś, 2016a) and yields, in particular, $e_2 = \pi$.

In the same time, Maxwell's formalism is based on the dipole at infinity, i.e. the total charge density over γ. The Eisenstein limit $n \to \infty$ presupposes the extra, conditionally convergent part of the dipole moment $\mathcal{X}(f)$, which has to be subtracted as it is made in (A.53) in the third order approximation in f.

Thus, extension of Maxwell's formalism to high concentrations requires a subtle mathematical treatment of the conditionally convergent series. It turns out that the analytical formulas from Chapters 3 and 4 for the effective constants of 2D elastic composites are actually based on Maxwell's type formalism. This implies that we take $e_2 = e_3^{(1)} = 0$. This remark shows a computationally simple implementation of Maxwell's formalism despite its subtle mathematical justification.

A.5 Mathematical pseudo-language, transformation rule sequences and rule-based programming

The Mathematical Pseudo-Language, described in details by Cohen (2002), covers basic operations common for Computer Algebra Systems (CAS). In our considerations the following operators are used:

$Kind(u)$	– type of u (e.g. $Kind(3)$ gives $integer$);
$Operand(u, i)$	– returns ith operand of the expression u;
$Map(F, u)$	– returns the expression u with operands of form $F(Operand(u, i))$;
$Length(u)$	– returns number of operands of the expression u;
$Delete(x, L)$	– returns a new list with all instances of x removed from list L;
$Operand\_list(u)$	– returns the list operands of a compound expression u;
$Construct(f, L)$	– returns an expression with main operator f and elements of list L as operands;
$"Operator"(u)$	– unevaluated form of expression $Operator(u)$.

One can describe symbolic operation or entire algorithm with help of *Transformation Rule Sequence*. The notation

$$expr_1 \quad \longmapsto \quad expr_2,$$

is a rule of symbolic transformation of expression matches the form of $expr_1$ with corresponding $expr_2$ form. Transformation rule or sequence of transformation rules (TRS) can be easily translated into MPL using pattern–matching procedures. Some CAS have the capability of *Rule–Based Programming* (RBP), i.e. direct implementing a TRS (see the implementation of the operator **next** in Section A.6).

In this paper we use transformation rules mostly to express operations modifying structural sums in expressions. For example, rule (2.9) modifies every structural sum in a given expression. On the other hand (2.12) applies to a particular index of structural sums in a given expression.

A model example of describing transformation rules is considered as the algorithm of *Derivative* operator using TRS and its implementation using RBP in *Mathematica* and *Maple* pattern-matching languages in Cohen (2002, pp. 182–183). For a pattern–matching algorithm example using MPL see the $Linear\_form$ operator in Cohen (2002, page 180).

A.6 Implementations in mathematica

A.6.1 Implementation of computations of coefficients A_q

We proceed to realization of the Algorithms 2.1 and 2.2 in *Mathematica* using rule-based programming. The implementation is presented as Algorithm A.1. The corresponding example in Algorithm A.2 computes coefficient A_5 (cf. example in Subsection 2.2.1.2).

Algorithm A.1 Implementation of Algorithms 2.1 and 2.2 in *Mathematica*.

```
A[1] = ρ e₂;
A[2] = ρ² e₂,₂;
A[q_] := next[A[q-1]]

next[expr_Plus] := next /@ expr
next[p_. e_x_,y_,z__] := p ρ e₂,x,y,z - p ───── e₁₊ₓ,₁₊ᵧ,z
                                           -1+x
```

$$A[1] = \rho\, e_2;$$
$$A[2] = \rho^2\, e_{2,2};$$
$$A[q\_] := \mathrm{next}[A[q-1]]$$

$$\mathrm{next}[expr\_Plus] := \mathrm{next}\ /@\ expr$$
$$\mathrm{next}[p\_.\ e_{x\_,y\_,z\_\_}] := p\,\rho\, e_{2,x,y,z} - p\,\frac{y}{-1+x}\, e_{1+x,1+y,z}$$

Algorithm A.2 Example call for $q = 5$.

```
A[5]
-4 ρ² e₅,₅ + 3 ρ³ e₂,₄,₄ + 6 ρ³ e₃,₄,₃ + 3 ρ³ e₄,₄,₂ -
2 ρ⁴ e₂,₂,₃,₃ - 2 ρ⁴ e₂,₃,₃,₂ - 2 ρ⁴ e₃,₃,₂,₂ + ρ⁵ e₂,₂,₂,₂
```

$$A[5]$$
$$-4\,\rho^2\, e_{5,5} + 3\,\rho^3\, e_{2,4,4} + 6\,\rho^3\, e_{3,4,3} + 3\,\rho^3\, e_{4,4,2} -$$
$$2\,\rho^4\, e_{2,2,3,3} - 2\,\rho^4\, e_{2,3,3,2} - 2\,\rho^4\, e_{3,3,2,2} + \rho^5\, e_{2,2,2,2}$$

A.6.2 Implementation of mirror terms reduction

One can also implement mirror terms reduction presented in Subsection 2.2.2.1 in *Mathematica*. Algorithm A.3 is the modification of the Algorithm A.1 ensuring sorted operands in coefficient A_q. Then we proceed with implementation of the *Less* (by overloading built-in procedure) and *Binary_search* operators (see Algorithm A.4). Finally, Algorithm A.5 defines operator for reducing mirror terms in A_q.

Algorithm A.3 Code for modified Algorithm A.1.

```
Unprotect[Re];
Re /: Re[c_. ρ^k_] = Re[c] ρ^k;

Attributes[Plus] =
    Complement[Attributes[Plus], {Orderless}];
A[1] = ρ e₂;
A[2] = ρ² e₂,₂;
A[q_] := A[q] = next[1][#] + next[2][#] &@A[q-1]

next[rule_][expr_Plus] := next[rule] /@ expr
next[1][p_. e_x_,y_,z__] := p ρ e₂,x,y,z
next[2][p_. e_x_,y_,z__] := -p ───── e_x+1,y+1,z
                                x-1
```

$$\mathrm{next}[1][p\_.\ e_{x\_,y\_,z\_\_}] := p\,\rho\, e_{2,x,y,z}$$
$$\mathrm{next}[2][p\_.\ e_{x\_,y\_,z\_\_}] := -p\,\frac{y}{x-1}\, e_{x+1,y+1,z}$$

For example, let us compute coefficient A_6 (Algorithm A.6). One can also check, with the code in Algorithm A.7, how many terms are reduced in higher order coefficients. Following results are also presented in Table 2.1.

Algorithm A.4 Implementation of *Binary_search* operator and overloaded *Less* operator in *Mathematica*.

```
Unprotect[Less];
Less[L1_List, L2_List] :=
 Module[{i, imax = Min[Length@L1, Length@L2]},
  i = 1;
  While[L1[[i]] === L2[[i]] ,
   If[i == imax, Return[Length@L1 < Length@L2]];
   i++];
  L1[[i]] < L2[[i]]]
Less[c1_. eₘ₁__ , c2_. eₘ₂__] := {m1} < {m2}

BinarySearch[e_, expr_, a1_, b1_] :=
 Module[{a = a1, b = b1,  mid},
  While[a ≤ b ,
   mid = Floor[(a + b)/2];
   If[e === expr[[mid]], Return[mid]];
   If[e < expr[[mid]], b = mid - 1, a = mid + 1]];
  Return[-1]]
```

Algorithm A.5 *Mirror_reduce* implemented in *Mathematica*.

```
Unprotect[Length];
Length[c_. eₘ_] := Length[{m}]

MirrorReduce[Aq_] :=
 Module[{B = List @@ Aq, k = 1, esum, i},
  While[k < Length[B],
   esum = B[[k]] /. eₘ_ :→ eSequence@@Reverse[{m}];
   If[Not[esum === B[[k]]],
    i = BinarySearch[esum, B, k, Length[B]];
    B[[k]] = 2 If[EvenQ[Length@B[[k]]], Re, Identity][B[[k]]];
    B = Delete[B, i]];
   k++];
  Plus @@ B]
```

Algorithm A.6 Example call for $q = 6$ (cf. example in the preceding section).

```
x = A[6]
```

$\rho^6 \, e_{2,2,2,2,2,2} - 2\,\rho^5\, e_{2,2,2,3,3} - 2\,\rho^5\, e_{2,2,3,3,2} +$
$3\,\rho^4\, e_{2,2,4,4} - 2\,\rho^5\, e_{2,3,3,2,2} + 6\,\rho^4\, e_{2,3,4,3} + 3\,\rho^4\, e_{2,4,4,2} -$
$4\,\rho^3\, e_{2,5,5} - 2\,\rho^5\, e_{3,3,2,2,2} + 4\,\rho^4\, e_{3,3,3,3} + 6\,\rho^4\, e_{3,4,3,2} -$
$12\,\rho^3\, e_{3,5,4} + 3\,\rho^4\, e_{4,4,2,2} - 12\,\rho^3\, e_{4,5,3} - 4\,\rho^3\, e_{5,5,2} + 5\,\rho^2\, e_{6,6}$

```
MirrorReduce[x]
```

$\rho^6 \, e_{2,2,2,2,2,2} - 4\,\rho^5\, e_{2,2,2,3,3} - 4\,\rho^5\, e_{2,2,3,3,2} +$
$6\,\rho^4\, \mathrm{Re}[e_{2,2,4,4}] + 12\,\rho^4\, \mathrm{Re}[e_{2,3,4,3}] + 3\,\rho^4\, e_{2,4,4,2} -$
$8\,\rho^3\, e_{2,5,5} + 4\,\rho^4\, e_{3,3,3,3} - 24\,\rho^3\, e_{3,5,4} + 5\,\rho^2\, e_{6,6}$

Algorithm A.7 Impact of the reduction on number of terms of A_q for $q =$ 2, 3, 4, ..., 18 (cf. Table 2.1).

```
{x = A[#]; Length@x, Length@MirrorReduce[x]} & /@
Range[2, 18]

{{2, 2}, {2, 2}, {4, 3}, {8, 6}, {16, 10}, {32, 20},
 {64, 36}, {128, 72}, {256, 136}, {512, 272},
 {1024, 528}, {2048, 1056}, {4096, 2080}, {8192, 4160},
 {16 384, 8256}, {32 768, 16 512}, {65 536, 32 896}}
```

References

Ackerson, B.J., Sorensen, C.M., Mockler, R.C., O'Sullivan, W.J., 1975. Scattering experiments on critical fluids. Phys. Rev. Lett. 34, 1371–1374.

Ackerson, B.J., Straty, G.C., 1978. Rayleigh scattering from methane. J. Chem. Phys. 69, 1207–1212.

Adler, P.M., 1992. Porous Media. Geometry and Transport. Butterworth-Heinemann, New York.

Adler, P.M., Malevich, V.V., Mityushev, V.V., 2013. Nonlinear correction to Darcy's law for channels with wavy walls. Acta Mech. 224, 1823–1848.

Adler, R.J., Taylor, J.E., 2008. Random Fields and Stochastic Geometry. Springer-Verlag, Berlin.

Adler, P.M., Thovert, J.F., 1999. Fractures and Fracture Networks. Kluwer, New York.

Adler, P.M., Thovert, J.F., Mourzenko, V.V., 2012. Fractured Porous Media. Oxford University Press, Oxford.

Akhiezer, N.I., 1990. Elements of the Theory of Elliptic Functions. American Mathematical Society, Providence, RI.

Ammari, H., Fitzpatrick, B., Kang, H., Ruiz, M., Yu, S., Zhang, H., 2018a. Mathematical and Computational Methods in Photonics and Phononics. American Mathematical Society.

Ammari, H., Kang, H., Lee, H., 2018b. Layer Potential Techniques in Spectral Analysis. American Mathematical Society.

Andrianov, I.V., Danishevskyy, V.V., Kalamkarov, A.L., 2010. Analysis of the effective conductivity of composite materials in the entire range of volume fractions of inclusions up to the percolation threshold. Composites, Part B, Eng. 41, 503–507.

Andrianov, I., Mityushev, V., 2018. Exact and "exact" formulae in the theory of composites. In: Drygaś, P., Rogosin, S. (Eds.), Modern Problems in Applied Analysis. Birkhäuser, Basel, pp. 15–34.

Andrianov, I.V., Starushenko, G.A., Danishevskyy, V.V., Tokarzewski, S., 1999. Homogenization procedure and Padé approximants for the effective heat conductivity of composite materials with cylindrical inclusions having square cross-section. Proc. R. Soc. A 455, 3401–3413.

Atkinson, S., Stillinger, F.H., Torquato, S., 2014. Existence of isostatic, maximally random jammed monodisperse hard-disk packings. Proc. Natl. Acad. Sci. USA 111, 18436–18441.

Awrejcewicz, J., Andrianov, I.V., Danishevskyy, V.V., 2018. Asymptotical Mechanics of Composites: Modelling Composites Without FEM. Springer Nature, Switzerland AG.

Baker, G.A., Graves-Moris, P., 1996. Padé Approximants. Cambridge University, Cambridge.

Banks, T., Torres, T.J., 2013. Two point Padé approximants and duality. arXiv:1307.3689v2 [hep-th].

Barber, D., 2012. Bayesian Reasoning and Machine Learning. Cambridge University Press.

Barbero, E.J., 2008. Finite Element Analysis of Composite Materials. CRC Press, Boca Raton.

Barnea, E., Mizrahi, J., 1973. A generalized approach to the fluid dynamics of particulate systems. Chem. Eng. J. 5, 171–189.

Bardzokas, D.I., Filshtinsky, M.L., Filshtinsky, L.A., 2007. Mathematical Methods in Electro-Magneto-Elasticity. Springer-Verlag, Berlin, Heidelberg.

Batchelor, G.K., 1972. Sedimentation in a dilute dispersion of spheres. J. Fluid Mech. 52, 245–268.

Batchelor, G.K., 1976. Brownian diffusion of particles with hydrodynamic interaction. J. Fluid Mech. 74, 1–29.

Batchelor, G.K., 1977. The effect of Brownian motion on the bulk stress in a suspension of spherical particles. J. Fluid Mech. 83, 97–117.

Batchelor, G.K., Green, J.T., 1972. The determination of the bulk stress in a suspension of spherical to order c^2. J. Fluid Mech. 56, 401–427.

Bauer, T., Hofling, F., Munk, T., Frey, E., Franosch, T., 2010. The localization transition of the two-dimensional Lorenz model. Eur. Phys. J. Spec. Top. 189, 103–118.

Baxter, R.J., 1968. Percus–Yevick equation for hard spheres with surface adhesion. J. Chem. Phys. 49, 2770–2774.

Beenakker, C.W.J., 1984. The effective viscosity of a concentrated suspension of spheres (and its relation to diffusion). Physica A 128, 48–81.

Belzons, M., Blanc, R., Boumlot, J.L., Camoin, C., 1981. Viscosite d'une suspension diluee et bidimensionnelle de spheres. C.R. Acad. Sci. Paris 292, 939–944.

Bender, C.M., Boettcher, S., 1994. Determination of $f(\infty)$ from the asymptotic series for $f(x)$ about $x = 0$. J. Math. Phys. 35, 1914–1921.

Bender, C.M., Vinson, J.P., 1996. Summation of power series by continuous exponentials. J. Math. Phys. 37, 4103–4119.

Bender, C.M., Wu, T.T., 1969. Anharmonic oscillator. Phys. Rev. 184, 1231–1260.

Benguigui, L., 1984. Experimental study of the elastic properties of a percolating system. Phys. Rev. Lett. 53, 2028–2030.

Berdichevsky, V.L., 1983. Variational Principles of Continuum Mechanics. Moscow, Nauka. In Russian.

Bergman, D.J., Stroud, D., 1992. Physical properties of macroscopically inhomogeneous media. In: von Ehrenreich, H., Turnbull, D. (Eds.), Solid State Phys., vol. 46. Academic Press Inc., pp. 147–269.

Bergman, D.J., 2004. Exact relations between critical exponents for elastic stiffness and electrical conductivity of percolating networks. In: Bergman, D.J., Inan, E. (Eds.), Continuum Models and Discrete Systems. In: NATO Science Series (Series II: Mathematics, Physics and Chemistry), vol. 158. Springer, Dordrecht, pp. 351–357.

Berkowitz, B., 2002. Characterizing flow and transport in fractured geological media: a review. Adv. Water Resour. 25, 861–884.

Berlyand, L., Mityushev, V., 2005. Increase and decrease of the effective conductivity of two phase composites due to polydispersity. J. Stat. Phys. 118, 481–509.

Bertin, E., 2015. An equation of state for active matter. Physics 8, 44.

Bhattacharjee, J.K., Mirzaev, S.Z., Kaatze, U., 2012. Does the viscosity exponent derive from ultrasonic attenuation spectra? Int. J. Thermophys. 33, 469–483.

Bicerano, J., Douglas, J.F., Brune, D.A., 1999. Model for the viscosity of particle dispersions. Polym. Rev. 39, 561–642.

Bielski, W., Wojnar, R., 2018. Stokes flow through a tube with wavy wall. In: Awrejcewicz, J. (Ed.), Dynamical Systems in Theoretical Perspective. DSTA 2017. In: Springer Proceedings in Mathematics & Statistics. Springer, Cham, pp. 379–390.

Bigon, D., Serkov, S.K., Movchan, A.B., Valentini, M., 1998. Asymptotic models of dilute composites with imperfectly bonded inclusions. Int. J. Solids Struct. 35, 3239–3258.

Bishop, C.M., 2006. Pattern Recognition and Machine Learning. Springer-Verlag, New York.

Bonneville, R., 2016. A semi empirical compact equation of state for hard sphere fluids at any density. Fluid Phase Equilib. 421, 9–15.

Bradshaw, R., Behnel, S., Seljebotn, D.S., Ewing, G., et al., 2011. Cython: the best of both worlds. Comput. Sci. Eng. 13, 31–39.

Brady, J.F., 1984. The Einstein viscosity correction in n dimensions. Int. J. Multiph. Flow 10, 113–114.

Brady, J.F., Durlofsky, L.J., 1988. The sedimentation rate of disordered suspensions. Phys. Fluids 31, 717–727.

Brady, J.F., 1993. The rheological behavior of concentrated colloidal dispersions. J. Chem. Phys. 99, 567–581.

Buchberger, B., et al., 1985. Symbolic computation (an editorial). J. Symb. Comput. 1, 1–6.

Burstyn, H.C., Sengers, J.V., Bhattacharjee, J.K., Ferrell, R.A., 1983. Dynamic scaling function for critical fluctuations in classical fluids. Phys. Rev. A 28, 1567–1578.

Buzzaccaro, S., Tripodi, A., Rusconi, R., Vigolo, D., Piazza, R., 2008. Kinetics of sedimentation in colloidal suspensions. J. Phys. Condens. Matter 20, 494219.

Caracciolo, S., Guttmann, A.J., Jensen, I., Pelissetto, A., Rogers, A.N., Sokal, A.D., 2005. Correction-to-scaling exponents for two-dimensional self-avoiding walks. J. Stat. Phys. 120, 1037–1100.

Caracciolo, S., Mognetti, B.M., Pelissetto, A., 2008. Two-parameter model predictions and θ-point crossover for linear polymer solutions. J. Chem. Phys. 128, 065104.

Carnahan, N.F., Starling, K., 1969. Equation of state for nonattracting rigid spheres. J. Chem. Phys. 51, 635–636.

Chao, C.K., Shen, M.H., 1993. Thermal problem of curvilinear cracks in bonded dissimilar materials. J. Appl. Phys. 73, 7129–7137.

Chatterjee, S.K., 2008. Crystallography and the World of Symmetry. Springer-Verlag, Berlin.

Chaudhary, K., Cardenas, B.M., Deng, W., Bennett, P.C., 2013. Pore geometry effects on intra-pore viscous to inertial flows and on effective hydraulic parameters. Water Resour. Res. 49, 1149–1162.

Chen, S., Acrivos, A., 1978. The effective elastic moduli of composite materials containing spherical inclusions at non-dilute concentrations. Int. J. Solids Struct. 14, 349–364.

Chen, P.Y., Smith, M.J.A., McPhedran, R.C., 2018. Evaluation and regularization of phase-modulated Eisenstein series and application to double Schlömilch-type sums. J. Math. Phys. 59, 072902.

Cherkaev, A., 2000. Variational Methods for Structural Optimization. Springer Verlag, NY.

Cherkaev, A., 2009. Bounds for effective properties of multimaterial two-dimensional conducting composites. Mech. Mater. 41, 411–433.

Chiu, S.N., Stoyan, D., Kendall, W.S., Mecke, J., 2013. Stochastic Geometry and Its Applications, 3rd ed. Wiley.

Cichocki, B., Ekiel-Jezewska, M.L., Szymczak, P., Wajnryb, E., 2002. Three-particle contribution to sedimentation and collective diffusion in hard-sphere suspensions. J. Chem. Phys. 117, 1231–1241.

Cichocki, B., Ekiel-Jezewska, M.L., Wajnryb, E., 1999. Lubrication corrections for three-particle contribution to short-time self-diffusion coefficients in colloidal dispersions. J. Chem. Phys. 111, 3265–3273.

Cichocki, B., Ekiel-Jezewska, M.L., Wajnryb, E., 2003. Three-particle contribution to effective viscosity of hard-sphere suspensions. J. Chem. Phys. 119, 606–619.

Clerc, J.P., Giraud, G., Laugie, J.M., Luck, J.M., 1990. The electrical conductivity of binary disordered systems, percolation clusters, fractals and related models. Adv. Phys. 39, 191–309.

Clercx, H.J.H., Schram, P.P.L.M., 1992. Three particle hydrodynamic interactions in suspensions. J. Chem. Phys. 96, 3137–3151.

Clisby, N., 2010. Accurate estimate of the critical exponent for self-avoiding walks via a fast implementation of the pivot algorithm. Phys. Rev. Lett. 104, 055702.

Clisby, N., 2017. Virial coefficients for D-dimensional, hard spheres. http://clisby.net/research/publications/clisby_hs_anu04.pdf.

Clisby, N., McCoy, B.M., 2005. New results for virial coefficients of hard spheres in D dimensions. Pramana J. Phys. 64, 775–783.

Clisby, N., McCoy, B.M., 2006. Ninth and tenth order virial coefficients for hard spheres in D dimensions. J. Stat. Phys. 122, 15–57.

Cohen, A.P., Dorosz, S., Schofield, A.B., Schilling, T., Sloutskin, E., 2016. Structural transition in a fluid of spheroids: a low-density vestige of jamming. Phys. Rev. Lett. 116, 098001.

Cohen, S., 2002. Computer Algebra and Symbolic Computation: Elementary Algorithms. A.K. Peters.

Czapla, R., 2018a. Random sets of stadiums in square and collective behavior of bacteria. IEEE/ACM Trans. Comput. Biol. Bioinform. 15, 251–256.

Czapla, R., 2018b. Stochastic Simulations of Geometric Objects on the Plane and Their Applications to RVE. PhD Thesis. Silesian University of Technology, Gliwice.

Czapla, R., Mityushev, V., 2017. A criterion of collective behavior of bacteria. Math. Biosci. Eng. 14, 277–287.

Czapla, R., Mityushev, V., Nawalaniec, W., 2010. Macroscopic conductivity of curvilinear channels. In: Jaworska, L. (Ed.), Int. Conf. Engineering, Education and Computer Science. Pedagogical University, Krakow.

Czapla, R., Nawalaniec, W., Mityushev, V., 2012a. Effective conductivity of random two-dimensional composites with circular non-overlapping inclusions. Compos. Mater. Sci. 63, 118–126.

Czapla, R., Nawalaniec, W., Mityushev, V., 2012b. Simulation of representative volume elements for random 2D composites with circular non-overlapping inclusions. Theor. Appl. Inform. 24, 227–242.

Czaplinski, T., Drygaś, P., Gluzman, S., Mityushev, V., Nawalaniec, W., Zietek, G., 2018. Elastic properties of a unidirectional composite reinforced with hexagonal array of fibers. Arch. Mech. 70, 1–33.

D'Adamo, G., Pelissetto, A., 2017. Polymer models with optimal good-solvent behavior. J. Phys. Condens. Matter 29, 435104.

Daintith, J., Wright, E., 2008. A Dictionary of Computing. Oxford University Press.

Darnton, N., Turner, L., Breuer, K., Berg, H.C., 2004. Moving fluid with bacterial carpets. Biophys. J. 86, 1863–1870.

Davis, K.E., Russel, W.B., Glantschnig, W.J., 1989. Disorder-to-order transition in settling suspensions of colloidal silica: X-ray measurements. Science 245, 507–510.

Davis, P.M., Knopoff, L., 2008. The elastic modulus, percolation and disaggregation of strongly interacting intersecting antiplane cracks. Proc. Natl. Acad. Sci. USA 106, 12634–12639.

De Gennes, P.G., 1977. Theoretical methods of polymer statistics. Riv. Nuovo Cimento 7, 363–378.

Deptuck, D., Harrison, J.P., Zawadzki, P., 1985. Measurement of elasticity and conductivity of a three – dimensional percolation system. Phys. Rev. Lett. 54, 913–916.

Des Cloizeaux, J., Conte, R., Jannink, G., 1985. Swelling of an isolated chain in a solvent. J. Phys. Lett. 46, L.595–L.600.

Devireddy, S.B.R., Biswas, S., 2014. Effect of fiber geometry and representative volume element on elastic and thermal properties of unidirectional fiber-reinforced composites. J. Composites, 629175.

Dolgih, V.N., Filshtinsky, L.A., 1980. Anti-plane shear problem for a composite with defects. Izvestia AN SSSR: Solid Mechanics 4, 103–110.

Doyeux, V., Priem, S., Jibuti, L., Farutin, A., Ismail, M., Peyla, P., 2016. Effective viscosity of two-dimensional suspensions: confinement effects. Phys. Rev. Fluids 1, 043301.

Drummond, J.E., Tahir, M.I., 1984. Laminar viscous flow through regular arrays of parallel solid cylinders. Int. J. Multiph. Flow 10, 515–540.

Drygaś, P., 2016a. Generalized Eisenstein functions. J. Math. Anal. Appl. 444, 1321–1331.

Drygaś, P., 2016b. Functional–differential equations in a class of analytic functions and its application to elastic composites. Complex Var. Elliptic Equ. 61, 1145–1156.

Drygaś, P., Mityushev, V., 2016. Effective elastic properties of random two-dimensional composites. Int. J. Solids Struct. 97–98, 543–553.

Drygaś, P., Mityushev, V., 2017. Contrast expansion method for elastic incompressible fibrous composites. Adv. Math. Phys., 4780928.

Drygaś, P., Mityushev, V., 2018. Lattice sums for polyanalytic functions. arXiv:1807.10116 [org/pdf].

Drygaś, P., Gluzman, S., Mityushev, V., Nawalaniec, W., 2017. Effective elastic constants of hexagonal array of soft fibers. Compos. Mater. Sci. 139, 395–405.

Durand, M., Sadoc, J.F., Weaire, D., 2004. Maximum electrical conductivity of a network of uniform wires: the Lemlich law as an upper bound. Proc. R. Soc. A 460, 1269–1285.

Durand, M., 2007. Low-Density Cellular Materials With Optimal Conductivity and Bulk Modulus. Congrès Français de Mécanique, Grenoble, France, pp. 1–6.

Efros, A.L., Shklovskii, B.I., 1976. Critical behaviour of conductivity and dielectric constant near the metal-non-metal transition threshold. Phys. Status Solidi B 76, 475–485.

Einstein, A., 1906. A new determination of the molecular dimensions. Ann. Phys. 16, 289–306. Erratum in Ann. Phys. 34, 1911 591.

Eischen, J.W., Torquato, S., 1993. Determining elastic behavior of composites by the boundary element method. J. Appl. Phys. 74, 159–170.

Erpenbeck, J.J., Luban, M., 1985. Monte-Carlo-molecular dynamics. Phys. Rev. A 32, 2920–2922.

Eshelby, J.D., 1957. The determination of elastic field of ellipsoidal inclusion and related problems. Proc. R. Soc. A 241, 376–396.

Feitosa, M., Marze, S., Saint-Jalmes, A., Durian, D.J., 2005. Electrical conductivity of dispersions: from dry foams to dilute suspensions. J. Phys. Condens. Matter 17, 6301–6305.

Feng, S., Halperin, B., Sen, P.N., 1987. Transport properties of continuum systems near the percolation threshold. Phys. Rev. B 35, 197–214.

Filishtinsky, L.A., 1964. Stresses and displacements in an elastic sheet weakened by a doubly periodic set of equal circular holes. J. Appl. Math. Mech. 28, 530–543.

Filshtinsky, L.A., Mityushev, V., 2014. Mathematical models of elastic and piezoelectric fields in two-dimensional composites. In: Pardalos, P., Rassias, T. (Eds.), Mathematics Without Boundaries. Springer, New York, pp. 217–262.

Fisher, I.Z., 1964. Statistical Theory of Liquids. University of Chicago Press, Chicago.

Flaherty, J.F., Keller, J.B., 1973. Elastic behavior of composite media. Commun. Pure Appl. Math. XXVI, 565–580.

Franke, N.A., Acrivos, A., 1967. On the viscosity of a concentrated suspension of solid spheres. Chem. Eng. Sci. 22, 847–853.

Frenkel, D., 1987. Velocity auto-correlation functions in a 2d lattice Lorentz gas: comparison of theory and computer simulation. Phys. Lett. 121, 385–389.

Frenkel, J., 1946. Kinetic Theory of Liquids. Dover Publications, NY.

Froba, A.P., Leipertz, A., 2005. Diffusion measurements in fluids by dynamic light scattering. Diffus. Fundam. 2, 63.1–63.25.

Gakhov, F.D., 1966. Boundary Value Problems. Dover Publ., New York.

Garboczi, E.J., Thorpe, M.F., DeVries, M.S., Day, A.R., 1991. Universal conductivity curve for a plane containing random holes. Phys. Rev. A 43, 6473–6482.

Gebart, B.R., 1992. Permeability of unidirectional reinforcements for RTM. J. Compos. Mater. 26, 1100–1133.

Gilleland, W.T., Torquato, S., Russel, W.B., 2011. New bounds on the sedimentation velocity for hard, charged and adhesive hard-sphere colloids. J. Fluid Mech. 667, 403–425.

Ginot, F., Theurkau, I., Levis, D., Ybert, C., Bocquet, L., Berthier, L., Cottin-Bizonne, C., 2015. Nonequilibrium equation of state in suspensions of active colloids. Phys. Rev. X 5, 011004.

Gluzman, S., Karpeev, D.A., 2017. Perturbative expansions and critical phenomena in random structured media. In: Drygaś, P., Rogosin, S. (Eds.), Modern Problems in Applied Analysis. Birkhäuser, pp. 117–134.

Gluzman, S., Karpeev, D.A., Berlyand, L.V., 2013. Effective viscosity of puller-like microswimmers: a renormalization approach. J. R. Soc. Interface 10, 20130720.

Gluzman, S., Mityushev, V., 2015. Series, index and threshold for random 2D composite. Arch. Mech. 67, 75–93.

Gluzman, S., Mityushev, V., Nawalaniec, W., Starushenko, G.A., 2016a. Effective conductivity and critical properties of a hexagonal array of superconducting cylinders. In: Pardalos, P., Rassias, T. (Eds.), Contributions in Mathematics and Engineering. In Honor of Constantin Caratheodory. Springer, pp. 255–297.

Gluzman, S., Mityushev, V., Nawalaniec, W., Sokal, G., 2016b. Random composite: stirred or shaken? Arch. Mech. 68, 229–241.

Gluzman, S., Mityushev, V., Nawalaniec, W., 2017. Computational Analysis of Structured Media. Elsevier.

Gluzman, S., Yukalov, V.I., 1997. Algebraic self-similar renormalization in theory of critical phenomena. Phys. Rev. E 55, 3983–3999.

Gluzman, S., Yukalov, V.I., 1998. Unified approach to crossover phenomena. Phys. Rev. E 58, 4197–4209.

Gluzman, S., Yukalov, V.I., Sornette, D., 2003. Self-similar factor approximants. Phys. Rev. E 67, 026109.

Gluzman, S., Yukalov, V.I., 2014. Extrapolation of perturbation theory expansions by self-similar approximants. Eur. J. Appl. Math. 25, 595–628.

Gluzman, S., Yukalov, V.I., 2015. Effective summation and interpolation of series by self-similar approximants. Mathematics 3, 510–526.

Gluzman, S., Yukalov, V.I., 2016. Self-similarly corrected Padé approximants for indeterminate problem. Eur. Phys. J. Plus 131, 340–361.

Gluzman, S., Yukalov, V.I., 2017a. Additive self-similar approximants. J. Math. Chem. 55, 607–622.

Gluzman, S., Yukalov, V.I., 2017b. Critical indices from self-similar root approximants. Eur. Phys. J. Plus 132, 535.

Golden, K.M., 2009. Climate change and the mathematics of transport in sea ice. Science 56, 562–584.

Golden, K.M., Ackley, S.F., Lytle, V.I., 1998. The percolation phase transition in sea ice. Science 282, 2238–2241.

Gonchar, A.A., 2011. Rational approximation of analytic functions. Proc. Steklov Inst. Math. 272 (Suppl. 2), S44–S57.

Grassberger, P., 1999. Conductivity exponent and backbone dimension in 2d percolation. Physica A 262, 251–263.

Greaves, G.N., Greer, A.L., Lakes, R.S., Rouxe, T., 2011. Poisson's ratio and modern materials. Nat. Mater. 10, 823–837.

Gregoire, G., Chate, H., Tu, Y., 2001. Active and passive particles: modeling beads in a bacterial bath. Phys. Rev. E 64, 011902.

Grigolyuk, E.I., Filishtinskii, L.A., 1970. Perforated Plates and Shells. Moscow, Nauka. In Russian.

Grigolyuk, E.I., Filishtinskii, L.A., 1992. Periodical Piece–Homogeneous Elastic Structures. Moscow, Nauka. In Russian.

Grigolyuk, E.I., Filishtinskii, L.A., 1994. Regular Piece-Homogeneous Structures With Defects. Fiziko-Matematicheskaja Literatura, Moscow. In Russian.

Grosberg, A.Y., Khokhlov, A.R., 1994. Statistical Physics of Macromolecules. AIP Press, Woodbury, NY, USA.

Guazzelli, E., Morris, J.F., 2011. A Physical Introduction to Suspension Dynamics. Cambridge University Press.

Guerrero, A.O., Bassi, A.B.M.S., 2008. On Pade approximants to virial series. J. Chem. Phys. 129, 044509.

Guinovart-Diaz, R., Bravo-Castillero, R.J., Rodriguez-Ramos, R., Sabina, F.J., 2001. Closed-form expressions for the effective coefficients of fibre-reinforced composite with transversely isotropic constituents-I. Elastic and hexagonal symmetry. J. Mech. Phys. Solids 49, 1445–1462.

Haines, B.M., Mazzucato, A.L., 2012. A proof of Einstein's effective viscosity for a dilute suspension of spheres. SIAM J. Math. Anal. 44, 2120–2145.

Hale, R.S., Bonnecaze, R.T., Hidrovo, C.H., 2014. Optimization of capillary flow through square micropillar arrays. Int. J. Multiph. Flow 58, 39–51.

Hashin, Z., 1962. The elastic moduli of heterogeneous materials. J. Appl. Mech. 29, 143–150.

Hashin, Z., 1983. Analysis of composite materials – a survey. J. Appl. Mech. 50, 481–505.

Hasimoto, H., 1959. On the periodic fundamental solutions of the Stokes equations and their applications to viscous flow past a cubic array of cylinders. J. Fluid Mech. 5, 317–328.

Hayakawa, H., Ichiki, K., 1995. Statistical theory of sedimentation of disordered suspensions. Phys. Rev. E 51, R 3815–R 3818.

He, H.X., Hamer, C.J., Oitmaa, J., 1990. High-temperature series expansions for the (2+1)-dimensional Ising model. J. Phys. A 23, 1775–1787.

Helsing, J., 1995. An integral equation method for elastostatics of periodic composites. J. Mech. Phys. Solids 6, 815–828.

Hill, R., 1964. Theory of mechanical properties of fibre-strengthened materials: I. Elastic behaviour. J. Mech. Phys. Solids 12, 199–212.

Hioe, F.T., McMillen, D., Montroll, E.W., 1978. Quantum theory of anharmonic oscillators. Phys. Rep. 43, 305–335.

Ho, D.T., Park, S.D., Kwon, S.Y., Park, K., Kim, S.Y., 2014. Negative Poisson's ratios in metal nanoplates. Nat. Commun. 5, 3255.

Hofling, F., Franosch, T., Frey, E., 2006. Localization transition of the three-dimensional Lorenz model and continuum percolation. Phys. Rev. Lett. 96, 165901.

Hohenberg, P.C., Galperin, B.I., 1977. Theory of dynamic critical phenomena. Rev. Mod. Phys. 49, 435–476.

Howse, J.R., Jones, R.A.L., Ryan, A.J., Gough, T., Vafabakhsh, R., Golestanian, R., 2007. Self-motile colloidal particles: from directed propulsion to random walk. Phys. Rev. Lett. 99, 048102.

Huang, J., 1999. Integral representations of harmonic lattice sums. J. Math. Phys. 40, 5240–5246.

Hurwitz, A., Courant, R., 1964. Vorlesungen Über allgemeine funktionentheorie und elliptische funktionen. Springer-Verlag, Berlin.

Illian, J., Penttinen, A., Stoyan, H., Stoyan, D., 2008. Statistical Analysis and Modelling of Spatial Point Patterns. John Wiley.

Jasiuk, I., Chen, J., Thorpe, M.F., 1994. Elastic moduli of two dimensional materials with polygonal and elliptical holes. Appl. Mech. Rev. 47 (1S), S18–S28.

Jeffrey, D.J., 1973. Conduction through a random suspension of spheres. Proc. R. Soc. Lond. A 338, 355–367.

Jikov, V.V., Kozlov, S.M., Olejnik, O.A., 1994. Homogenization of Differential Operators and Integral Functionals. Springer, Berlin.

Jones, E., Oliphant, T., Peterson, P., et al., 2001. SciPy: open source scientific tools for Python. http://www.scipy.org/. (Accessed 11 February 2019).

Jun, S., Jasiuk, I., 1993. Elastic moduli of two-dimensional composites with sliding inclusions – a comparison of effective medium theories. Int. J. Solids Struct. 30, 2501–2523.

Kamien, R.D., Liu, A.J., 2007. Why is random close packing reproducible? Phys. Rev. Lett. 99, 155501.

Karakin, A.V., 1985. Analytical solution to some plane problems of convection in mantle. Izvestiya AN SSSR, Phys. Solid Earth 2, 16–25.

Kastening, B., 2006. Fluctuation pressure of a fluid membrane between walls through six loops. Phys. Rev. E 73, 011101.

Kawasaki, K., 1970. Kinetic equations and time correlation functions of critical fluctuations. Ann. Phys. 61, 1–56.

Kawasaki, T., Coslovich, D., Ikeda, A., Berthier, L., 2015. Diverging viscosity and soft granular rheology in non-Brownian suspensions. Phys. Rev. E 91, 012203.

Keller, J.B., 1963. Conductivity of a medium containing a dense array of perfectly conducting spheres or cylinders or nonconducting cylinders. J. Appl. Phys. 34, 991–993.

Keller, J.B., 1964. A theorem on the conductivity of a composite medium. J. Math. Phys. 5, 548–549.

Kirkpatrick, S., 1973. Percolation and conduction. Rev. Mod. Phys. 45, 574–588.

Kleinert, H., 2006. Path Integrals in Quantum Mechanics, Statistics, Polymer Physics and Financial Markets. World Scientific, Singapore.

Koch, D.L., Brady, J.F., 1988. Anomalous diffusion in heterogeneous porous media. Phys. Fluids 31, 965–973.

Kostko, A.F., Anisimov, M.A., Sengers, J.V., 2004. Probing structural relaxation in complex fluids by critical fluctuations. J. Exp. Theor. Phys. Lett. 79, 117–120.

Krauth, W., 2006. Statistical Mechanics. Algorithms and Computations. Oxford University Press.

Krieger, I.M., Dougherty, T.J., 1959. Concentration dependence of the viscosity of suspensions. Trans. Soc. Rheol. 3, 137–152.

Kurtyka, P., Rylko, N., 2013. Structure analysis of the modified cast metal matrix composites by use of the RVE theory. Arch. Metall. Mater. 58, 357–360.

Kurtyka, P., Rylko, N., 2017. Quantitative analysis of the particles distributions in reinforced composites. Compos. Struct. 182, 412–419.

Kushch, V.I., Sevostianov, I., 2016. Maxwell homogenization scheme as a rigorous method of micromechanics: application to effective conductivity of a composite with spheroidal particles. Int. J. Eng. Sci. 98, 36–50.

Ladd, A.J.C., 1990. Hydrodynamic transport coefficients of random dispersions of hard spheres. J. Chem. Phys. 93, 3484–3494.

Lakes, R., 1987. Foam structures with a negative Poisson's ratio. Science 235, 1038–1040.

Lanzani, L., Shen, Z., 2004. On the robin boundary condition for Laplace's equation in Lipschitz domains. Commun. Partial Differ. Equ. 29, 91–109.

Lekhnitskii, S.G., 1981. Theory of Elasticity of an Anisotropic Elastic Body. Mir Publishers, Moscow.

Lemlich, R., 1978. Theory for limiting conductivity of polyhedral foam at low-density. J. Colloid Interface Sci. 64, 107–110.

Lemlich, R., 1985. Semi-theoretical equation to relate conductivity to volumetric foam density. Ind. Eng. Chem. Process Des. Dev. 24, 686–687.

Li, B., Madras, N., Sokal, A.D., 1995. Critical exponents, hyperscaling, and universal amplitude ratios for two- and three-dimensional self-avoiding walks. J. Stat. Phys. 80, 661–754.

Li, J., Ostling, M., 2015. Conductivity scaling in supercritical percolation of nanoparticles – not a power law. Nanoscale 7, 3424–3428.

Liu, H., 2006. A very accurate hard sphere equation of state over the entire stable and metastable region. arXiv:cond-matt/0605392.

Loeser, J.G., Zhen, Z., Kais, S., Herschbach, D.R., 1991. Dimensional interpolation of hard sphere virial coefficients. J. Chem. Phys. 95, 4525–4544.

Lopez de Haro, M., Santos, A., Bravo Yuste, S., 1998. A student-oriented derivation of a reliable equation of state for a hard-disc fluid. Eur. J. Phys. 19, 281–286.

Lopez de Haro, M., Santos, A., Bravo Yuste, S., 2008. Simple equation of state for hard disks on the hyperbolic plane. J. Chem. Phys. 129, 116101-1–116101-2.

Losert, W., Bocquet, L., Lubensky, T.C., Gollub, J.P., 2000. Particle dynamics in sheared granular matter. Phys. Rev. Lett. 85, 1428–1431.

Maestre, M.A.G., Santos, A., Robles, M., Lopez de Haro, M., 2011. On the relation between virial coefficients and the close-packing of hard disks and hard spheres. J. Chem. Phys. 134, 084502.

Maier, B., Rädler, J.O., 1999. Conformation and self-diffusion of single DNA molecules confined to two dimensions. Phys. Rev. Lett. 82, 1911–1914.

Majewski, M., Kursa, M., Holobut, P., Kowalczyk-Gajewska, K., 2017. Micromechanical and numerical analysis of packing and size effects in elastic particulate composites. Composites, Part B, Eng. 124, 158–174.

Malevich, A.E., Mityushev, V.V., Adler, P.M., 2006. Stokes flow through a channel with wavy walls. Acta Mech. 182, 151–182.

Malevich, A.E., Mityushev, V.V., Adler, P.M., 2008. Couette flow in channels with wavy walls. Acta Mech. 197, 247–283.

Markov, K.Z., 1999. Elementary micro mechanics of heterogeneous media. In: Markov, K., Preziosi, L. (Eds.), Heterogeneous Media: Micromechanics Modeling Methods and Simulations, pp. 1–162.

Marsland, S., 2014. Machine Learning: An Algorithmic Perspective, second edition. Chapman and Hall/CRC.

Maxwell, J.C., 1873. Electricity and Magnetism, a Treatise on Electricity and Magnetism, vol. 1, 1st edn. Clarendon Press, Oxford, UK.

Mazur, P., Geigenmiller, U., 1987. A simple formula for the short-time self-diffusion coefficient in concentrated suspensions. Physica A 146, 657–661.

McCoy, B.M., 2001. Do hard spheres have natural boundaries? arXiv:cond-mat/0103556v1.

McKenzie, D.R., McPhedran, R.C., Derrick, G.H., 1978. The conductivity of lattices of spheres II. Proc. R. Soc. A 362, 211–232.

McPhedran, R.C., McKenzie, D.R., 1978. The conductivity of lattices of spheres I. The simple cubic lattice. Proc. R. Soc. A 359, 45–63.

McPhedran, R.C., Movchan, A.B., 1994. The Rayleigh multipole method for linear elasticity. J. Mech. Phys. Solids 42, 711–727.

Mertens, S., Moore, C., 2012. Continuum percolation thresholds in two dimensions. Phys. Rev. E 86, 061109.

Messina, R., Aljawhari, S., Bécu, L., Schockmel, J., Lumay, G., Vandewalle, N., 2015. Quantitatively mimicking wet colloidal suspensions with dry granular media. Sci. Rep. 5, 10348.

Mikhlin, S.G., 1964. Integral Equations. Pergamon Press, Oxford Etc.

Milton, G., 2002. The Theory of Composites. Cambridge University Press.

Mishuris, G.S., Kuhn, G., 2002. Asymptotics of elastic field near the tip of interface crack under nonclassical transmission conditions. In: Karihaloo, B.L. (Ed.), IUTAM Symposium on Analytical and Computational Fracture Mechanics of Non-Homogeneous Materials. Springer, pp. 57–67.

Mishuris, G.S., Movchan, A.B., Bigoni, D., 2012. Dynamics of a fault steadily propagating within a structural interface. Multiscale Model. Simul. 10, 936–953.

Mityushev, V., 1993. Plane problem for the steady heat conduction of material with circular inclusions. Arch. Mech. 45, 211–215.

Mityushev, V., 1997a. Transport properties of finite and infinite composite materials and Rayleigh's sum. Arch. Mech. 49, 345–358.

Mityushev, V., 1997b. Transport properties of regular array of cylinders. Z. Angew. Math. Mech. 77 (2), 115–120.

Mityushev, V., 1999. Transport properties of two-dimensional composite materials with circular inclusions. Proc. R. Soc. A 455, 2513–2528.

Mityushev, V., 2000. Thermoelastic plane problem for material with circular inclusions. Arch. Mech. 52, 915–932.

Mityushev, V., 2006. Representative cell in mechanics of composites and generalized Eisenstein–Rayleigh sums. Complex Var. Elliptic Equ. 51, 1033–1045.

Mityushev, V., 2011. Riemann-Hilbert problems for multiply connected domains and circular slit maps. Comput. Methods Funct. Theory 11, 575–590.

Mityushev, V., 2015. Random 2D composites and the generalized method of Schwarz. Adv. Math. Phys., 535128.

Mityushev, V., 2018. Cluster method in composites and its convergence. Appl. Math. Lett. 77, 44–48.

Mityushev, V., Adler, P.M., 2002a. Longitudinal permeability of arrays of circular cylinders I. A single cylinder in the unit cell. Z. Angew. Math. Mech. 82, 335–345.

Mityushev, V., Adler, P.M., 2002b. Longitudinal permeability of arrays of circular cylinders II. An arbitrary distribution of cylinders inside the unit cell. Z. Angew. Math. Phys. 53, 486–517.

Mityushev, V., Adler, P.M., 2006. Darcy flow around a two-dimensional lens. J. Phys. A 39, 3545–3560.

Mityushev, V., Drygaś, P., 2019. Effective properties of fibrous composites and cluster convergence. Multiscale Model. Simul. 17, 696–715.

Mityushev, V., Nawalaniec, W., 2015. Basic sums and their random dynamic changes in description of microstructure of 2D composites. Compos. Mater. Sci. 97, 64–74.

Mityushev, V., Nawalaniec, W., 2019. Effective conductivity of a random suspension of highly conducting spherical particles. Appl. Math. Model. 72, 230–246.

Mityushev, V., Rogosin, S., 2000. Constructive methods for linear and nonlinear boundary value problems for analytic functions. In: Theory and Applications. In: Monographs and Surveys in Pure and Applied Mathematics. Chapman & Hall / CRC, Boca Raton etc.

Mityushev, V., Rylko, N., 2010. Mathematical model of electrokinetic phenomena in two-dimensional channels. In: Jaworska, L. (Ed.), Int. Conf. Engineering, Education and Computer Science. Pedagogical University, Krakow, pp. 5–20.

Mityushev, V., Rylko, N., 2012. Optimal distribution of the nonoverlapping conducting disks. Multiscale Model. Simul. 10, 180–190.

Mityushev, V., Rylko, N., 2013. Maxwell's approach to effective conductivity and its limitations. Q. J. Mech. Appl. Math. 66, 241–251.

Mityushev, V., Nawalaniec, W., Rylko, N., 2018a. Introduction to Mathematical Modeling and Computer Simulations. CRC–Taylor & Francis, Boca Raton.

Mityushev, V., Rylko, N., Bryla, M., 2018b. Conductivity of two-dimensional composites with randomly distributed elliptical inclusions. Z. Angew. Math. Mech. 98 (4), 512–516.

Mityushev, V., Nawalaniec, W., Nosov, D., Pesetskaya, E., 2019. Schwarz's alternating method in a matrix form and its applications to composites. Applied Mathematics and Computation 356, 144–156.

Modes, C.D., Kamien, R.D., 2007. Hard disks on the hyperbolic plane. Phys. Rev. Lett. 99, 235701.

Modes, C.D., Kamien, R.D., 2008. Geometrical frustration in two dimensions: idealizations and realizations of a hard disc fluid in negative curvature. Phys. Rev. E 77, 041125.

Moffatt, H.K., 1964. Viscous and resistive eddies near a sharp corner. J. Fluid Mech. 18, 1–18.

Mourzenko, V.V., Thovert, J.F., Adler, P.M., 2011. Conductivity of isotropic and anisotropic fracture networks, from the percolation threshold to very large densities. Phys. Rev. E 84, 036307.

Movchan, A.B., Nicorovici, N.A., McPhedran, R.C., 1997. Green's tensors and lattice sums for elastostatics and elastodynamics. Proc. R. Soc. A 453, 643–662.

Mulero, A., Cachadina, I., Solana, J.R., 2009. The equation of state of the hard-disc fluid revisited. Mol. Phys. 107, 1457–1465.

Murat, M., Marianer, S., Bergman, D.J., 1986. A transfer matrix study of conductivity and permeability exponents in continuum percolation. J. Phys. A 19, L275–L279.

Murty, M., Devi, V.S., 2011. Pattern Recognition. An Algorithmic Approach.

Muskhelishvili, N.I., 1966. Some Mathematical Problems of the Plane Theory of Elasticity. Nauka, Moscow.

Muthukumar, M., Nickel, B.G., 1984. Perturbation theory for a polymer chain with excluded volume interaction. J. Chem. Phys. 80, 5839–5850.

Muthukumar, M., Nickel, B.G., 1987. Expansion of a polymer chain with excluded volume interaction. J. Chem. Phys. 86, 460–476.

Nagy, F., Duxbury, P., 2002. Permeability and conductivity of platelet-reinforced membranes and composites. Phys. Rev. E 66, 020802.

Natanson, V.Ya., 1935. On the stresses in a stretched plate weakened by identical holes located in chessboard arrangement. Mat. Sb. 45 (5), 616–636. In Russian.

Nawalaniec, W., 2015. Random non-overlapping walks of disks on the plane. In: Mityushev, V.V., Ruzhansky, M. (Eds.), Current Trends in Analysis and Its Applications: Proceedings of the 9th ISAAC Congress. Springer.

Nawalaniec, W., 2016. Algorithms for computing symbolic representations of basic e-sums and their application to composites. J. Symb. Comput. 74, 328–345.

Nawalaniec, W., 2017. Efficient computation of basic sums for random polydispersed composites. Comput. Appl. Math. 37 (2), 2237–2259.

Nawalaniec, W., 2019a. Basicsums: a Python package for computing structural sums and the effective conductivity of random composites. J. Open Source Software 4 (39), 1327. https://doi.org/10.21105/joss.01327.

Nawalaniec, W., 2019b. The basicsums Python package documentation. https://basicsums.bitbucket.io. (Accessed 11 February 2019).

Nawalaniec, W., 2019c. Classifying and analysis of random composites using structural sums feature vector. Proc. R. Soc. A 475, 2225.

Nielsen, L.F., 2005. Composite Materials Properties as Influenced by Phase Geometry. Springer Verlag, Berlin, Heidelberg.

Nieuwenhuizen, Th.M., van Velthoven, P.F.J., Ernst, M.H., 1986. Diffusion and long-time tails in a two-dimensional site-percolation model. Phys. Rev. Lett. 57, 2477–2480.

Novak, I.L., Kraikivski, P., Slepchenko, B.M., 2009. Diffusion in cytoplasm: effects of excluded volume due to internal membranes and cytoskeletal structures. Biophys. J. 97, 758–767.

O'Brein, R.W., 1979. A method for the calculation of the effective transport properties of suspensions of interacting particles. J. Fluid Mech. 91, 17–39.

O' Hern, C.S., Silbert, L.E., Liu, A.J., Nagel, S.R., 2003. Jamming at zero temperature and zero applied stress: the epitome of disorder. Phys. Rev. E 68, 011306.

O' Hern, C.S., Silbert, L.E., Liu, A.J., Nagel, S.R., 2006. Reply to comment on jamming at zero temperature and zero applied stress: the epitome of disorder. Phys. Rev. E 70, 043302.

Olsson, P., Teitel, S., 2007. Critical scaling of shear viscosity at the jamming transition. Phys. Rev. Lett. 99, 178001.

Ostoja-Starzewski, M., 2008. Microstructural Randomness and Scaling in Mechanics of Materials. Chapman & Hall/CRC/Taylor & Francis.

Palacci, J., Cottin-Bizonne, C., Ybert, C., Bocquet, L., 2010. Sedimentation and effective temperature of active colloidal suspensions. Phys. Rev. Lett. 105, 088304.

Paladin, G., Peliti, L., 1982. Fixed dimensional computation of critical transport properties of fluids. J. Phys. Lett. 43, L.15–L.20.

Parisi, G., Zamponi, F., 2010. Mean field theory of hard sphere glasses and jamming. Rev. Mod. Phys. 82, 789–845.

Paule, P., Buchberger, B., Kartashova, L., Kauers, M., Schneider, C., Winkler, F., 2009. Algorithms in symbolic computation. In: Buchberger, B., et al. (Eds.), Hagenberg Research. Springer, Berlin, Heidelberg, pp. 5–62.

Paulin, S.E., Ackerson, B.J., 1990. Observation of a phase transition in the sedimentation velocity of hard spheres. Phys. Rev. Lett. 64, 2663–2666;
Phys. Rev. Lett. E 65, 1990 668.

Pearson, D.S., Shikata, T., 1994. Viscoelastic behavior of concentrated spherical suspensions. J. Rheol. 38, 601–616.

Pedregosa, F., et al., 2011. Scikit-learn: machine learning in Python. J. Mach. Learn. Res. 12, 2825–2830.

Pelisetto, A., Vicari, E., 2002. Critical phenomena and renormalization group theory. Phys. Rep. 368, 549–727.

Perrins, W.T., McKenzie, D.R., McPhedran, R.C., 1979. Transport properties of regular array of cylinders. Proc. R. Soc. A 369, 207–225.

Penrose, O., 1995. Metastable decay rates, asymptotic expansions, and analytic continuation of thermodynamic functions. J. Stat. Phys. 78, 267–283.

Pesetskaya, E., Czapla, R., Mityushev, V., 2018. An analytical formula for the effective conductivity of 2D domains with cracks of high density. Appl. Math. Model. 53, 214–222.

Phelan, R., Weaire, D., Peters, E.A.J.F., Verbist, G., 1996. The conductivity of a foam. J. Phys. Condens. Matter 8, L475–L482.

Piccolroaz, A., Mishuris, G., Movchan, A., Movchan, N., 2012. Perturbation analysis of Mode III interfacial cracks advancing in a dilute heterogeneous material. Int. J. Solids Struct. 49, 244–255.

Poladian, L., 1988a. Asymptotic behaviour of the effective dielectric constants of composite materials. Proc. R. Soc. A 426, 343–359.

Poladian, L., 1988b. General theory of electrical images in sphere pairs. Q. J. Appl. Math. 41, 395–417.

Poladian, L., McPhedran, R.C., 1986. Effective transport properties of periodic composite materials. Proc. R. Soc. A 408, 45–59.

Post, A.J., Glandt, D.E., 1986. Cluster concentrations and virial coefficients for adhesive particles. J. Chem. Phys. 84, 4585–4594.

Pozrikidis, C., 1987. Creeping flow in two-dimensional channel. J. Fluid Mech. 180, 495–514.

Rayleigh, 1892. On the influence of obstacles arranged in rectangular order upon the properties of medium. Philos. Mag. Ser. 34, 481–502.

Rasoulzadeh, M., Panfilov, M., 2018. Asymptotic solution to the viscous/inertial flow in wavy channels with permeable walls. Phys. Fluids 30, 106604.

Regner, B.M., Vucinic, D., Domnisoru, C., Bartol, T.M., Hetzer, M.W., Tartakovsky, D.M., Sejnowski, T.J., 2013. Anomalous diffusion of single particles in cytoplasm. Biophys. J. 104, 1652–1660.

Rintoul, M.D., Torquato, S., 1996. Metastability and crystallization in hard-sphere systems. Phys. Rev. Lett. 77, 4198–4201.

Rohrmann, R.D., Santos, A., 2014. Equation of state of sticky-hard-sphere fluids in the chemical-potential route. Phys. Rev. E 89, 042121.

Ryan, S.D., Mityushev, V., Vinokur, V.V., Berlyand, L., 2015. Rayleigh approximation to ground state of the Bose and Coulomb glasses. Sci. Rep. 5, 7821.

Rylko, N., 2000. Transport properties of a rectangular array of highly conducting cylinders. J. Eng. Math. 38, 1–12.

Rylko, N., 2014. Representative volume element in 2D for disks and in 3D for balls. J. Mech. Mater. Struct. 9 (4), 427–439.

Rylko, N., Krzaczek, B., Mityushev, V., 2013. Conductivity of fibre composites with fractures on the boundary of inclusions. Multiscale Model. Simul. 11, 152–161.

Samuel, M.A., Li, G., 1994. Estimating perturbative coefficients in quantum field theory and the ortho-positronium decay rate discrepancy. Phys. Lett. B 331, 114–118.

Sangani, A.S., Acrivos, A., 1982. Slow flow past periodic arrays of cylinders with application to heat transfer. Int. J. Multiph. Flow 8, 193–206.

Sangani, A.S., Acrivos, A., 1983. The effective conductivity of a periodic array of spheres. Proc. R. Soc. A 386, 263–275.

Sangani, A.S., Yao, C., 1988. Transport processes in random arrays of cylinders. II. Viscous flow. Phys. Fluids 31, 2435–2444.

Santos, A., 2014. Playing with marbles: structural and thermodynamic properties of hard-sphere systems. In: Cichocki, B., Napiorkowski, M., Piasecki, J. (Eds.), 5th Warsaw School of Statistical Physics. Warsaw University Press, Warsaw, pp. 203–296.

Santos, A., Lopez de Haro, M., Bravo Yuste, S., 1995. An accurate and simple equation of state for hard disks. J. Chem. Phys. 103, 4622–4625.

Santos, A., Yuste, S.B., Lopez de Haro, M., 2011. Inferring the equation of state of a metastable hard-sphere fluid from the equation of state of a hard-sphere mixture at high densities. J. Chem. Phys. 135, 181102.

Santos, A., Yuste, S.B., Lopez de Haro, M., Odriozola, G., Ogarko, V., 2014. Simple effective rule to estimate the jamming packing fraction of polydisperse hard spheres. Phys. Rev. E 89, 040302(R).

Santos, A., Yuste, S.B., Lopez de Haro, M., Ogarko, V., 2017. Equation of state of polydisperse hard-disk mixtures in the high-density regime. Phys. Rev. E 96, 062603.

Schnyder, S.K., 2014. Anomalous Transport in Heterogenous Media. Inaugural-Dissertation, Dusseldorf. pp. 1–141.

Scholle, M., 2004. Creeping Couette flow over an undulated plate. Arch. Appl. Mech. 73, 823–840.

Scholle, M., Haas, A., Aksel, N., Wilson, M.C.T., Thompson, H.M., Gaskell, P.H., 2009. Eddy genesis and manipulation in plane laminar shear flow. Phys. Fluids 21 (7), 073602.

Scholle, M., Marner, F., 2018. Couette flow with geometrically induced unsteady effects. PAMM 18, 1–2.

Segre, P.N., Pusey, P.N., 1997. Dynamics and scaling in hard-sphere colloidal suspensions. Physica A 235, 9–18.

Selvadurai, A.P.S., Nikopour, H., 2012. Transverse elasticity of a unidirectionally reinforced composite with an irregular fibre arrangement: experiments, theory and computations. Compos. Struct. 94, 1973–1981.

Sengers, J.V., Perkins, R.A., Huber, M.L., Le Neeinde, B., 2009. Thermal diffusivity of H_2O near the critical point. Int. J. Thermophys. 30, 1453–1465.

Shannon, S.R., Choy, T.C., Fleming, R.J., 1996a. An improved perturbation approach to the 2D Edwards polymer: corrections to scaling. J. Chem. Phys. 10, 8951–8957.

Shannon, S.R., Choy, T.C., Fleming, R.J., 1996b. Corrections to scaling in 2-dimensional polymer statistics. Phys. Rev. B 53, 2175–2178.

Skinner, T.O.E., Schnyder, S.K., Aarts, D.G.A.L., Horbach, J., Dullens, R.P.A., 2013. Localization dynamics of fluids in random confinement. Phys. Rev. Lett. 111, 128301.

Speedy, R.J., 1994. On the reproducibility of glasses. J. Chem. Phys. 100, 6684–6691.

Stanley, H.E., 1971. Introduction to Phase Transitions and Critical Phenomena. Oxford University Press.

Stevenson, P.M., 2016. The effective exponent $\gamma(Q)$ and the slope of the β-function. Phys. Lett. B 761, 428–430.

Stumpf, M.P.H., Porter, M.A., 2012. Critical truths about power laws. Science 335, 665–666.

Szász, D., 2000. Boltzmann's ergodic hypothesis, a conjecture for centuries? In: Szász, D., Bunimovich, L.A., et al. (Eds.), Hard Ball Systems and the Lorentz Gas. Springer, Berlin, pp. 421–446.

Takatori, S.C., Brady, J.F., 2015. Towards a thermodynamics of active matter. Phys. Rev. E 91, 32117.

Tamayol, A., Bahrami, M., 2010a. Parallel flow through ordered fibers: an analytical approach. J. Fluids Eng. 132, 114502-1.

Tamayol, A., Bahrami, M., 2010b. Transverse permeability of fibrous porous media. In: Proceedings (CD) of the 3rd International Conference on Porous Media and Its Applications in Science and Engineering ICPM3. Montecatini, Italy.

Telega, J.J., 2004. Stochastic homogenization: convexity and nonconvexity. In: Castañeda, P.P., Telega, J.J., Gambin, B. (Eds.), Nonlinear Homogenization and Its Applications to Composites, Polycrystals and Smart Materials. In: NATO Science Series. Kluwer Academic Publishers, Dordrecht, pp. 305–346.

Telega, J.J., Wojnar, R., 1998. Flow of conductive fluids through poroelastic media with piezoelectric properties. J. Theor. Appl. Mech. 36 (3), 775–794.

Telega, J.J., Wojnar, R., 2000. Flow of electrolyte through porous piezoelectric medium: macroscopic equations. C. R. Acad. Sci., Sér. 2, Méc. Phys. Chim. Astron. 328 (3), 225–230.

Telega, J.J., Wojnar, R., 2007. Electrokinetics in random piezoelectric porous media. Bull. Pol. Acad. Sci., Tech. Sci. 55, 125–128.

Tian, J., Gui, Y., Mulero, A., 2010a. New virial equation of state for hard-disk fluids. Phys. Chem. Chem. Phys. 12, 13597–13602.

Tian, J., Gui, Y., Mulero, A., 2010b. New closed virial equation of state for hard-sphere fluids. J. Phys. Chem. B 114, 13399–13402.

Tian, J., Jiang, H., Gui, Y., Mulero, A., 2009. Equation of state for hard-sphere fluids offering accurate virial coefficients. Phys. Chem. Chem. Phys. 11, 11213–11218.

Tikhonov, A.N., Samarskii, A.A., 1977. Equations of Mathematical Physics, 5th edition. Nauka, Moscow. Translation of 1st edition Pergamon Press, Oxford, 1963.

Torquato, S., 2002. Random Heterogeneous Materials: Microstructure and Macroscopic Properties. Springer-Verlag, New York.

Torquato, S., 2018. Perspective: basic understanding of condensed phases of matter via packing models. J. Chem. Phys. 149, 020901.

Torquato, S., Stillinger, F.H., 2010. Jammed hard-particle packings: from Kepler to Bernal and beyond. Rev. Mod. Phys. 82, 2634–2672.

Torquato, S., Truskett, T.M., Debenedetti, P.G., 2000. Is random close packing of spheres well defined? Phys. Rev. Lett. 84, 2064–2067.

Valle, F., Favre, M., De Los Rios, P., Rosa, A., Dietler, G., 2005. Scaling exponents and probability distributions of DNA end-to-end distance. Phys. Rev. Lett. 95, 158105.

van der Walt, S., Colbert, S.C., Varoquaux, G., 2011. The numPy array: a structure for efficient numerical computation. Comput. Sci. Eng. 13, 22–30.

Vilanove, R., Poupinet, D., Rondelez, F., 1988. A critical look at measurements of the ν exponent for polymer chains in two dimensions. Macromolecules 21, 2880–2887.

Vilanove, V., Rondelez, F., 1980. Scaling description of two-dimensional chain conformations in polymer monolayers. Phys. Rev. Lett. 45, 1502–1505.

Vilchevskaya, E., Sevostianov, E.I., 2015. Overall thermal conductivity of a fiber reinforced composite with partially debonded inhomogeneities. Int. J. Eng. Sci. 98, 99–109.

von zur Gathen, J., Gerhard, J., 2013. Modern Computer Algebra, 3 edition. Cambridge University Press.

Wajnryb, E., Dahler, J.S., 1997. The Newtonian viscosity of a moderately dense suspensions. In: Prigogine, I., Rice, S.A. (Eds.), Adv. Chem. Phys., vol. 102. Wiley, New York, pp. 193–313.

Wall, H.S., 1948. Analytic Theory of Continued Fractions. Chelsea Publishing Company, New York.

Wall, P., 1997. A comparison of homogenization, Hashin–Shtrikman bounds and the Halpin–Tsai equations. Appl. Math. 42, 245–257.

Wang, M., Brady, J.F., 2015. Constant stress and pressure rheology of colloidal suspensions. Phys. Rev. Lett. 115, 158301.

Weil, A., 1999. Elliptic Functions According to Eisenstein and Kronecker. Springer-Verlag, Berlin.

Wheatley, J., 2013. Calculation of high-order virial coefficients with applications to hard and soft spheres. Phys. Rev. Lett. 110, 200601.

Whitesides, G.M., Grzybowski, B., 2002. Modeling self-assembly across scales: the unifying perspective of smart minimal particles. Science 295, 2418–2421.

Wisniak, J., 2003. Heike Kamerlingh-the virial equation of state. Indian J. Chem. Technol. 10, 564–572.

Wojnar, R., 1995. Thermoelasticity and homogenization. J. Theor. Appl. Mech. 33, 323–335.

Wojnar, R., 1997. Homogenization of piezoelectric solid and thermodynamics. Rep. Math. Phys. 40, 585–598.

Wojnar, R., Bielski, W., 2015. Laminar flow past the bottom with obstacles – a suspension approximation. Bull. Pol. Acad. Sci., Tech. Sci. 63, 685–695.

Wojnar, R., Bielski, W., 2018. Gravity driven flow past the bottom with small waviness. In: Drygaś, P., Rogosin, S. (Eds.), Modern Problems in Applied Analysis. In: Trends in Mathematics. Birkhäuser, pp. 181–202.

Wojnar, R., Bytner, S., Galka, A., 1999. Effective properties of elastic composites subject to thermal fields. In: Hetnarski (Ed.), Thermal Stresses V, pp. 257–466.

Wolovick, M.J., Moore, J.C., 2018. Stopping the flood: could we use targeted geoengineering to mitigate sea level rise? Cryosphere 12, 2955–2967.

Wu, Y., Yi, N., et al., 2015. Three-dimensionally bonded spongy graphene material with super compressive elasticity and near-zero Poisson's ratio. Nat. Commun. 6, 6141.

Wu, X.L., Libchaber, A., 2000. Particle diffusion in a quasi-two-dimensional bacterial bath. Phys. Rev. Lett. 84, 3017–3020.

Wu, G.W., Sadus, R.J., 2005. Hard sphere compressibility factors for equation of state development. AIChE J. 51, 309–313.

Yakubovich, S., Drygaś, P., Mityushev, V., 2016. Closed-form evaluation of 2D static lattice sums. Proc R. Soc. A 472 (2195), 20160662.

Yazdchi, K., Srivastava, S., Luding, S., 2011. Microstructural effects on the permeability of periodic fibrous porous media. Int. J. Multiph. Flow 37, 956–966.

Yelash, L.V., Kraska, T., 2001. A generic equation of state for the hard-sphere fluid incorporating the high density limit. Phys. Chem. Chem. Phys. 3, 3114–3118.

Yukalov, V.I., Gluzman, S., 1997a. Critical indices as limits of control functions. Phys. Rev. Lett. 79, 333–336.

Yukalov, V.I., Gluzman, S., 1997b. Self-similar bootstrap of divergent series. Phys. Rev. E 55, 6552–6570.

Yukalov, V.I., Gluzman, S., 1998. Self-similar exponential approximants. Phys. Rev. E 58, 1359–1382.

Yukalov, V.I., Gluzman, S., 1999a. Self-similar crossover in statistical physics. Physica A 273, 401–415.

Yukalov, V.I., Gluzman, S., 1999b. Weighted fixed points in self-similar analysis of time series. Int. J. Mod. Phys. B 13, 1463–1476.

Zhang, K., 2016. On the concept of static structure factor. arXiv:1606.03610.

Zick, A.A., Homsy, G.M., 1982. Stokes flow through periodic arrays of spheres. J. Fluid Mech. 115, 13–26.

Ziff, R.M., Torquato, S., 2017. Percolation of disordered jammed sphere packings. J. Phys. A, Math. Theor. 50, 085001.

Ziman, J.M., 1979. Models of Disorder. Cambridge University Press, Cambridge.

Zimmerman, R.W., 1996. Effective conductivity of a two-dimensional medium containing elliptical inhomogeneities. Proc. R. Soc. A 452, 1713–1727.

Zohdi, T.I., Wriggers, P., 2008. An Introduction to Computational Mechanics. Springer-Verlag, Berlin.

Zuzovsky, M., Brenner, H., 1977. Effective conductivities of composite materials composed of cubic arrangements of spherical particles embedded in an isotropic matrix. J. Appl. Math. Phys. 28, 979–992.

Index